High Pressure Chemistry and Biochemistry

NATO ASI Series

Advanced Science Institutes Series

A series presenting the results of activities sponsored by the NATO Science Committee, which aims at the dissemination of advanced scientific and technological knowledge, with a view to strengthening links between scientific communities.

The series is published by an international board of publishers in conjunction with the NATO Scientific Affairs Division

A Life Sciences	Plenum Publishing Corporation
B Physics	London and New York
C Mathematical	D. Reidel Publishing Company
and Physical Sciences	Dordrecht, Boston, Lancaster and Tokyo
D Behavioural and Social Sciences	Martinus Nijhoff Publishers
E Engineering and	Dordrecht, Boston and Lancaster
Materials Sciences	
F Computer and Systems Sciences	Springer-Verlag
G Ecological Sciences	Berlin, Heidelberg, New York, London,
	Paris, and Tokyo

Series C: Mathematical and Physical Sciences Vol. 197

High Pressure Chemistry and Biochemistry

edited by

R. van Eldik

Institute for Physical and Theoretical Chemistry. ·
Johann Wolfgang Goethe University, Frankfurt am Main, F.R.G.

and

J. Jonas

Department of Chemistry,
University of Illinois, Urbana, Illinois, U.S.A.

D. Reidel Publishing Company

Dordrecht / Boston / Lancaster / Tokyo

Published in cooperation with NATO Scientific Affairs Division

Proceedings of the NATO Advanced Study Institute on
Advances in High Pressure Studies of Chemical and Biochemical Systems
Corfu, Greece
September 28-October 11, 1986

Library of Congress Cataloging in Publication Data

NATO Advanced Study Institute on High Pressure Studies of Chemical and Biochemical
    Systems (1986: Kerkyra, Corfu)
    High pressure chemistry and biochemistry.

    (NATO ASI series. Series C, Mathematical and physical sciences; vol. 197)
    "Proceedings of the NATO Advanced Study Institute on High Pressure Studies of
Chemical and Biochemical Systems, Corfu, Greece, September 28–October 11, 1986"—
T.p. verso.
    "Published in cooperation with NATO Scientific Affairs Division."
    Includes Index.
    1. Chemistry, Physical and theoretical—Congresses. 2. Biological chemistry—Con-
gresses. 3. High pressure (Science)—Congresses. I. Eldik, Rudi van. II. Jonas, J.
III. North Atlantic Treaty Organization. Scientific Affairs Division. IV. Title. V. Series:
NATO ASI series. Series C, Mathematical and physical sciences; vol. 197.
    QD450.N38   1986   541.3   87–4645
    ISBN 90–277–2457–1

Published by D. Reidel Publishing Company
P.O. Box 17, 3300 AA Dordrecht, Holland

Sold and distributed in the U.S.A. and Canada
by Kluwer Academic Publishers,
101 Philip Drive, Assinippi Park, Norwell, MA 02061, U.S.A.

In all other countries, sold and distributed
by Kluwer Academic Publishers Group,
P.O. Box 322, 3300 AH Dordrecht, Holland

D. Reidel Publishing Company is a member of the Kluwer Academic Publishers Group

Printed in The Netherlands

TABLE OF CONTENTS

PREFACE

It was the objective of the ASI on "Advances in High Pressure Studies of Chemical and Biochemical Systems" to present the current status of such studies and to emphasize the advances achieved during the nine years since the previous ASI on "High Pressure Chemistry". These advances are partly due to the improved instrumentation enabling static and dynamic measurements at pressures several orders of magnitude higher than before, and partly due to the more general availability of high pressure equipment. This has led to a remarkable development in various areas of physics and chemistry, and especially in biochemistry.

Throughout the presentation of this Advanced Study Institute the emphasis fell on the teaching character of such a summer school, and the contributions in this volume are of such a nature. Following a general introduction to modern high pressure research, a series of chapters on theoretical and experimental studies of gases, fluids and solids at high temperatures and pressures are presented with special emphasis on the physical aspects involved. Instrumentation used in such studies, viz. shock compression, NMR spectroscopy, laser scattering, x-ray and neutron scattering, and vibrational spectroscopy are treated in detail. The subsequent chapters are devoted to the application of high pressure techniques in the broad areas of organic, inorganic and biochemistry. The formal lectures were supplemented by 29 contributed papers, for which a list of titles is included.

Following the successful completion of this ASI, we the editors gratefully acknowledge the very effective cooperation of the organizing committee, and wish to thank all the lecturers and contributors to this volume for the great care they took in preparing the lectures and manuscripts. On behalf of all the lecturers and participants, we wish to express our sincere appreciation for the generous financial support offered by the Scientific Affairs Division of the North Atlantic Treaty Organization.

Frankfurt am Main   R. van Eldik
Urbana   J. Jonas
December 1986

SECTION I:

GENERAL AND PHYSICAL ASPECTS

MODERN ASPECTS OF HIGH PRESSURE CHEMISTRY

H. C. Drickamer
School of Chemical Sciences, Department of Physics
and Materials Research Laboratory
University of Illinois
Urbana, IL 61801 USA

ABSTRACT. Modern aspects of high pressure chemistry and closely related disciplines are discussed. Atomic and molecular phenomena are emphasized.

Modern high pressure research spans so many disciplines from geology through chemistry and physics to biochemistry and molecular biology that no one could possibly have read a significant share of the current literature, let alone be able to present the results in a brief paper. This outline emphasizes topics broadly of interest to chemists with a brief excursion into aspects of physics of some concern to chemists and some mention of aspects of biology which impinge on chemistry.

There have been important developments in static high pressure techniques, especially in the use of new materials, most notably in the diamond anvil cell. However, the principles of all modern high pressure equipment can be traced directly to P.W. Bridgman and progress has been, in a sense, evolutionary. Nevertheless, the pressure range has been expanded dramatically through clever use of diamonds and through the availability of optical, x-ray and electronic devices which permit the use of very small samples. The use of shock compression for very high pressure experimentation is largely a development of the past forty years and it has evolved considerably in that period. To date its impact has been more extensive in physics than in chemistry and I shall omit its discussion of it here; this decision is admittedly in some degree arbitrary.

The two most generally significant developments involve: (1) the use of novel and hitherto unavailable tools to measure macroscopic properties with greater detail and precision and in regions of temperature and pressure not previously accessible earlier and (2) the study of the molecular, atomic and electronic properties of matter. The recognition that pressure has much to offer in understanding interactions on the atomic scale has indeed been revolutionary. The implementation of these ideas has been greatly enhanced by improvements in various forms of optical spectroscopy, in nuclear magnetic resonance, and in x-ray and neutron diffraction.

1

R. van Eldik and J. Jonas (eds.), High Pressure Chemistry and Biochemistry, 1–7.
© *1987 by D. Reidel Publishing Company.*

In parallel with these experimental developments has come a deeper understanding and theoretical insight at the molecular level, as well as new techniques such as molecular dynamics. The extraction of quantitative results from theory and the ability to establish and use much more realistic models for atomic and molecular interactions has been made possible in large part by high speed computers. Because of my own limitations and biases, the emphasis in this review will be largely on experimental developments.

It is, from some aspects, convenient to classify the consequences of modern high pressure research in three categories:

(1) Characterization

Under this heading one might include the characterization of electronic or vibrational excitations according to their pressure shifts, the characterization of types of chemical reactions according to the pressure coefficients of their rates, the characterization of structures according to their compressibilities, the characterization of protein types according to their resistance to pressure denaturation etc.

(2) Testing of Theories

In this rubric one could include listing of theories of electronic phenomena such as insulator-metal transitions in solids or supercritical fluids, theories which predict relative displacement of electronic or vibrational energy levels ("pressure tuning"), theories predicting structural changes in crystals, theories of reaction rates in condensed phases, theories of electron transfer or energy transfer phenomena, theories of molecular dynamics in fluids etc.

(3) Transformations

Most obviously this category would include structural phase transitions, but it is much broader. It includes a wide variety of electronic transitions with optical, electrical, magnetic or chemical consequences; it includes new types of chemical reactions or better reaction paths with improved yields, it includes changes of conformation of proteins both reversible and irreversible; and transitions from fluid to crystalline or amorphous phases, glass transitions etc.

The examples listed in these three categories are clearly illustrative rather than exhaustive, and we shall hear much more about these and other examples in the course of the Institute proceedings.

Chemistry-Physics Interaction

One aspect of high pressure physics which impinges strongly on chemistry is the measurement of the equation of state. This is, of course, perhaps the oldest branch of high pressure science. Modern developments for fluids include the extension to greater ranges of temperature and pressure and to fluid mixtures in the supercritical region. For solids the use of high pressure x-ray diffraction, in part in conjunction with synchrotron radiation has extended tremendously both the range and precision of results. One of the most general advances in the past four decades has been the ability of theory to predict and to correlate p-v-t relationships especially in solids, over very wide ranges of density.

A second aspect of modern high pressure physics which is of interest to chemists is the study of phase transitions, particularly in solids. Long ago Bridgman showed that first order phase transitions were common in solids. The modern aspects involve, in the first place, the determination of the structure of the high pressure phases formed. A more dramatic departure has been the possibility of studying the mechanism of phase transitions via Raman spectra of soft modes. In general, lattice vibrations become stiffer with increasing compression. However, when a phase transition is imminent one vibrational mode softens, i.e. shifts to lower frequency, and the lattice rearranges by movement along this mode. It has also been possible to predict phase transitions by calculation of the free energy associated with different lattices at different densities. These calculations have been largely applied to Si, Ge and zincblende or wurtzite structures.

Significant pressure studies have also been made on freezing, on transitions involving amorphous phases, and on phase transitions near critical points. There is a broad class of transformations where the degree of cooperativity is such they have neither the characteristics of first order transition, nor of chemical equilibria. These are of particular interest.

The field of high pressure magnetism is a third area in which physicists and chemists have had parallel interests. Physicists have been largely concerned with paramagnetic-ferromagnetic-antiferromagnetic transitions in metals and alloys and in the behavior of spin glasses. chemists have emphasized changes of spin state in insulating compounds of transition metals and, to some extent, rare earths. In these studies Mossbauer resonance and NMR have proved to be very powerful tools.

Magnetic transitions are a special case of the broad field of electronic transitions, where, again, chemistry and physics meet. Physicists have emphasized insulator-semiconductor-metal transitions and transitions to the superconducting state. In chemistry the more relevant electronic transitions are neutral to ionic transitions in electron donor-acceptor complexes, photochromic-thermochromic transitions, transitions to chemically reactive states, etc.

High Pressure Chemistry

I have somewhat arbitrarily divided the discussion of chemistry into fluids and solids, although many phenomena take place in either phase. Under solids I include both crystalline and amorphous materials.

In modern high pressure work on fluids the gas and liquid phases form a continuum - in fact, much of the interesting research is in the supercritical region. Perhaps the most general distinction is that the relative importance of the repulsive parameter becomes greater as one goes from supercritical fluids to highly compressed liquids. One of the most extensively studied phenomena is the solubility of solids and gases like H_2, N_2, O_2 and CH_4 in supercritical fluids including metals. Chemistry in this region also has unusual aspects. One can support a flame in a solution of CH_4 in water with O_2 injection at 1000 bars. A particularly elegant set of experiments involves the insulator-metal transition in supercritical mercury and alkali metals. These experiments provide the most direct and clear cut test of theories of metal-insulator transitions such as that suggested by Mott. Transport coefficients in the supercritical region vary rapidly and provide sensitive tests for theories. E.g., many years ago large anomalies were observed in the thermal diffusion coefficient in supercritical fluids.

Compressed liquids at lower temperatures also provide important tests of theories and new challenges for theory. Measurements of diffusion, rotational relaxation, vibrational dephasing etc. have formed the basis for modern theories of motion in liquids. The basic feature that arises is that one must extract the temperature coefficient at constant density for transport phenomena in order to relate theory and experiment.

Electronic and molecular spectroscopy as a function of pressure have been very important in demonstrating the significance of polarizability in intermolecular interactions and separating the relative importance of van der Waals interactions, solvation and related effects for various solvents. These phenomena will be discussed in some detail in later lectures.

An important aspect of high pressure studies on fluids in the ability to make large variations in physical parameters (e.g. viscosity, density, dielectric constant etc.) at constant temperature. This ability frequently provides a unique test of relationships which may be more apparent than real when extracted from studies of temperature or solvent effects at ambient pressure.

It is frequently possible to freeze a liquid at, say 25°C by applying a modest pressure (a few kilobars or tens of kilobars). This provides a discontinuity at constant temperature in properties such as static dielectric constant, density, viscosity, etc. and a unique test of the relationship of these variables to reaction rates, rearrangements, radiative and non-radiative rates of luminescence decay, electron and energy transfer process etc. E.g., the static dielectric constant of a polar liquid such as acetonitrile may change across the freezing point by an order of magnitude, with only a very modest change in the high frequency dielectric constant.

The study of chemical equilibria and chemical reaction rates in solution is one of the oldest parts of high pressure research, but it is still one of the intense activity with new aspects arising.

Basically, one can write:

$$K_x = e^{-\overline{\Delta G}/RT} \tag{1}$$

$$\frac{\partial \ln K_x}{\partial p}\Big)_T = -\frac{\overline{\Delta V}}{RT} \tag{2}$$

where K_x is defined in terms of mole fractions and $\overline{\Delta V}$ is the difference in partial molar volumes of products and reactants. Similarly from Eyring reaction rate theory, the rate coefficient k is given by:

$$k = \frac{k_B T}{h} e^{-\Delta G^{\ddagger}/RT} \tag{3}$$

and $$\frac{\partial \ln k}{\partial p}\Big)_T = -\frac{\Delta V^{\ddagger}}{RT}$$

where $\Delta V^{\ddagger}$ is the difference in volume between the system in the transition state (activated complex) and the reactants. The measurement of rates and equilibria permits one to extract $\overline{\Delta V}$ and $\Delta V^{\ddagger}$ and from these quantities plus some chemical intuition one can frequently select among suggested reaction mechanisms. This has been and is a valuable tool.

The basic assumption of Eyring's equation is that the decay of the transition state into products takes place irreversibly along a single reaction coordinate. In 1940 Kramers pointed out that there are circumstances under which this picture may be oversimplified. At very low viscosity of solvent the decay may not be irreversible while at high viscosity one may have to take into account solvent rearrangement. There have been a number of theoretical extensions of Kramer's argument and a few critical experiments. The resolution of this problem remains an important aspect of modern high pressure kinetics in solution.

One should point out that pressure in the range of 10's of kilobars has proven to be a valuable tool in synthesis, e.g. in overcoming steric hindrance. In addition to thermal chemistry at high pressure there has been significant recent effort in the area of high pressure photochemistry. A remarkable number of materials seem to react photochemically in solution much more rapidly under a few kilobars pressure than they do at one atmosphere. Sorting out the products can be a complex task, but promises to be a field of considerable interest.

Chemical interest in the solid state, at high pressure and otherwise, is growing rapidly. The study of electronic phenomena in solids is a very broad and significant field and is covered in detail in a further lecture. The effects of compression on electronic excitations and emissions on changing steric effects, on radiative and non-radiative rates, on electron transfer processes and on the balance between intramolecular and intermolecular interactions is the subject of intensive and fruitful investigation. The use of polymeric or glassy media has added a new dimension to the work.

Reactivity in crystalline solids provides a form of steric control not present in liquid solution. Symmetry effects also control the kinds of excitations which are permitted. Interactions between lattice vibrations and molecular vibrations lattice modes change rapidly in energy with pressure.

The forms of reactivity most common in solids are isomerization and polymerization, both of which have been studied, but not extensively. The effect of solid state structural phase transitions on chemical reactivity in solids is an area which is virtually untouched. Glass transitions in polymers, the sol-gel process and similar rearrangements have been shown to have very large pressure coefficients, but systematic studies are in their early stages.

High Pressure Biochemistry

Perhaps the fastest growing area of high pressure research involves application to biochemical and biophysical problems. The lectures of Heremans, Weber and Wong will cover these phenomena far more thoroughly than is possible in this overview. Only a couple of aspects are considered here.

It is possible to induce changes of conformation of proteins in solution with modest pressures. These can conveniently be deduced from changes in fluorescence energy, intensity, lifetime or polarization. The active sites whose neighborhood one studies can be either fluorescent amino acids like tryptophane or suitable ligands. For single stranded proteins the changes of conformation may be reversible up to ~ eight kilobars. More complex multiple stranded proteins usually undergo irreversible changes in conformation in the range of two to three kilobars. One should note that reversibility of conformation as detected by, say, fluorescence polarization and by biological activity do not necessarily occur in the same time period. Minor pressure induced changes in conformation near critical sites, such as the heme group in myoglobin can drastically effect reaction rates with O_2 and CO. Important information can be gathered for light scattering, NMR and Raman and infrared spectroscopy as well as from electronic spectra. One may expect many important developments in this field in the next few years.

Since the deepest parts of the ocean are at ~ one kilobar interesting observations can be made of pressure effects on more highly organized biological systems. There exist barophyllic bacteria which are stable at say 600 bars, but not at significantly lower or higher pressures. A comparison of the structure and conformation of some of their proteins with similar proteins from animals existing at one atmosphere would be of interest.

High Pressure Geochemistry

The field of geochemistry is not concerned specifically in this institute although several of the papers have relevance to this area. In addition to pressure induced crystal structure changes which have been the bread and butter of earlier high pressure geochemistry, there are several newer aspects. A number of elements can occur in unusual oxidation state and miscibility patterns. At high temperature and modest pressure geothermal and hydrothermal syntheses are important processes both in nature and artifically to make useful crystals. Migration of ions under extreme conditions is of interest. One anticipates a considerable broadening of topics under the geochemistry rubric in future years.

Note on References

The references attached to the papers on various specific topics at the conference provide an access to the literature which is both broader and deeper than would be possible to attach to a survey of this kind. For topics not covered at this conference one can find references in the papers from AIRAPT and related conferences.

Acknowledgement

The author would like to acknowledge financial support from the Division of Materials Science of the Department of Energy under Contract DE-AC02-76ER01198.

PHYSICS OF DENSE FLUIDS

Marvin Ross
Lawrence Livermore National Laboratory
Physics Department, H-Division
P.O. Box 808
Livermore, Ca 94550, USA

ABSTRACT.

We review recent advances in the theory of dense fluids and of the
application of these methods to the study of shock compressed liquids.

I.   INTRODUCTION

In the present article we examine the properties of liquids and fluids
at high pressure and temperature with special emphasis on theoretical
and shockwave studies. The extremely high pressures and temperatures
generated in shockwave experiments on liquids provides the theorist
with a unique opportunity to study intermolecular forces, electronic
and chemical properties, and melting at extreme conditions.

In a shockwave experiment one measures the shock and particle
velocities and from these the pressure and density of the final state
can be obtained directly. For some materials it is also possible to
measure temperature and optical properties. But for the most part
detailed information about atomic and molecular processes must come
from theoretical studies. It is the responsibility of the theoretician
to interpret the experimental results in terms of microscopic physics
and to extract additional information and insight.

This article is organized into three main sections. Section 2 is
a review of some of the theoretical and computational methods that are
useful in the study of dense fluids. In section 3 we introduce the
fundamental relations of shock physics. Section 4 will be devoted to
examining the application of liquid theory to the interpretation of
shock data.

2.   THEORY OF FLUIDS AT HIGH PRESSURE

The most important advances in the theory of fluids have come from
computer simulations and the development of computationally fast
approximate models. For a number of well-defined potentials,

9

R. van Eldik and J. Jonas (eds.), High Pressure Chemistry and Biochemistry, 9–49.
© 1987 by D. Reidel Publishing Company.

simulations carried out by Monte Carlo and molecular dynamics methods
have been used to determine the pair distribution and thermodynamic
and transport properties over a wide range of conditions.   An
extensive set of data exists for the hard sphere, Lennard-Jones and
inverse-power potentials.

Computer simulations should be viewed as experiments producing
exact data which complement laboratory experiments. For the case of
strongly coupled degenerate plasmas with densities comparable to those
found in high-energy astrophysics and laser-fusion-compression
experiments, computer simulations are the only 'experimental' data.
Computer simulation results have been extremely valuable in the
development and testing of approximate methods such as variational
fluid theory and the hypernetted chain method. These methods are now
'work horses' for applications to simple fluids and dense plasmas.
For a detailed review of liquid theory, computer methods and results,
see Barker and Henderson[1] and Hansen and McDonald.[2] Although
computer simulations provide a sound basis for statistical theories,
they typically deal with idealized systems. The physically interesting
atomic and electronic processes must be introduced separately with
approximate models.

But in addition to accurate statistical mechanical theories the
determination of real fluid properties requires a detailed description
of the intermolecular forces. From a rigorous point of view this part
of the problem is in the least satisfactory state. Almost all useful
intermolecular potentials are at least semiempirical having been
obtained in some way from experimental information.

In this section we first discuss the computer simulations which
provide "exact" results but which are slow and expensive and then
examine several approximate methods which are, computationally fast
and have been tested for accuracy against the computer simulations.
We close this section with some remarks regarding intermolecular
potentials.

2.1 Computer Simulations

The two principal computer simulation methods are Monte Carlo and
molecular dynamics. In the Monte Carlo method, random sampling
techniques are used in which particles are moved and an ensemble of
possible configurations is generated. Thermodynamic properties are
obtained by averaging the properties of these configurations. In
molecular dynamics the ensemble of configurations is obtained by
directly integrating the complete set of Newton's equations of
motion. Despite these fundamental differences the methods have
several common features. In both, N molecules are confined to a box
of volume V at temperature T. Computer size and speed limits N to
several hundred particles. To minimize surface effects periodic
boundary conditions are employed. This consists of filling three
dimensional space by replications of the original cell in which the
molecules in each cell occupy the same relative positions.

The initial or starting configuration may be ordered or random.
The potential energy or force exerted on a particle is computed by

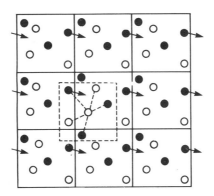

Fig 2.1. A two dimensional periodic system. Each particle in the central box interacts with its nearest neighbors in the same or in a neighboring cell as defined by the dashed lines. If a particle moves outside the central cell, then its counterpart in an adjacent cell moves in through the opposite boundary. One can assess the validity of the periodic boundary conditions by repeating the simulations using different numbers of particles.

summing over the interactions with nearest neighbors in the same and adjoining cells.

In the Monte Carlo method a particle, chosen at random, is moved from position (x,y,z) to $(x, + \delta x, y + \delta y, z + \delta z)$ by randomly choosing δx, δy, and δz. The change in potential energy δU is computed and if the move results in a decrease in energy the new configuration is accepted and its properties are included in the ensemble averaging. If the energy increases the computer selects a random number between 0 and 1 and compares it with $\exp(-\delta U/kT)$. The move is allowed if the random number is the smaller, otherwise the move is rejected and the previous configuration counted again. The properties of the system are obtained by taking the ensemble average over all configurations.

In the molecular dynamics technique the velocities and positions for all N interacting particles are determined by solving Newtons equations. The molecular dynamics technique has the advantage that it can be used for studying non-equilibrium processes provided the characteristic relaxation times are significantly shorter than the computer time ($\sim 10^{-11}$ seconds). The equilibrium properties determined by Monte Carlo and molecular dynamics are in excellent agreement. The free energy and entropy can be determined for a state of interest by a thermodynamic integration along a path connecting to one in which the free energy is known. For example by integrating from the gas along a reversible path to the liquid.

Constant pressure molecular dynamics and Monte Carlo simulations as opposed to the constant volume methods described above, have been increasingly used to study high pressure solid-solid phase transitions.[3] The new method allows the system to change both its

volume and shape in response to an imbalance in the forces and permits
the detailed study of the mechanics of a transition.

Some Monte Carlo Results

The first molecular dynamics calculations were made for hard-sphere
systems. This potential is

$$\phi(r) = \infty, \ r \leq d,$$

and

$$\phi(r) = 0, \ r > d. \tag{2.1}$$

where d is the hard sphere diameter. Thus the dynamics breaks down
into a series of "billiard ball" collisions following which molecules
move in straight lines. The pressure data in the fluid (Fig. 2.2) may
be expressed as

$$\frac{\beta P}{\rho} = (1 + \eta + \eta^2 - \eta^3) / (1 - \eta)^3 \quad , \tag{2.2}$$

where $\eta = (\pi/6)\rho d^3$, $\beta = 1/k_B T$, k_B is the Boltzmann's constant and ρ
is the number of particles per unit volume. η is the hard sphere
packing fraction. The excess Helmholtz free energy per molecule, A,
can be obtained from

$$\frac{\beta A}{N} = \int_0^\rho \{\frac{\beta P}{\rho'} - 1\} \frac{d\rho'}{\rho'} + \log \rho - 1 \tag{2.3}$$

which leads to

$$\frac{A_{HS}}{NkT} = \frac{4\eta - 3\eta^2}{(1 - \eta)^2} \tag{2.4}$$

This expression is referred to as the Carnahan-Starling[4] hard sphere

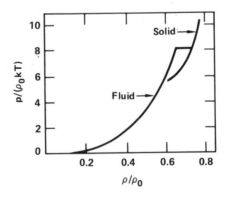

Fig. 2.2. Equation of state of hard sphere system. The quantity
ρ_0 is the close-packed density.

free energy. Although the hard sphere system is an idealized model it is a very useful prototype and widely used as a reference system for fluids. More realistic systems have softer repulsions. Extensive computer results also exist for systems fo particles interacting through the softer inverse power potential, $\phi(r) = \varepsilon(r^*/r)^n$ for n = 12, 9, 6, 4 and 1 and the more realistic Lennard–Jones (LJ) potential, $\phi(r) = \varepsilon[(r^*/r)^{12} - 2(r^*/r)^6]$. Simple analytic expressions have been fitted to these "experimental" data which permit rapid evaluation of thermodynamic properties.[5]

2.2 Fluid Variational Theory

A number of approximate methods have been developed for calculating liquid properties. They have invariably been derived from a perturbation theory in which the properties of the reference system is well known. The most widely used reference system is the hard sphere fluid. At high pressures many of these standard perturbation theories have serious short comings. In this section we limit ourselves to those which have proven useful at high pressures as validated by testing against computer simulations and successful applications to the study of dense liquids. The most useful treatment of high pressure liquids has been with variational theory.

In the variational formulation of perturbation theory the Helmholtz free energy A of an interacting system is approximated by the use of the Gibbs–Bogolyubov inequality:

$$A \leq A_0 + \langle \phi - \phi_0 \rangle_0 \quad . \tag{2.5}$$

This equation states that the free energy of the interacting system is bounded above by the free energy of a reference system (A_0) plus the difference in potential energy between the actual system and the reference, averaged over all configurations of the reference system.

The reference system of virtually unanimous choice has been the hard sphere. This system represents the limiting case of particles interacting by a repulsive potential and has been well characterized by the systematic molecular dynamic computer studies. Analytic expressions for $A_{HS}(d,\rho)$ and the pair distribution function $g_{HS}(r/d,\rho)$ have been obtained, as a function of ρ and d from computer data. Closed-form analytic approximations to g_{HS} have also been derived from solutions of the Percus–Yevick equations. For more details see reference 1. Consequently, the problem becomes to a large measure how to choose the hard-sphere diameter. It is chiefly the manner in which this choice is made that distinguishes the various perturbation theories.

Mansoori and Canfield were the first to use hard spheres as a reference system in a variational theory. The excess Helmholtz free energy of the fluid reduces to

$$A \leq A_{HS} + \frac{N}{2} \rho \int_0^\infty g_{HS}(r, \eta) \, \phi(r) d^3r \quad , \tag{2.6}$$

where A_{HS} and g_{HS} are the free energy and pair distribution

function of the reference system. Whereas hard spheres have many
attractive features, they incorporate a repulsion that tends to be
unrealistically severe at high density and hence limits their accuracy.

Soft Sphere Variational Theory[6]

A more realistic system is the inverse 12th-power or soft sphere (SS)
reference potential. For this system, the excess Helmholtz free
energy may be written as the inequality

$$A \leq A_{12}(\lambda) + \frac{N}{2} \rho \int_0^\infty g_{12}(r, \lambda) \left[\phi(r) - \phi_{12}(r)\right] d^3r \qquad (2.7)$$

where λ is some characteristic parameter of the reference system which
minimizes the right side of Eq. 2.7. Judging from the usefulness of
the hard-sphere packing fraction η as a universal fluid scaling
parameter, it is tempting to relate the parameter λ to η and express
the reference-state properties as functions of η. We may approximate
A_{12} in terms of hard sphere theory by using the equation

$$A_{12} \leq A_{HS} + \frac{N}{2} \rho \int_d^\infty g_{PY}(r, \eta) \phi_{12}(r) d^3r \qquad . \qquad (2.8)$$

where g_{PY} is the Percus-Yevick approximation to the hard sphere pair
distribution function.

Listed in Table 2.1 are Helmholtz free energies for the inverse
12th-power system as determined from the Monte Carlo calculations of
Hoover et al.[7] and Hansen.[8] These are shown in columns 2 and 3,

TABLE 2.1. Comparison of inverse 12th-power excess
Helmholtz free energies from Monte Carlo calculations (MC),
fitted with soft sphere theory using Eq. (2.8), and from
hard sphere theory using Eq. 2.9.

$\rho(\epsilon/kT)^{1/4}$	MC[a]	MC[b]	Eq. (2.9)	Eq. (2.8)
0.1	0.40	0.40	0 40	0.44
0.2	0.91	0.91	0.90	0.99
0.3	1.53	1.53	1.52	1.67
0.4	2.32	2.33	2.32	2.53
0.5	3.33	3.34	3.32	3.60
0.6	4.60	4.61	4.60	4.93
0.7	6.20	6.21	6.20	6.60
0.8	8.19	8.21	8.20	8.67

[a]   Results from Hoover et al., Ref. 7.
[b]   Results from Hansen, Ref. 8.

respectively. In column 5 are results computed by the hard sphere variational method using Eq. (2.8). The agreement is of the order of 8%. We now determine a function of η which, when added to the RHS of Eq. (2.8) and included in the variational procedure, will result in computed properties that agree exactly with the Monte Carlo results. $F_{12}(\eta) = - (\eta^4/2 + \eta^2 + \eta/2)$ is such a function. The free energy calculated with Eq. 2.9 is shown in column 4 of Table 2.1. Thus the free-energy function which reproduces the Monte Carlo inverse 12th-power results exactly in terms of hard-sphere parameters, is

$$A_{12} \leq A_{HS} + \frac{N}{2} \rho \int_d^\infty g_{PY}(r, \eta) \, \phi_{12}(r) \, d^3r + F_{12}(\eta) \, NkT \quad . \quad (2.9)$$

We now replace $A_{12}(\lambda)$ in Eq. (2.7) by $A_{12}(\eta)$, and the full expression for A becomes

$$A \leq A_{HS} + \frac{N}{2} \rho \int_d^\infty g_{PY}(r, \eta) \, \phi_{12}(r) \, d^3r$$

$$+ F_{12}(\eta) \, NkT + \frac{N}{2} \rho \int_0^\infty g_{12}(r, \lambda) \, [\phi(r) - \phi_{12}(r)] \, d^3r$$

$$(2.10)$$

It contains terms with $g_{12}(r, \lambda)$ and $g_{PY}(r, \eta)$. We approximate g_{12} with g_{PY} and by combining the second and fourth terms, Eq. (2.10) is rewritten as

$$A \leq A_{HS} + \frac{\rho N}{2} \int_d^\infty g_{PY}(r, \eta) \, \phi(r) \, d^3r + F_{12}(\eta) \, NkT, \quad (2.11)$$

This equation is formally identical to the original hard-sphere variational formulation except for the additional term, $F_{12}(\eta)$. This term modifies the hard sphere system in such a manner as to convert it to an inverse twelfth power reference preserving the usefulness of η and the analyticity of g_{HS}. The η is chosen to have the value that minimizes the Helmholtz free energy. The pressure (P) and excess internal energy (U) are obtained by taking the derivatives of A:

$$\beta P/\rho = 1 + \rho \left(\frac{\partial \beta A}{\partial \rho} \right)_T \quad , \quad (2.12)$$

$$\beta U = \beta \left(\frac{\partial \beta A}{\partial \beta} \right)_\rho \quad . \quad (2.13)$$

Young and Rogers[9] have constructed a thermodynamically consistent inverse 12th power fluid variational theory (I12) which uses an accurate fit to the reference fluid free energy and tabulated accurate reference fluid radial distribution functions computed from a new integral equation. In reduced variables, where $x = r/a$, $a = (3/4\pi\rho)^{1/3}$

Table 2.2. Comparison of theoretical (I12 and SS) and simulation (EX) reduced total pressures and reduced excess internal energies for the Lennard-Jones fluid.[9] $T^* = kT/\varepsilon$ and $\rho^* = \rho\, r^{*3}/\sqrt{2}$.

			PV/NkT			U/NkT	
T^*	ρ^*	EX	I12	SS	EX	I12	SS
100	0.2	1.221	1.225	1.216	0.036	0.037	0.034
	0.5	1.675	1.682	1.669	0.115	0.117	0.111
	1.00	2.95	2.98	2.96	0.361	0.366	0.358
	2.0	9.50	9.51	9.57	1.767	1.772	1.779
	2.5	16.29	16.45	16.55	3.304	3.350	3.365
20	0.2	1.270	1.272	1.258	-0.005	-0.002	-0.010
	0.5	1.930	1.947	1.922	0.026	0.034	0.018
	1.333	7.999	8.031	8.089	0.942	0.960	0.958
	1.765	16.68	16.66	16.75	2.65	2.65	2.66
5	0.2	1.169	1.177	1.149	-0.202	-0.178	-0.200
	0.5	1.867	1.859	1.822	-0.474	-0.448	-0.488
	1.00	6.336	6.437	6.491	-0.456	-0.422	-0.448
	1.279	13.44	13.37	13.46	0.435	0.430	0.413
2.74	0.1	0.97	1.00	0.97	-0.223	-0.174	-0.199
	0.3	1.04	1.07	1.01	-0.650	-0.580	-0.630
	0.55	1.65	1.62	1.60	-1.172	-1.122	-1.181
	1.00	7.37	7.27	7.37	-1.525	-1.525	-1.559
	1.10	10.17	10.23	10.30	-1.351	-1.313	-1.352
1.35	0.1	0.72	0.76	0.75	-0.578	-0.383	-0.423
	0.3	0.35	0.35	0.27	-1.548	-1.284	-1.374
	0.5	0.30	0.18	0.14	-2.496	-2.327	-2.430
	0.9	4.58	4.52	4.60	-4.192	-4.144	-4.220
	0.95	6.32	6.00	6.08	-4.230	-4.239	-4.311
0.810	0.801	0.057	-0.058	0.073	-7.068	-6.963	-7.074
	0.8839	1.946	2.106	2.122	-7.707	-7.580	-7.687
0.72	0.8350	-0.080	-0.182	0.019	-8.400	-8.285	-8.391
	0.9158	2.248	2.614	2.687	-9.079	-8.909	-9.031

and $z = (Nr^{*3}/\sqrt{2}\, V)(\varepsilon/kT)^{1/4}$, the variational inequality becomes

$$\frac{A}{NkT} \leq \frac{A_{12}(z)}{NkT} + \frac{3}{2} \int_0^\infty \left[\frac{\phi(r)}{kT} - \frac{\Gamma}{x^{12}} \right] g_{12}(x, z)\, x^2\, dx, \qquad (2.14)$$

where $\Gamma = 4\,\sqrt{2}\pi z/3)$ is the variational parameter used to minimize the right-hand side of Eq. (2.14). Values of $g_{12}(x, z)$ for arbitrary z are determined by using cubic spline interpolation in z space. Pressure and energy are then determined by numerical differentiation of the free energy with respect to volume and temperature, respectively.

RESULTS OF CALCULATIONS

Calculations for the thermodynamic properties of the fluid theories
were carried out and the results were compared with computer
experiments for two potentials--the Lennard-Jones (LJ)

$$\phi(r) = \epsilon \left[\left(\frac{r^\star}{r} \right)^{12} - 2 \left(\frac{r^\star}{r} \right)^6 \right] , \qquad (2.15)$$

and the exponential-six (exp-6)

$$\phi(r) = \epsilon \left[\left(\frac{6}{\alpha - 6} \right) \exp[\alpha(1 - r/r^\star)] - \left(\frac{\alpha}{\alpha - 6} \right) \left(\frac{r^\star}{r} \right)^6 \right] . \qquad (2.16)$$

The results for the Lennard-Jones potential are shown in Table
2.2 and are compared with the Monte Carlo simulations here. Similar
agreement was found using the exponential-six using $\alpha = 13.5$.[9]
The results show that the soft sphere and inverse-12th
variational theories are in good overall agreement with computer
simulations of various model fluids. The agreement between theory and
simulation is excellent at high pressure near the melting curve. SS
theory, which is really an approximation to the I12 theory, useful as
a practical method for calculating fluid properties. An advantage of
the SS theory is the use of an analytic $g_{HS}(r, \eta)$ subroutine,
which avoids the problem of interpolation errors with tabular
functions. For most applications the marginal improvement of I12
theory may not justify the additional computational complexity.
Although perturbation theories are fast and convenient and give
good results for thermodynamic properties they do not predict accurate
structural properties. The integral equations such as the modified
hypernetted chain (HNC) equation are more complicated computationally
but as a result of several recent advances they can now be made to
yield thermodynamic and structural properties in excellent agreement
with computer simulations. Integral equations are also more directly
applicable to mixtures and to ionic systems than are perturbation
theories. A complete discussion the integral equations are beyond the
scope of this review, but a good comparison of this method and
variational has been reported by Talbot et al.[10]

2.3 Intermolecular Potentials at High Density.

Among the least satisfactory features of liquid physics has been the
difficulty in obtaining accurate intermolecular forces from first
principles. Typically the forces calculated between pairs of
molecules are too stiff when used at high densities because they
neglect many-body interactions. The case of H_2 illustrates this well.
Figure 2.3 compares the H_2-H_2 pair potential determined from
quantum-mechanical calculations with that derived from shock data.
For many dense liquids a useful semiempirical potential is the
exp-six (Eq. 2.16), where α, $r^\star$ and ϵ are adjustable parameters,
typically obtained by fitting to experimental data, which determine

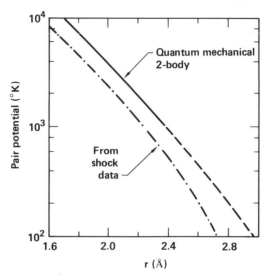

Fig. 2.3. Many-body effects in H_2-H_2 potential. Comparison of
the hydrogen pair potential obtained from ab initio H_2-H_2
calculations with one obtained from shock data.

the repulsive stiffness, the position of the potential minimum and its
depth. The exponential term is a more realistic characterization of
interatomic repulsive forces than the inverse twelfth power
(Lennard-Jones) potential. At very small separations the attraction
may become unrealistically large. This can be avoided by multiplying
the attractive term by a damping function.

A good example of such a procedure is to be found in the widely
used Aziz potential for helium.[11] The form of the potential is

$$V(r) = \epsilon V^\star(x)$$

$$V^\star(x) = A \exp(-\alpha x) - \left(\frac{C_6}{x^6} + \frac{C_8}{x^8} + \frac{C_{10}}{x^{10}} \right) F(x) \quad , \qquad (2.17)$$

where

$$F(x) = \exp\left[- \left(\frac{D}{x} - 1 \right)^2 \right] \text{ for } x < D$$

$$= 1 \qquad\qquad\qquad \text{ for } x \geq D$$

and $D = 1.28$ and $x = r/r_m$.

The function $F(x)$, which is considered to be universal for all
spherical systems, was obtained by fitting a potential of this form to
the accurately known potential curve of the $^3\Sigma_u^+$ state of H_2. This
fixed the value of parameter D.

The input data for this potential are the SCF HF calculations of the repulsive interaction, C_6, C_8 and C_{10} are the multipole dispersion coefficients. The parameters ϵ/k = 10.8 K and r_m = 2.98 Å result from constraints on the value of the reduced potential and its slope at its minimum. The value of α may be determined from scattering data or fitted to Hartree Fock results at small r.

A considerable simplification in the choice of potential parameters for shock-compressed molecular fluids has emerged from the recognition that the repulsive pair potentials approximately obey the corresponding-states scaling principle.[12] That is, for two fluids the ratios of the depths of their potential wells (ϵ) are in the same ratio as their critical temperatures (T), and the cubes of their characteristic length scales (r*) are in proportion to their critical volumes (V_c). As a result of having fitted parameters to the Ar potential, it has been possible to predict within experimental error the Hugoniot curves for most molecular liquids. This means that the repulsive forces which dominate the high-temperature properties have similar functional dependencies for most fluids. An exception is hydrogen, which has a much softer repulsive force than other closed-shell molecules.

3.   SHOCK COMPRESSION

A shock wave is a disturbance propagating at supersonic speed in a material, preceded by an extremely rapid rise in pressure, density and temperature. The general reader often associates shock waves with explosions and other uncontrolled and irreversible processes. Although shock waves are irreversible, the process is well understood and can be controlled to produce a desired response. Shock compression studies obtain high pressures by introducing a rapid impulse through the detonation of a high explosive, the impact of a high-speed projectile or the absorption of an intense pulse of radiation. A shock wave not sustained loses energy through viscous dissipation and reduces to a sound wave (e.g., thunder). High-speed optical and electronic methods are necessary to measure certain dynamic variables which determine pressure, density and energy. In shock-wave experiments, the passage time of the shock is short compared ot the disassembly or 'fly away' time of the sample. As a result the attainable pressures for a given material are limited only by the energy density supplied by the driver. Chemical explosives have been used to obtain pressures up to 1 Mbar in liquids and up to 13 Mbar in metals, with accompanying temperatures of tens of thousands of degrees Kelvin. Final pressures ranging from 20-158 Mbar have been reported using underground nuclear explosions. In inertial confinement studies, pellets of liquid deuterium are subjected briefly to laser-driven dynamic pressures of about 1000 Mbar and temperatures in excess of 10^7 K. An excellent introduction to shock wave physics is to be found in Zeldovich and Raizer.[13]

3.1 Introduction to the Dynamics of Shockwaves

Consider a fluid or solid at rest with constant density and pressure, ρ_0, Po bounded on the left by a plane piston in a cylinder of area A.

Let us assume that at t_0 the piston is set in motion with a constant velocity u_p. This motion compresses the material before it and a disturbance with velocity propagates in the direction. The first infinitesimal compression at the piston face results in the propagation of a sound wave with velocity C. However, subsequent compressions at the piston face take place with the material at higher densities and result in higher local sound speeds. This produces a train of waves in which the first is at the speed of sound in the undisturbed material and the last, closest to the piston face, is supersonic. Because the last wave can catch but not pass the first, all the waves eventually coalesce into a single, steep steady wave front across which exists a sharp discontinuity in pressure, density, and temperature. The width of the discontinuity is generally a few molecular mean-free-path lengths.

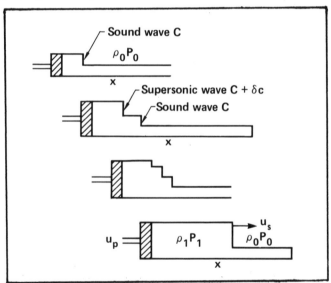

Fig. 3.1. Formation of a one-dimensional planar shock wave. Shown by pressure-density-distance p-x plots at successive times.

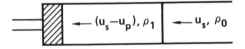

Fig. 3.2. Steady shock wave as viewed by an observer on the shock front. The fluid streams toward the shock front at a velocity u_s and rushes to the left away at the velocity u_s-u_p.

We can now apply the general laws of conservation of mass, momentum and energy to determine the pressure, density and velocity of the disturbance.

A useful coordinate system is shown in Fig. 3.2 in which the observer is moving with the shockfront.

During a period δt a mass of $\rho_0 u_s A \delta t$ passes through the shockfront. Conservation of mass requires that

$$\rho_0 u_s A \delta t = \rho_1 (u_s - u_p) A \delta t$$

$$\rho_0 u_s = \rho_1 (u_s - u_p) \tag{3.1}$$

This leads to an expression for the density change

$$\frac{\rho_0}{\rho_1} = 1 - \frac{u_p}{u_s} \quad . \tag{3.2}$$

The pressure jump across the shock may be determined from the conservation of momentum. To the observer the momentum flow from the unshocked fluid into the shockfront is

$$(\rho_0 u_s A \delta t) \, u_s \quad .$$

The momentum flow away from the shockfront is

$$\rho_1 (u_s - u_p) A \delta t \, (u_s - u_p) \quad ,$$

or using Eq. 3.1

$$\rho_0 u_s A \delta t \, (u_s - u_p) \quad .$$

The change in momentum must equal the difference in forces across the front $(P_1 - P_0) A \delta t$. Thus conservation of momentum leads to an expression for the pressure change.

$$P_1 - P_0 = \rho_0 u_s u_p \quad . \tag{3.3}$$

Similarly the law of conservation of energy.

$$\rho_1 (u_s - u_p) \left[E_1 + \frac{P_1}{\rho_1} + \frac{(u_s - u_p)^2}{2} \right] = \rho_0 u_s \left[E_0 + \frac{P_0}{\rho_0} + \frac{u_s^2}{2} \right] .$$

leads to the energy equation.

$$E_1 - E_0 = \frac{1}{2} (P_1 + P_0) (V_0 - V) \quad , \tag{3.4}$$

in which we introduced the specific volumes $V_0 = 1/\rho_0$ and $V_1 = 1/\rho_1$ in place of density. Equations 3.2–3.4 are referred to as the Hugoniot equations.

The derivation of these equations considers the shockfront as a

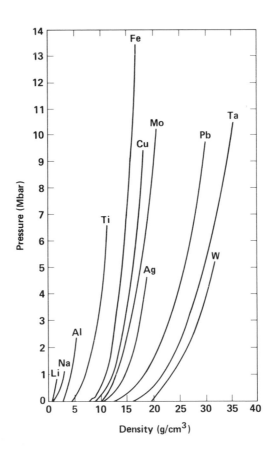

Fig. 3.3. Experimental Hugoniots for metals.

discontinuity but says nothing of its width. Only the equilibrium
properties on either side are considered in the flow properties.

The Hugoniot equations represent the locus of all final states
that can be reached by shock-compressing a material from a given
initial state; the resultant curve of pressure against volume is known
as a Hugoniot. If the initial state is known, then the final-state
properties may be determined from a measurement of any two of these
five properties (u_s, u_p, P, ρ, E). Of these, the shock and particle
velocities are the most commonly measured. A discussion of
experimental methods is beyond the scope of the present review. For
the interested reader a complete description of the shock wave
techniques for liquids has been given by Nellis and Mitchell.[14]

Shock-wave measurements have been carried out on most elements on
a very large number of compounds. The most extensive single source of
all published Hugoniot data up to 1977 is the LLNL Compendium of Shock
Wave Data by van Thiel et al.[15] Marsh has made available all of the

Los Alamos data including previously unreported measurements.[16] These
two collections include data for elements, alloys, rocks, minerals and
compounds, plastics and synthetics, woods, assorted liquids, aqueous
solutions and high explosives. Al'tshuler et al. have reported a
critical analysis of shock measurements for most of the metals.[17]
     The assumptions made in deriving the Hugoniot equations were
one-dimensional steady motion, thermodynamic equilibrium immediately
ahead of an behind the shock front, and negligible material strength.
The first assumption can be met experimentally. The second will hold
if the relaxation times are much less than the resolution of the
detectors ~ 10 ns. The third is justified for pressures appreciably
greater than the yield strength of the material. Typical yield
strengths in metals are 20 kbar and much less for most materials. In
the case of liquids, which have no yield strength, this creates no
problem. The temperature cannot be obtained from the Hugoniot
equations, but must be derived from an equation of state. In a few
cases it has been determined from a direct measurement of the emitted
radiation.
     Several equations of general interest follow from the
conservation laws. In place of energy we introduce the enthalpy $H = E + PV$. We can write the Hugoniot equation in the equivalent form

$$H_1 - H_0 = \frac{1}{2} (P_1 - P_0) (V_0 + V_1) \quad . \tag{3.5}$$

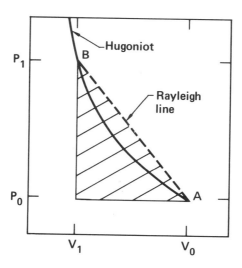

Fig. 3.4. The crosshatched area represents the piston kinetic energy
and is the total energy transmitted by the piston to a unit mass
initially at rest.

     Eqs: 3.2 and 3.3 can be combined to obtain explicit expressions
for the shock velocity

$$u_s^2 = V_o^2 \left(\frac{P_1 - P_o}{V_1 - V_o} \right) \tag{3.6}$$

and particle velocity

$$u_p^2 = (V_o - V_1)(P_1 - P_o) \tag{3.7}$$

These expressions are particularly useful for clarifying the relationship between shockwaves and thermodynamic properties. Consider the dashed line AB in Fig. 3.4 connecting the initial Hugoniot point $P_o V_o$ to P_1, V_1. The slope $(P_1 - P_o)/(V_1 - V_o)$, referred to as the Rayleigh line, is related to the square of the shock velocity according to Eq. 3.6.

If the material is initially at rest the kinetic energy acquired by compression of the piston

$$\frac{u_p^2}{2} = \frac{1}{2}(P_1 - P_o)(V_o - V_1) \tag{3.8}$$

is equal to the crosshatched area. If $P_1 \gg P_o$ then from Eq. 3.4 the area is approximately equal to the increase in energy,

$$E_1 - E_o = \frac{1}{2} P_1 (V_o - V_1)$$

and

$$E_1 - E_o = \frac{u_p^2}{2} \quad . \tag{3.9}$$

It can be shown that in the weak limit $(P_1 \to P_o)$ shock compression is isentropic. Hence Eq. 3.6 reduces to

$$u_s = V_o \sqrt{\frac{P_1 - P_o}{V_1 - V_o}} \to V_o \sqrt{-\left(\frac{\partial P}{\partial V}\right)_S} = C \quad ,$$

and the initial shock velocity approaches the sound speed (C).

The entropy of the compressed fluid increases as the shock pressure increases. This results in a large increase in temperature and demonstrates the irreversible nature of the shock process. Figure 3.5 compares the calculated temperatures and pressures for isentropically and shock compressed liquid hydrogen.

3.2 Structure of a Shock Front in a Liquid

In the derivation of the Hugoniot equations the shock front has been treated as a discontinuous surface. In fact this region must be likened to a transition zone in which molecules are rushing in with a

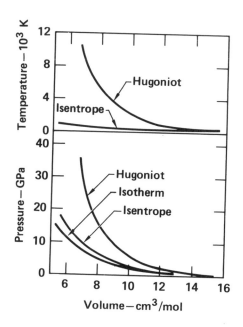

Fig. 3.5. The Hugoniot, isotherm and isentrope of molecular hydrogen
illustrate the large pressure and temperature rise that accompanies
shock compression. Liquid hydrogen V_0 = 28.4 cm^3/mole.

Boltzmann distribution at the initial temperature and rushing out at a
Boltzmann distribution at the shock temperature. The passage is
irreversible and accompanied by a large entropy change. Through this
transition zone there occur dissipative processes associated with
viscosity and thermal conductivity. Experimental methods are not
capable of directly probing the structure of shock front. Thus to
properly describe the microscopic nature of shock compression it is
necessary to study the process using molecular dynamics.

Klimenko and Dremin[18] started with a 2592 molecule of equilibrium
liquid argon initially at V_0 = 36cm^3/mole, T_0 = 131 K, and P_0 = 145
atm. The atoms interact through a Lennard Jones potential. To
produce a shockwave a layer of particles at one end of the cell is set
in motion with a velocity u_p in the x-direction. This piston
compresses material before it and propagates a shock wave down the
cell. After a time interval the process reaches a steady motion and
the density temperature and pressure profiles shown by the solid
curves in Fig. 3.6 were calculated.

The material is compressed rapidly to a width of about 10 Å (3
atomic diameters) within a time of about 10^{-13} sec. This time is
comparable to the translational relaxation time and prevents complete
equilibrium in the front. This results in significant overheating of
the temperature T_x in the direction of propagation (not shown). T_z
(shown) is the temperature parallel to the front.

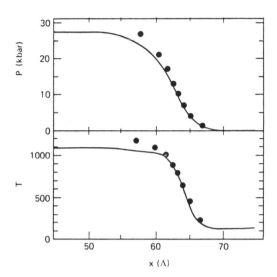

Fig. 3.6. Pressure and temperature profiles argon taken from Refs. 18 and 19. The full curves are the results from atomistic molecular dynamics. The dots are the solution of the Navier-Stokes equations. Pressure in kilobars, and temperatures in Kelvins are given as functions of distance (angstroms). The full curves include, from the Navier-Stokes viewpoint, the bulk-viscosity contribution to the mean pressure. The data correspond to a shock velocity of 2.6 km/sec.

In contrast to molecular dynamics, continuum mechanics ignores atomic structure and describes material properties in terms of continuous gradients of mass momentum and energy. Properties such as viscosity and thermal conductivity are assumed rather than derived and are determined from gradients of the velocity and temperature. Macroscopic theory should apply best to conditions where gradients are small. Hoover[19] and coworkers[20] have shown that despite these apparent limitations calculations of the shock profile using the continuum Navier-Stokes equations agree well with those of molecular dynamics simulations.

Just why the agreement is as good as it is over such a wide range of fluid conditions has not been explained.[20] Theoretical efforts to go beyond the Navier-Stokes level have not been completed yet.

4.   SHOCK COMPRESSION OF SIMPLE LIQUIDS

Theoretical and experimental studies have been carried out on a wide range of shock compressed liquids. Space limitations require us to restrict the discussion to a few materials which best illustrate some particularly unique or interesting physics. Because of their simplicity it is appropriate to begin with a discussion of the rare

gas liquids. We follow these with discussions of successively more
complicated molecules.

4.1 Argon and Xenon

The static diamond cell and shockwave techniques for studying the
properties of matter at very high pressure have so little in common
that it may not be apparent to the nonspecialist that the two methods
can be used to generate equation of state data that are directly
comparable. Recently diamond-window cell measurements for argon have
been pursued to 800 kbar at 298 K. These measurements provide an
excellent opportunity to study their intermolecular forces and to
compare shock and static methods for a simple system.[21]

A Comparison of Static and Shock Measurements

In the diamond-window static high-pressure cell the faces of two
opposing diamond anvils are squeezed together and isothermally
compress a sample of a few microns in size to megabar pressures. The
sample density is determined from a direct x-ray measurement of the
unit cell volume. Pressure may be determined routinely to high
accuracy by measuring the density of a secondary standard such as NaCl
or by recording the shift of the R_1 ruby fluorescence line. The
pressure dependence of the ruby line has been calibrated against
isotherms reduced from shock data.
    In contrast to static methods shockwave experiments create high
pressures by introducing a rapid impulse into a sample by the impact
of a high speed projectile. The pressure, energy and compression are
determined directly from the measurement of the shock and particle
velocity and a knowledge of the initial conditions. Determination of
the shock pressure is absolute and does not require a secondary
standard. This feature has made shock data valuable for independent
static pressure calibrations. In the case of liquid argon the final
temperatures are on the order of several tens of thousands of
degrees. The very different nature of the final states makes any
direct comparison heavily reliant on theoretical methods. In this
regard argon has several useful features. It is a relatively simple
inert gas atom with a large electron band gap. The forces between
atoms can be approximated by an effective pair-wise additive
interactions so that established methods in statistical mechanics
provide a convenient means for calculating thermodynamic properties.
    Shock compression experiments on liquid argon have been carried
out by several workers.[14,22] Nellis and Mitchell[14] using the two stage
light-gas gun at Livermore, achieved final pressures up to 900 kbar
and temperatures calculated to be approximately 30000 K. Recently
Grigoriev et al.,[23] have reported measurements of a liquid argon
Hugoniot to 670 kbar and the first shock temperature measurements on
this substance. In the range above 400 kbar the temperature is over
13000 K and is sufficiently high to cause an appreciable degree of
electronic excitation. Below this pressure, argon behaves as an
insulator and the only contribution to the thermodynamic properties

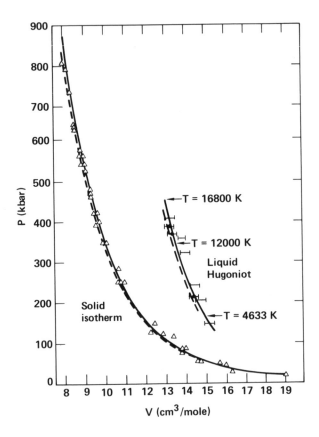

Fig. 4.1. A comparison of static (triangles), shock-wave (bars) and
theoretical (curves) results. Dotted bars (|–●–|) are shock data
from Ref. 14 and bars (|––|) are from Ref. 23. Solid curves were
calculated using the exp-6 with α = 13.2 and dashed curves with α =
13.0. The initial conditions of the shock-compressed liquid are at 87
K and 28.64 cm³/mol.

are from atom-atom interactions and atomic motion. It is these data,
below 400 kbar and uncomplicated by thermal electron excitation, that
provide a determination of the interatomic potential and a direct
comparison with the room temperature isotherm.

Monte Carlo calculations of the pressure and energy were made
using the exponential-six potential (Eq. 2.16) with parameters
ε/k=122K, r^* =3.85 Å and using two different values of α(13.2 and 13.0).

Starting from a set of initial conditions, a Hugoniot curve is
determined using an equation-of-state model (in this case the Monte
Carlo method) by first choosing a final state volume and then
iterating on the temperature until the calculated values of E and P
satisfy Eq. (3.4). The Hugoniots and isotherms calculated using these

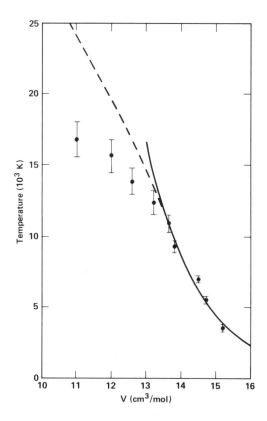

Fig. 4.2. A comparison of the measured and theoretical shock temperatures. The solid curve represents Monte Carlo results (α = 13.2) which do not include electronic excitations. The dashed curve includes electronic excitations. The bars are experimental data from Ref. 23.

potentials are shown in Fig. 4.1. T_0 and V_0 for liquid argon are 87 K and 28.64 cm^3/mol respectively. The best fit to the shock data of Nellis and Mitchell and to the diamond-window cell measurements is given by the exp-6 with an α = 13.2. The curves calculated using α = 13.0 are within the experimental precision but lie on the lower pressure side. The newly reported Soviet measurements[23] are in substantial agreement with the present results although their data is shifted on average to slightly higher pressures. The curves calculated using the two values of α provide a measure of the agreement between two very different sets of experimental data and of the uncertainty to which an effective potential can be determined.

An important feature of the shock process that provides a particularly sensitive probe of the interatomic forces at small separations is the high temperature that is generated along with the

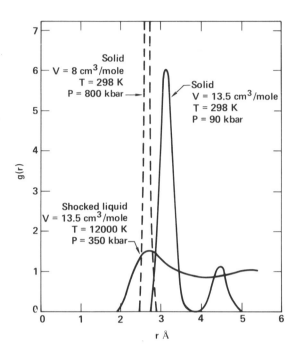

Fig. 4.3. A comparison of the calculated argon pair distribution
function, g(r) for the statically compressed solid and shock
compressed liquid.

high pressure. Grigoriev et al., determined the shock temperatures by
measuring the emitted radiation and comparing these values to a
standard pyrometer. The measured and calculated shock temperatures
are shown in Fig. 4.2. Shown are the Monte Carlo results for α =
13.2 which do not include electronic excitation (solid curve) and the
temperatures from an argon Hugoniot calculation which does include
these excitations (dashed curve). Above 13000 K (400 kbar) the
electronic excitation begins to absorb a significant amount of the
shock energy which leads to a lowering of the temperature. The
measured values fall significantly below those predicted by theory.
The Soviet authors found the same result and have explained this as
due to electron screening of the emitted shock radiation. Below 400
kbar where these effects are small the agreement between theory and
temperature measurements is excellent. In the next section we return
to consider the higher temperature partially ionized argon. The
overall agreement between shock and static data is excellent despite
the appearance of some excessive scatter in the diamond window results
between 100 and 200 kbar. The goodness of the agreement lends
credence to the validity of the ruby scale at these pressures to an
accuracy within 10%.
      The importance of temperature in the study of intermolecular

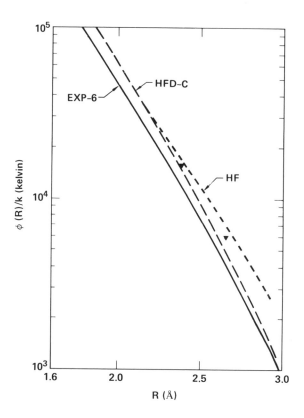

Fig. 4.4. Pair potentials φ(R)/k in units of degrees kelvin plotted versus interatomic separation. Curves are discussed and referenced in the text. HFD-C is a purely two-body potential and exp-6 is the best fit to both static and shock data. The difference between the two curves represents the softening due to many-body interactions. Also shown are Hartree-Fock (HF) results. The triangles represent configuration interaction (CI) results.

forces is well illustrated in Fig. 4.3 which shows the pair distribution in the fluid at 350 kbar and 12000 K, and in the 298 K solid at the same volume (13.5 cc/mole) and at the highest density reached in the diamond cell (8.0 cc/mole). These curves identify the range of separations probed in the various experiments. As expected, the angle-averaged pair distribution of atoms in the solid are sharply peaked at separations corresponding to nearest neighbor shells whereas the fluid distribution is smeared over a wide range of separations. The first maximum in the shocked fluid appears near 2.6 Å which is about the same position as in the solid at 8 cc/mole and 800 kbar. But, the atoms in the fluid have a much higher kinetic energy and as a result undergo collisions at much shorter separations which probe the

potential down to 2 Å. It is this feature that provides the
information needed for extrapolating the isotherm to much higher
compressions.

The exp-6 potential has been characterized as an effective
pair-potential because it includes in a phenomenological fashion the
effects of the many-body interactions while retaining the convenient
features of a two-body function. For this reason it is of interest to
compare the present results to those for a purely two-body potential.
Figure 4.4 compares our best fit, exp-6(α = 13.2), with the two-body
potential of Aziz and Chen[24] (HFD-C, their designation) which they
obtained by fitting to second virial coefficients, gas phase transport
data and to Hartree-Fock atom-atom calculations at small interatomic
separations. Also shown are the Hartree-Fock results of Gilbert and
Wahl[25] and Christianson et al.[26] Two CI calculations of Wadt[27] lie
between the Hartree-Fock and HFD-C curves. The difference observed
between the exp-6 and HFD-C must be attributed to an effective
softening of the short range repulsion by many body interactions.
These results are similar to those shown for hydrogen in Fig. 2.3 in
which the short range potential derived from shock data has also been
found to lie well below the two-body potential.

Electronic Energy Levels and Metallization

At pressures up to 400 kbar electronic thermal excitations in shock
compressed argon are negligible and this substance remains a simple
closed shell insulator fluid whose properties are determined by the
repulsive pair potential. Above this pressure the temperature is
greater than 12000 K and rising exponentially with compression. In
this higher temperature regime the properties are increasingly
dominated by electron thermal excitations from the closed 3p shell
into the 3d conduction band. A proper calculation of the Hugoniot
requires a knowledge of the density dependence of these energy levels.

The Wigner Seitz (WS) method has been used to calculate the

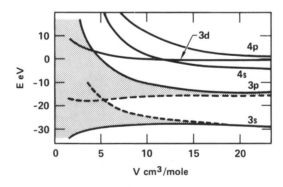

Fig. 4.5. Calculated Wigner-Seitz energy levels of argon with atomic
notations. Solid curves are solutions for k = 0. Dashed curves show
bandwidth based on the estimate of the energy at the maximum k value.

electronic energy levels for compressed argon as a function of
volume.[28] Although less rigorous than more modern methods the results
shown in Fig. 4.5 provide a simple illustration of the changes in
electronic structure with compression. At normal solid density the
lowest the 3d state is filled and the lowest excited state is the
4s-like conduction band. The effect of compression is to raise the
energy of the 3p and 4s relative to the 3d which at high compression
becomes the lowest excited state. The d state wavefunction is
relatively localized and unperturbed by compression. The calculations
predict that the 3d and 4s levels will cross at a volume of 4.5
cm³/mole. At this volume argon should become metallic. McMahan[21] has
carried out extensive electron band theory calculations which predict
argon will undergo a structural transition from fcc to hcp at a
pressure below 2.3 Mbar with a volume change less than 0.1%. The
energy differences between these phases are too small to permit a more
precise prediction. McMahan also predicts that hcp argon will become
metallic at 4.3 Mbar and fcc argon at 5.5 Mbar as the result of a band
gap closure. The transition from fcc to hcp, if it occurs, is not
expected to increase the pressure by more than 0.4% and should be
easily identified.

Hugoniot calculations were made for a model that treats the
electron thermal properties using semiconductor statistics, and
computes the fluid properties with a fluid variational theory
employing the exp-6 interatomic pair potential. The intermolecular
forces are assumed to be unaffected by the electronic excitation. The
total energy and pressure of the fluid are written

$$E(V, T) = E_{ins}(V, T) + \Delta E(V) N_e(V, T) + E_e(V, T), \qquad (4.1)$$

$$P(V, T) = P_{ins}(V,T) - \frac{\partial \Delta E}{\partial V} N_e(V, T) + P_e(V,T), \qquad (4.2)$$

where number of electrons N_e thermally excited to the conduction band
is determined using semiconductor theory.

$$N_e(T, V) = 2(g_v g_c)^{1/2} \left(\frac{2\pi kT}{h^2}\right)^{3/2} (m_v^* m_c^*)^{3/4}$$

$$x \exp\left[-\left(\frac{\Delta E(V)}{2kT}\right)\right] \frac{V}{N} . \qquad (4.3)$$

$\Delta E(V)$ is the volume-dependent gap. The E_e and P_e are the thermal
energy and pressure, respectively, of the free electrons and holes.
The effective masses, m_v^*, are taken to be m_e, the free-electron mass.
The band degeneracies are g_v and g_c. E_{ins} and P_{ins} are the atomic
properties treated as an insulating fluid and computed using the
fluid-variational theory.

Theoretical calculations are compared with experimental data[30] in
Fig. 4.6. The upper curve INS (for insulator) was computed using the
exp-6 pair potential and does not include electronic excitations. The

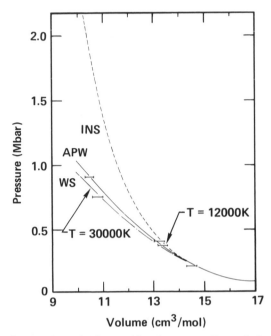

Fig. 4.6. Experimental and theoretical argon Hugoniot data plotted as pressure versus volume. V_0 = 28.6 cm^3/mole. Experimental results are shown with error bars. Theoretical curve labeled INS was computed using an intermolecular potential and does not include electronic excitations. The curves labeled WS and APW refer to Hugoniots calculated using conduction band gaps predicted by the Wigner-Seitz and augmented plane wave (APW) methods.

two lower curves include electronic thermal excitation calculated using an electron band gap. The maximum electronic excitation encountered in argon near 1 Mbar less than 0.12 electrons/atom, out of a possible 6 in the full p shell, at the highest temperature.

By absorbing some of the shock energy, the excited electrons act as thermal sinks, keeping the temperature down and lowering the pressure. The volume dependence of the energy gap also plays an important role. Because the 3p-3d gap is narrowing with decreasing volume, it makes a negative contribution to the total pressure in Eq. 4.2, lowering it further. These electronic thermal processes lead to the observed softening of the experimental Hugoniot and shock temperatures. Similar results have been observed in xenon which are due to thermal excitation of electrons from the 5p to 5d states.[31]

Xenon, which has a filled 5p band and empty 6s and 5d-conduction levels is of particular interest because it has the smallest conduction band gap and is the leading candidate for metallization among the rare gases. Band calculations predict that for an fcc lattice under compression, the energy levels of the d-like conduction

band will overlap those of the 5p core and xenon will become metallic
at a pressure above 1.3 Mbar.[31]
    CsI which is isoelectronic with Xe is expected to demonstrate -
similar electronic properties. Predictions have been made that CsI
will become metallic near 1 Mbar as a result of the closure of the 5p
to 5d gap. Williams and Jeanloz[32] measured the optical absorption
edges of CsI to 0.9 Mbar using a diamond anvil cell and Reichlin
et al.[33] measured the optical reflectivity to 1.7 Mbar. These
results lead to a best estimate of 1.1 ± 0.1 Mbar for the insulator to
metal transition.

4.3 Molecular Hydrogen

The equation of state (EOS) of hydrogen at very high density is of
considerable theoretical interest and is important for modeling of
giant planet interiors. Theoretical studies indicate that at
sufficiently high pressures, between 2 and 4 Mbar molecular hydrogen
will undergo a phase transition into a metallic state.
    Static measurements for molecular hydrogen have been made on the
solid up to 370 kbar by van Straaten et al.[34] and to 200 kbar by
Shimizu et al.[35] These recent studies, both using the diamond anvil
cell, represent an order of magnitude extension of the pressure range
over earlier work.
    Shockwave data has been reported for deuterium to 750 kbar and
seven times liquid density and for hydrogen to 110 kbar and three
times liquid density.[36] In these experiments temperatures up to
7000 K are achieved, thereby probing the potential to intermolecular
separations of about 1.5 Å. This is approximately the nearest
neighbor separation of molecules in the vicinity of the metallic
transition. However, the shock data provides little information about
the potential nearer the attractive well at low energies, and this
information is best obtained from static experiments.

Intermolecular potentials

    The experimental H_2 and D_2 Hugoniot data has been analyzed using
several different H_2-H_2 potentials. The first of these is the
potential of Silvera and Goldman (SG)[37] who used it to explain the
available static compression data. The SG potential $\phi_{SG}(r)$
consists of two terms, i.e.,

$$\phi_{SG}(r) = \phi_p(r) + \phi_t(r) \quad . \tag{4.4}$$

Silvera and Goldman associate $\phi_p(r)$ with the potential of an isolated
H_2-H_2 pair and $\phi_t(r)$ with the "average" contribution by the Axelrod-
Teller triple dipole intraction. They are given, respectively, by

$$\phi_p(r) = \exp(\alpha - \beta r - \gamma r^2) - \left(\frac{C_6}{r^6} + \frac{C_8}{r^8} + \frac{C_{10}}{r^{10}} \right) f(r) , \tag{4.5}$$

$$\phi_t(r) = \frac{C_9}{r^9} f(r) \quad .$$ (4.6)

where the modulation function f(r),

$$f(r) = \exp\left[-\left(\frac{1.28 \, r_m}{r} - 1\right)^2\right] \, , \quad r \leq 1.28 \, r_m \, ,$$

$$= 1 \, , \qquad\qquad\qquad r > 1.28_m \, r \quad , \qquad (4.7)$$

attenuates the long-range multipole terms at small r. The factor r_m is
the position of the attractive minimum of $\phi_p(r)$. The SG potential has
been found to reproduce both the melting properties and the static
compression data of liquid H_2 and D_2 at 75–300 K to 2 GPa. However, it
becomes less suitable to explain the high pressure static and the dynamic
data. To remedy this, $\phi_{SG}(r)$ was modified by softening it at small r.[38]

Fluid theory and Hugoniot calculations[38]

The thermodynamic properties were calculated using the Helmholtz free
energy (A) given by

$$A = 2.5 \, NkT + N[0.5 \, h\nu + kT \ln [1 - \exp(-\beta h\nu)]]$$

$$+ A_{int} + NkT \ln \rho + \text{constant},$$ (4.8)

The first term is due to the kinetic energies of translation and
rotation; the second term, within the bracket, represents the free
energy of a vibrating molecule with frequency $h\nu/k = 6340$ K and 4395 K
for H_2 and D_2, respectively. The term A_{int} is the intermolecular
potential energy contribution evaluated from fluid variational theory:

$$A_{int} = A_{HS}(\eta) + F(\eta) \, NkT$$

$$+ (\rho N/2) \int dr \, \phi \, (r) \, g_{HS}(r, \, \eta) + A_{QM} \quad , \qquad (4.9)$$

The last term A_{QM} is the first-order quantum mechanical correction to A
in the Wigner-Kirkwood expansion. It makes a negligible contribution
under the experimental shock-wave conditions except at the initial
states of liquid H_2 and D_2.

Figure 4.7 compares the experimental liquid D_2 Hugoniot with
theoretical calculations using the modified SG potential which agree
satisfactorily with the data over the entire range. The Hugoniot
calculated with the unmodified SG potential are too stiff. Molecular
dissociation was found to be negligible except on the double shocked
Hugoniot at 75 GPa where it leads to a 1% decrease in the volume which
is well within experimental error.

Calculated solid isotherms for hydrogen are shown in Table 4.1
and are in agreement with the diamond-anvil experimental data of van
Straaten.

TABLE 4.1.   5 K isotherms of H_2.

V (cm^3/mol)	Pressure (kbar)	
	Calculations	Experiment
10	22.6	22.7
9	35.1	34.6
8	56.2	54.1
7	89.0	87.8
6	153.	149.
5	274.	271.
4	529.	537.[a]
3	1173.	1120.[a]
2	3429.	3482.[a]
1.6	6082.	5892.[a]

[a] Extrapolated values.

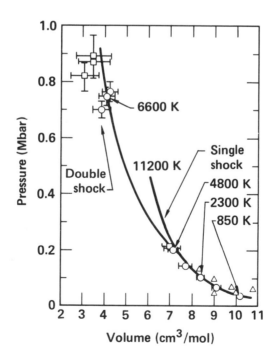

Fig. 4.7.   Single and double-shocked deuterium Hugoniots.

Metallic Transition

The possibility exists that hydrogen may become a metal by dissociation into a monatomic phase or, as in the case of iodine, first convert to an electrically conducting molecular phase as a result of energetically overlapping valence and conduction bands. The conducting molecular phase would then convert into a monatomic metal at some higher pressure. Electron band calculations predicted a band crossing in the molecule at 2.4 cm^3/mol,[39] or 210 GPa (2.1 Mbar) on our theoretical curve.

The molecular-to-metallic first-order phase transition was estimated using ϕ to obtain the properties of solid H_2. The transition pressure, predicted by equating the pressures and the Gibbs free energies, is estimated to be between 3.1 and 3.6 Mbar. These values suggest that an insulating molecular-to-metallic molecular transition will first occur followed at higher pressure by the diatomic to monatomic metal transitions.[38]

4.4 The Dissociation of Dense Liquid Nitrogen

The dissociation of molecular nitrogen has been the subject of several recent experimental and theoretical studies. McMahan and LeSar[40] predicted that the molecular solid at 0 K should dissociate to a monatomic phase near 0.8 Mbar. However subsequent diamond-anvil cell studies have shown no evidence for such a transition up to 1.3 Mbar. The failure to observe this transition has been attributed to a large energy barrier between the two phases. In a series of papers Nellis et al.[41] have reported shockwave results which indicate that liquid nitrogen begins to undergo a transformation at a pressure of 0.31 Mbar and a temperature of about 7000 K. The most likely explanation is that they have observed molecular dissociation.

In the simplest model, the one which neglects dissociation, the intermolecular potential is described by a spherical angle averaged interaction which neglects the diatomic structure of the molecule. The parameters were chosen to scale from the argon potential by the use of the law of corresponding states. The vibrational and electronic energy levels are taken to be the free molecule values.

The calculated principal Hugoniot shown in Fig. 4.10 (curve A) is in good agreement with the data of Nellis et al.[41] up to about 310 kbar. Above this pressure the theoretical curve continues to rise rapidly but fails to exhibit the large increase in compressibilty observed experimentally. Clearly this model is incomplete because it fails to incorporate dissociation.

Consider a reacting mixture of atoms and molecules. Let x be the fraction of molecules that have undergone dissociation into atoms. We write F the free energy per two atoms as;

$$F = (1 - x)\, F^0_{N_2} + x F^0_{2N} + F_{mix} + A_{int}, \qquad (4.10)$$

where $F^0_{N_2}$ and F^0_{2N} are respectively the free energies of the isolated

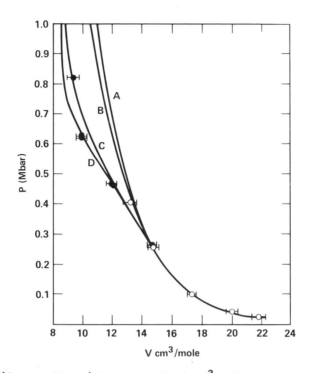

Fig. 4.8. Nitrogen Hugoniots, V_0 = 34.7 cm³/mole.
Theoretical curves: (A) with no dissociation; (B) including
dissociation with dissociation energy 9.76 eV; curves C and D include
dissociation with atomic phase binding energy.

molecule and of two atoms. F_{mix} is the free energy of mixing.
Minimization of F determines X. A_{int} is the potential energy of the
fluid calculated using fluid variational theory. The results (curve
B) differ only slightly from curve A which neglects dissociation.

There are a number of reasons why the present model might fail
even when the dissociation energy is included. One is that some of
the energy levels of the pure components may be density dependent.
But the most likely cause probably stems from the neglect of the
atomic phase binding energy. In the gas phase 9.76 ev are required to
break the chemical bond and move the atoms to infinite separation.
Clearly this value cannot be correct in the very dense fluid because
it does not account for the recombination of atoms into other chemical
states. This binding would return some of the energy expended in the
bond breaking and decrease the energy needed for dissociation. An
analogous effect known as 'ionization lowering' is observed in dense
plasmas where the interaction of the ionized electron with the
remaining particles leads to a lowering of the effective ionization
energy. We may refer to the molecular analog as 'dissociation
lowering.' As an example consider the case in which the molecule is

compressed to such a high density that the molecular bond distance
becomes comparable to the intermolecular separation. At this density
much less than 9.76 ev would be required to dissociate the molecule.
Thus we may conclude that the dissociation energy should decrease
continuously from the gas phase value to some much smaller figure at
high density.

We introduce this reasoning into our thermodynamic model by
adding to the free energy a volume dependent term E_b equal the binding
energy of the atomic phase per two atoms. It contributes to the total
free energy in proportion to the fraction of molecules dissociated.

$$F = (1-x)F^0_{N_2} + xF^0_{2N} + F_{mix} + A_{int} + xE_b \quad . \tag{4.11}$$

The volume dependence leads to an additional term in the pressure
equation:

$$P = RT/V(1+x) + P_{int} - \left(\frac{\partial E_b}{\partial V}\right)x \tag{4.12}$$

and to a volume dependent dissociation energy

$$D = 9.76 + E_b. \tag{4.13}$$

Since E_b has a negative value it lowers the dissociation energy
and leads to an increase in the dissociation fraction. E_b is
determined by fitting to the Hugoniot and is consistent with the
calculated cohesive energy for atomic nitrogen.

The introduction of a binding energy for the atomic phase results
in the softening of the Hugoniot shown as curve C (or D). The
calculated values of the dissociation energy and the fraction of
molecules dissociated are plotted in Fig. 4.9. Fig. 4.10 is an
overview of the nitrogen equation of state for the present model
showing the 0 K molecular and metallic isotherms and the Hugoniot.
Also shown are the results of McMahan who has calculated isotherms for
atomic nitrogen in several different crystal structures. The overall
agreement of the theoretical model with McMahan's results are
consistent with what one would expect from having determined
empirically the binding energy from expanded metal data. Physically
what is happening is that every time a molecule dissociates, the two
atoms are converted into metal atoms lying on a much lower isotherm.
Thus the dissociation process leads to a drop in pressure or
equivalently to a volume collapse. An analogous behavior is that
observed in shock compressed liquid argon and xenon. In these liquids
electrons are thermally excited from the top of a filled p-like
valence band to the bottom of an unfilled d-like conduction band in
which they have a lower pressure. The energy band gap separating
these two states is decreasing with decreasing volume and this
introduces a negative contribution to the pressure just as it does in
the present case. And with the same effect. The negative
contribution to the thermal pressure leads to the large increase in

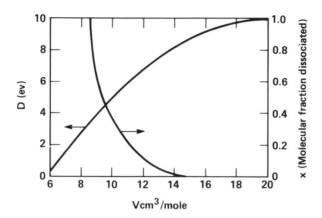

Fig. 4.9. Dissociation energies (D) and fractions of dissociated molecules (x) calculated along theoretical Hugoniot.

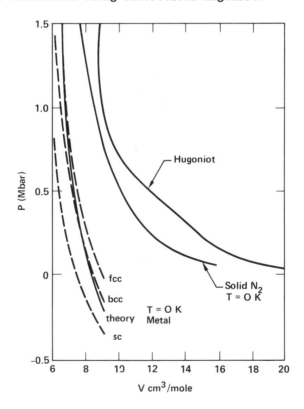

Fig. 4.10. Calculated molecular and atomic nitrogen equation of state (solid curves). The dashed curves represent the results of McMahan for the metal with different crystal structures.

the compressibility observed along the Hugoniot.

4.5 Polyatomic Molecules; Water and Ammonia.

The properties of water and ammonia have been studied with measurements

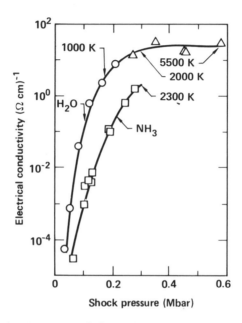

Fig. 4.11. Electrical conductivity of water and ammonia. From Ref. 43.

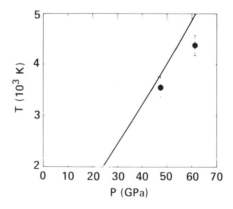

Fig. 4.12. NH$_3$ shock temperature plotted as a function of pressure. Solid line is a calculation using fluid variational theory. From Ref. 43.

of the Hugoniot, shock temperatures and electrical conductivities. The most interesting feature of work prior to 1979 were the early electrical conductivity measurements of Hamman and Linton[42] on water and aqueous solutions of KCl, KOH and HCL at shock pressures ranging from 70-133 kbar. By combining the measured ionic conductivities of these compounds it was estimated that the ionization of water becomes nearly complete at 200 kbar and that this transformation may be responsible for the change in the compressibility of the water Hugoniot above and below the 150-200 kbar region. The shock temperature and electrical measurements on water have been extended to 600 kbar and more recently these were made for ammonia.[43]

Comparison of the calculated temperatures with the experimental results for NH_3 in Fig. 4.12 shows that the model is in reasonably good agreement with the data. The small deviation of 10% at the highest pressure maybe an indication of an additional absorption mechanism at these high pressures. Previous electrical conductivity measurements on ammonia have shown that it becomes conductive in the range 70-280 GPa. The mechanism proposed to explain the high conductivity was molecular dissociation-ionization. The relatively small difference between the measured temperatures and those calculated using this non-dissociative model suggests that dissociation is not sufficient to affect the equation of state in a dramatic way. Including dissociation-ionization in the model would have the effect of lowering the calculated temperatures in the direction of better agreement with the experimental data.

Theoretical studies on water are complicated by strong electrostatic interactions and highly nonspherical potentials. Ree[44] has carried out a detailed study of shock compressed water using fluid perturbation theory and intermolecular potentials based on quantum mechanical ab-initio calculations. Of the several potentials examined Ree found best agreement with experiment was given by the Stillinger and Rahman ST2 potential. This is a non-spherical atom-atom potential which must be suitably averaged over all orientations of molecules 1 and 2 at fixed intermolecular separations.

4.5 Carbon Compounds

Chemical decomposition is known to occur in small molecules containing carbon and in hydrocarbons. Yakusheva et al.[45] have shown that the shock compression of many transparent and colourless hydrocarbons above about 100 kbar and 1000 K is accompanied by a break in the Hugoniot curve and a sharp increase in the light absorbance, resulting in a loss of transparency. These features suggest a rapid pyrolysis of the initial molecule and the formation of an opaque carbon condensate (i.e., tars, graphite). Liquid CS_2 exhibits similar behaviour when it is shock-compressed to approximately 60 kbar and statically compressed to about the same pressure. CO probably decomposes into carbon and CO_2.

Theoretical calculations of hydrocarbon Hugoniots, including methane, predict that shock heating induces C-H bondscission and favours the condensation of C atoms into elemental carbon.[46] The

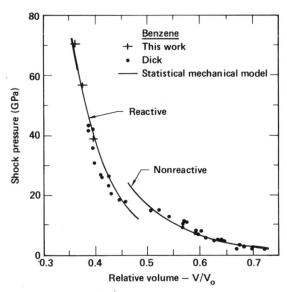

Fig. 4.13. Shock pressure vs relative volume for benzene. The curve
above 20 GPa was calculated assuming decomposition. The curve below
20 GPa was calculated assuming benzene retains its C_6H_6 molecular
structure. From Ref. 46.

formulation considers a chemical equilibrium involving C (diamond), C
(graphite), and H_2 plus nine other low molecular-weight species, CH_4,
C_2H_2, C_6H_6, C_2H_4, C_2H_6, C_3H_6, C_3H_8, C_4H_{10}, and C_5H_{12}. The
concentrations are determined by minimizing the Gibbs free energy.
The Hugoniot calculated for molecular benzene shown in Fig. 4.13
suggests the benzene molecule dissociates above a shock pressure of
130 kbar. Similar results have been obtained for a large number of
hydrocarbons.

4.6 The Structure of Dense Alkali Halide Melts

The unique characteristic of shockwave experiments is that they can be
used to explore states of matter at very high pressure and temperature
that are inaccessible by other techniques. This property makes it
valuable for studying melting at extreme conditions. The usefulness
of shock melting data is not that it simply represents more data but
that it greatly extends the range of conditions over which to test the
applicability of melting laws and concepts.
        When a substance melts at atmospheric pressure the added energy
does not lead to a rise in temperature until the process has been
completed. Under shock compression the pressure-temperature path
passes through the melting curve. This feature was first observed by
Kormer, et al.[47] (Fig. 4.12) in shock temperature measurements on
NaCl and KCl. Similar results for CsI, have recently been obtained by

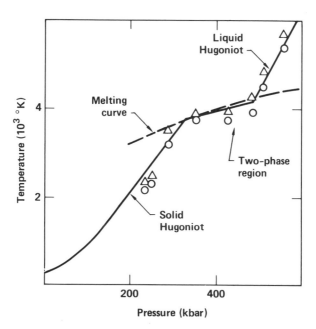

Fig. 4.14. Hugoniot temperature measurements and estimated melting curve for KCl. Copied after Kormer.

Radousky, et al.[48]

X-ray scattering and neutron diffraction experiments coupled with Monte Carlo and hypernetted chain (HNC) equation calculations have established that at atmospheric pressure alkali halide melts are characterized by a relatively open NaCl-like structure containing about 5 to 6 atoms in the nearest neighbor shell. The application of pressure is believed to result in a gradual increase in the coordination number. But very little is known experimentally.

The hypernetted chain equation is now widely used for calculating the properties of ionic fluids.[49] Calculations were made using an exponential six function to represent the pair-potential for the inert gas xenon and for CsI using an exponential-six with a coulomb term added.

$$\phi(r) = \phi_{exp6}(r) + \frac{z_1 z_2 e^2}{r}$$

where z_1 and z_2 are the ion charges. The xenon results provide a reference against which to judge the occurrence of a structural change in CsI to an inert gas-like structure.

Figure 4.15a shows the partial distribution functions of liquid CsI calculated at its normal melting temperature. The atomic separations in the figure are plotted in units of r/a where a is the mean ion sphere radius, or $a = (3/4\pi V/N)^{1/3}$. For these potentials $g_{++} =$

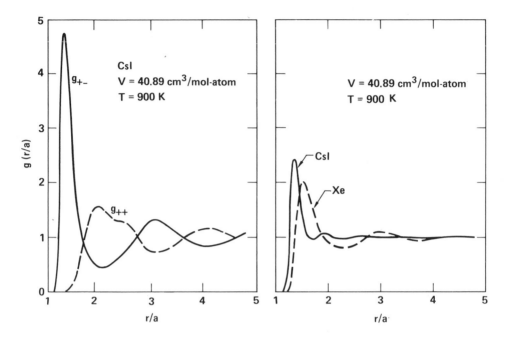

Fig. 4.15. (a) Partial distribution functions for CsI; (b)
Comparison of total distribution functions for CsI and Xe. P ~ 0
kbar.

$g_{-\ -}$. The figure exhibits the characteristic alkali halide arrangement
of alternating shells of unlike and like charge. Figure 4.15b compares
the total pair distribution function, $g(r) = (g_{++}(r) + g_{+\ -}(r))/2$, of
CsI with that for the xenon-like fluid (potential 1). The figure
shows two very different structures. The reader may have noted that
there is a shoulder in the first g_{++} peak near $(r/a) = 2.3$. This is
the start of a pressure induced splitting into two new peaks.
    As the pressure along the freezing line increases the splitting
of the g_{++} peak becomes more pronounced. Increasing the pressure to
about 300 kbar (3650 K) to near the observed shock freezing point
shifts the first g_{++} peak to inside the $g_{+\ -}$ first peak envelope
(Fig. 4.16a). As a result the total distribution functions of Xe and
CsI (Fig. 4.16b) are now virtually identical. Each ion has about 12
nearest neighbors, as in a close-packed system, of which seven are
oppositely charged and five have the same charge. Figure 4.16a
demonstrates that the oppositely charged neighbors on the average
approach each other more closely than do ions with the same charge.
But a considerable degree of interpenetration exists. At pressures up
to 700 kbar no important changes were observed.
    These results demonstrate that at sufficiently high density the
short range repulsive forces will be dominant over the long range
attraction. Near the pressure of 300 kbar, where the Hugoniot enters

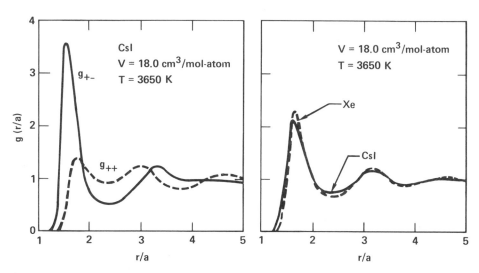

Fig. 4.16. (a) Partial distribution functions for CsI; (b) Comparison of total distribution functions for CsI and Xe. P ~ 300 kbar.

the fluid, the contribution of the exp-6 is an order of magnitude larger than that of the Coulomb term. Thus the properties are dominated by the strong repulsive forces and the liquid adopts a xenon or hard sphere-like structure. The application of pressure has the effect of "dialing down," or decoupling the influence of the coulomb forces. The results demonstrate the existence of a gradual pressure-induced shift in the structure of an alkali halide melt from an open arrangement to one characteristic of a simple non-ionic fluid. Tallon[50] has considered the possibility that the shapes of melting curves could be understood by a continuous pressure induced change in the melt to a more closely packed state. Tallon explains the curvature and projected occurrence of the maxima in terms of a continuous transition in the melt from the lower density six-coordinated state to a higher density eight-coordinated state at higher pressure. His conclusions are generally confirmed by the present results in the sense that we also believe that the fluid becomes more densely packed.

Recently several authors have suggested that liquid silicates and magmas undergo an increase in coordination number with increasing pressure.[51-52] Thus, it appears that the phenomena observed in alkali halides is not isolated but is a more general feature of ionic materials that has important consequences for geophysics.

ACKNOWLEDGEMENTS

Work performed under the auspices of the U.S. Department of
Energy by the Lawrence Livermore National Laboratory under contract
number W-7405-ENG-48.

REFERENCES

1. J. A. Barker and D. H. Henderson, Rev. Mod. Phys. **48**, 587 (1976).
2. J. P. Hansen and I. R. McDonald, 'Theory of Simple Liquids,'
    Academic Press, NY, 1976.
3. M. L. Klein, Ann. Rev. Phys. Chem. **36**, 525 (1985).
4. N. F. Carnahan and K. E. Starling, J. Chem. Phys. **51**, 635
    (1969).
5. F. H. Ree, J. Chem. Phys. **73**, 5401 (1980).
6. M. Ross, J. Chem. Phys. **71**, 1567 (1979).
7. W. G. Hoover, M. Ross, K. W. Johnson, D. Henderson, J. A. Barker
    and B. C. Brown, J. Chem. Phys. **52**, 4931 (1970).
8. J. P. Hansen, Phys. Rev. A **2**, 221 (1970).
9. D. A. Young and F. J. Rogers, J. Chem. Phys. **81**, 2789 (1984).
10. J. Talbot, J. L. Lebowitz, E. M. Waisman, D. Levesque and J. J.
    Weis, to be published, J. Chem. Phys. 1986.
11. R. A. Aziz, V. P. S. Nain, J. S. Carley, W. L. Taylor and G. T.
    McConville, J. Chem. Phys. **70**, 4330 (1979).
12. M. Ross and F. H. Ree, J. Chem. Phys. **73**, 6146 (1980).
13. Ya. B. Zeldovich and Yu. P. Raizer, Physics by Shock Waves and
    High Temperatures Hydrodynamic Phenomena, Vols. 1-2 (New York:
    Academic Press), 1966.
14. W. J. Nellis, and A. C. Mitchell, J. Chem. Phys. **73**, 6137, 1980.
15. M. Van Thiel, M. J. Shaner, and E. Salinas, Lawrence Livermore
    National Laboratory Report UCRL 50108, Vols. 1-3, Rev. 1, 1977.
16. S. P. Marsh, LASL Shock Data, (Berkeley: University of
    California Press, 1980).
17. L. V. Al'tshuler, A. A. Bakanova, I. P. Dudoladov, E. A. Dynin,
    R. F. Trunin, and B. A. Chekin, J. Appl. Mech. Tech. Phys. **22**,
    145, 1981.
18. V. Y. Klimenko and A. N. Dremin, in Detonatsiya, Chernogolovka,
    edited by O. N. Breusov et al. (Akad. Nauk, Moscow, SSSR, 1979),
    p. 79.
19. W. G. Hoover, Phys. Rev. Lett. **23**, 1531 (1979).
20. B. L. Holian, W. G. Hoover, B. Moran and G. K. Straub, Phys. Rev.
    A **22**, 2798 (1980).
21. M. Ross, H. K. Mao, P. M. Bell and J. A. Xu, to be published J.
    Chem. Phys. 1986.
22. M. van Thiel and B. J. Alder, J. Chem. Phys. **44**, 1056 (1966);
    R. D. Dick, R. H. Warnes, and J. Skalyo, Jr., J. Chem. Phys.
    53, 1648 (1970).
23. F. V. Grigoriev, S. B. Kormer, O. L. Mikhailova, M. A. Mochalov,
    and V. D. Urlin, Sov. Phys. JETP **61**, 751 (1985), [Zh. Eksp.
    Teor. Fiz. **88**, 1271 (1985)].

24. R. A. Aziz and H. H. Chen, J. Chem. Phys. **67**, 5719 (1977).
25. T. L. Gilbert and A. C. Wahl, J. Chem. Phys. **47**, 3425 (1967).
26. P. A. Christiansen, K. S. Pitzer, Y. S. Lee, J. H. Yates, W. C. Ermler and N. W. Winter, J. Chem. Phys. **75**, 5410 (1981).
27. W. R. Wadt, J. Chem. Phys. **68**, 402 (1978).
28. M. Ross, Phys. Rev. **171**, 777 (1968).
29. A. K. McMahan, Phys. Rev. B **33**, 5344, (1986).
30. M. Ross, W. J. Nellis, and A. C. Mitchell, Chem. Phys. Lett. **68**, 532 (1979).
31. M. Ross and A. K. McMahan, Phys. Rev. B **21**, 1658 (1980).
32. Q. Williams and R. Jeanloz, Phys Rev. Lett. **56**, 163 (1986).
33. R. Reichlin, M. Ross, S. Martin, K. A. Goettel, to be published in Phys. Rev. Lett. 1986.
34. J. Van Straaten, R. J. Wijngarden, and I. F. Silvera, Phys. Rev. Lett. **48**, 97 (1981).
35. H. Shimizu, E. M. Brody, H. K. Mao, and P. M. Bell, Phys. Rev. Lett. **47**, 128 (1981).
36. W. J. Nellis, M. Ross, A. C. Mitchell, M. van Thiel, D. A. Young, F. M. Ree and R. J. Trainor, Phys. Rev. A **27**, 608 (1983).
37. I. F. Silvera and V. V. Goldman, J. Chem. Phys. **69**, 4209 (1978).
38. M. Ross, F. H. Ree, and D. A. Young, J. Chem. Phys. **79**, 1487 (1983).
39. C. Friedli, and N. W. Ashcroft, Phys. Rev. B **16**, 662 (1977).
40. A. K. McMahan and R. LeSar, Phys. Rev. Lett. **54**, 1929 (1985).
41. W. J. Nellis and A. C. Mitchell, J. Chem. Phys. **73**, 6137 (1980); W. J. Nellis, N. C. Holmes, A. C. Mitchell, and M. van Thiel, Phys. Rev. Lett. **53**, 1661 (1984).
42. S. D. Hamann and M. Linton, Trans. Faraday Soc. **65**, 2186 (1969).
43. H. B. Radousky, A. C. Mitchell, W. J. Nellis and M. Ross, Proceedings of the APS Topical Conference on Shockwaves in Condensed Matter, Spokane, WA (1985).
44. F. H. Ree, J. Chem. Phys. **76**, 6287 (1982).
45. O. B. Yakusheva, V. V. Yakushev, A. N. Dremin, Russ. J. Phys. Chem. **51**, 973 (1977).
46. W. J. Nellis, F. H. Ree, R. J. Trainor, A. C. Mitchell and M. B. Boslough, J. Chem. Phys. **80**, 2789 (1984).
47. S. B. Kormer, Usp. Fiz. Nauk. **94**, 641 (1968) [Sov. Phys.-Usp. **11**, 229 (1968)].
48. H. B. Radousky, M. Ross, A. C. Mitchell, and W. J. Nellis, Phys. Rev. B **31**, 1457 (1985).
49. M. Ross and F. J. Rogers, Phys. Rev. B **31**, 463 (1985).
50. J. L. Tallon Phys. Lett. **72A**, 150 (1979).
51. C. A. Angel, P. A. Cheeseman, and S. Tomaddon, Science **218**, 885 (1982).
52. I. Kushiro, in Physics of Magmatic Processes, edited by R. B. Hargraves (Princeton University Press, Princeton, New Jersey, 1980), pp. 93-120.

PROPERTIES OF GASES UNDER HIGH COMPRESSION

B. Le Neindre
L.I.M.H.P. - C.N.R.S.
Université Paris-Nord
93430 Villetaneuse
France

ABSTRACT. Some experimental results of thermophysical properties of gases under high compression are compared with different computer simulations. A special emphasis is placed on the fluid-solid transition and on the approach to this transition on the gas side. The thermophysical properties which have been analyzed are the following : equations of state, compressibility coefficients, heat capacities, velocity of sound, dielectric constant, viscosity, thermal conductivity, phase equilibria. We will see that under high compression, the most of these properties vary monotonously with respect to temperature and pressure. A comparison is given with gases under low or moderated compression which present large anomalies of some of their thermophysical properties.

1. INTRODUCTION

Several quantities can be defined to characterize the compression of gases. We will recall some of them.

The compressibility factor pV is the product of the pressure p and the volume V and the virial coefficients are the coefficients in the expansion of pV in powers of the density 1/V or the pressure P. The density expansion is the more fundamental of the two.

$$PV = RT \left[1 + B_{(v)}/V + C_{(v)}/V^2 + ... \right] \qquad (1)$$

For gases at moderate density, the consecutive coefficients can be related to interactions between pairs, triplets, etc... of molecules.

The compression factor Z is the ratio

$$Z = PVM/RT \qquad (2)$$

If we consider a system of a given number of molecules at a constant temperature, an external force acting on the system generates an increase of the pressure and the compression factor but a decrease of the volume and the compressibility coefficients.

Compressibility coefficients are defined as follows. The coefficient of isothermal compressibility κ is given by :

51

R. van Eldik and J. Jonas (eds.), High Pressure Chemistry and Biochemistry, 51–92.

$$\kappa = -(1/V)(\partial V/\partial P)_T \tag{3}$$

In an adiabatic change, the coefficient of adiabatic compressibility is given by

$$\chi = -(1/V)(\partial V/\partial P)_S \tag{4}$$

The ratio of isothermal and adiabatic compressibilities is related to the ratio of heat capacities at constant pressure and volume

$$(\kappa/\chi) = (C_p/C_v) = \gamma \tag{5}$$

If a given number of molecules are in a box of a constant volume, for instance the critical volume, a small change in temperature generates large density fluctuations and the interaction between long range forces becomes predominant a large increase in the isothermal compressibility follows.

We will examine ranges of the phase diagram where the compression factor is large and we will see how other thermophysical properties vary.

In the three phase diagram shown in fig. 1, the pressure temperature range of this study is the hatched part which is bounded by the critical isotherm, the critical isobar and the fluid-solid transition. However the main interesting ranges are the approach of the fluid-solid transition for which no pretransitional changes can be seen and the approach of the critical point which presents large

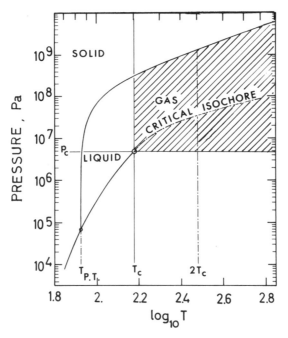

Figure 1. Three phase diagram of a substance.

enhancement in some thermophysical properties.

2. EXPERIMENTAL RESULTS ALONG THE FLUID SOLID TRANSITION.

There is only a limited number of gases for which experiments have been conducted at high pressure for temperature larger than the critical temperature. Among these gases, noble gases are of special interest because they allow to develop a corresponding state treatment and a direct comparison with computer calculations. Properties of molecular gases like hydrogen, deuterium, nitrogen, methane are important for their application for the modelling of the planets of our solar system.

2.1. The melting curve of helium

Melting properties of He were reported by Mills et al. (1). They have measured the melting pressure P_m, the fluid volume V_f, the solid volume V_s and the volume change on melting $\Delta V_m = V_f - V_s$. These melting properties were added to those measured by Grilly and Mills (2) from 1 to 3.5 kbar and by Crawford and Daniels (3) from 1 to 10 kbar to fit a Simon's equation.

$$P_m \text{ (kbar)} = -0.008112 + 0.01691 \ T_m^{1.5550} \tag{6}$$

in the melting pressure range from 1 to 20 kbar and the melting temperature range from 14 to 97 K. Values of the volume change on melting were used in a least squares fit to obtain

$$\Delta V_m \text{ (cm}^3/\text{mole)} = 0.6640 \ (P_m + 0.1604)^{-0.3569} \tag{7}$$

The parameters of equation 6 were modified to include the value $P_m = 115$ kbar at $T_m = 297$ K obtained by Pinceaux and Besson (4) and the set of data measured by Loubeyre (5) from 222.15 K ($P_m = 76.6$ kbar) to 366.4 K ($P_m = 158.2$ kbar). This author has reported the following Simon type equation for the melting pressure

$$P_m \text{(kbar)} = -0.03186 + 0.01783 \ T_m^{1.5417} \tag{8}$$

The relative volume change on melting $\Delta V_m/V_s$ was also estimated from refractive index measurements at 266, 281 and 344 K. At 266 and 281 K there is a good agreement between these values and extrapolated data of Mills et al. (1). However at 344 K, the value of ref. 5 is much smaller. The author (5) has explained this difference by the existence of a triple point at 300 K.

2.2. The melting curve or argon

Precise measurements of the $P_m - T_m$ melting curve of argon were performed by Hardy et al (6) at pressures ranging from 0.58 – 11.41 kbar and temperatures from 97.8 – 272.9 K.
There data were fitted with the following Simon type equation

$$P_m \text{ (bar)} = -2293.252 + 2.67348 \ T_m^{1.52299} \tag{9}$$

Liebenberg et al. (7) have determined the melting pressure P_m = 13.24 kbar at 295.7 K with a volume change on melting of 1.11 cm^3/mole. Recently the melting curve of argon has been measured up to 717 K and 60 kbar using a new interferometric technique in a diamond anvil cell by Zha et al. (8).

2.3. The melting curve of molecular hydrogen

Recent measurements of melting points of H_2 were carried on by Diatschenko et al (9) up to 368 K and 77 kbar. These measurements agree well with previous results obtained at lower pressure (10-14) but they disagree with a room temperature point measured in a diamond anvil cell (15). These data were least squares fitted to the following modified Simon equation.

$$P_m(kbar) = -0.5149 + 1.702 \ 10^{-3} \ (T_m + 9.689)^{1.8077} \tag{10}$$

with a standard deviation of 0.20 kbar.

Volume changes on melting ΔV_m decrease with increasing temperature as shown by a least-squares fit of data points up to 20 Kbar.

$$\Delta V_m \ (cm^3/mole) = 9.355 \ (T_m - 6.720)^{-0.6200} \tag{11}$$

At the calculated pressure of 2.5 Mbar for the possible transition from a molecular to a metallic phase the molar volume change is finite and has the extrapolated value ΔV_m = 0.07 cm^3/mole.

2.4. The melting curve of molecular deuterium

Measurements of melting properties of molecular deuterium were also reported by Diatschenko et al. (9). Pressure melting data were least squares fitted to the following Simon equation

$$P_m(kbar) = -0.5187 + 3.436 \times 10^{-3} \ T_m^{\ 1.691} \tag{12}$$

with a standard deviation of 0.16 kbar up to 40 kbar. A good agreement was found with literature data except for one point (16).

The volume change on melting was fitted in function of the melting temperature between 75 and 156 K by

$$\Delta V_m(cm^3/mole) = 38.49 \ (T_m + 0.1265)^{-0.9093} \tag{13}$$

2.5. The melting curve of nitrogen

Melting properties were measured by Mills et al. (17) up to 20 kbar. Combining their measurements with previous ones (18-20) they relate the melting pressure P_m to the melting temperature T_m by

$$P_m(kbar) = -1.592 + 0.0009598 \ T_m^{\ 1.7895} \tag{14}$$

and give the volume change on melting by

$$\Delta V_m(cm^3/mole) = 2.244 \ (P_m + 0.831)^{-0.5707} \tag{15}$$

3.   THEORETICAL INTERPRETATION OF THE FLUID SOLID TRANSITION

In recents years there has been a revival of interest in the phenomenon of melting. Several attempts were made to predict melting properties on theoretical grounds.

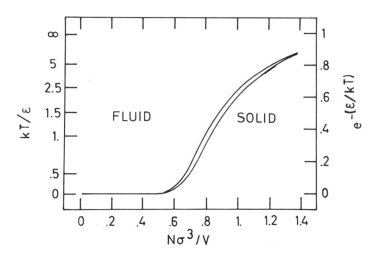

Figure 2. Phase diagram of a soft sphere fluid.

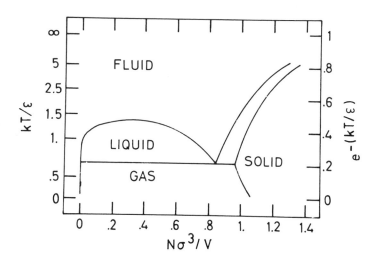

Figure 3. Phase diagram of a Lennard-Jones fluid.

Phase diagrams of simulated fluids by molecular dynamics or Monte-Carlo methods are characteristic of intermolecular potentials as these methods limit the small number of interactive particles of the many body problem to an increase of the density parameter. Considering the phase diagram of a first fluid (fig. 2) made of soft spheres of diameter σ, in additive binary interactions with a soft repulsive potential (n=12) selected among homogeneous potential of the type

$$\Phi_{ij} = \varepsilon \ (\sigma/r_{ij})^n \tag{16}$$

The second fluid differs of the first one by the addition of an attractive part $(r_{ij})^{-6}$ to the interaction potential which is described by a Lennard-Jones potential.

$$\Phi_{ij} = 4\varepsilon \ [(\sigma/r_{ij})^{12} - (\sigma/r_{ij})^6] \tag{17}$$

The phase diagram is shown in figure 3.

There is a fundamental difference between figures 2 and 3, as the Lennard-Jones fluid is the only one to show a liquid-vapor equilibrium line.

This equilibrium line ends with a critical point having the following reduced coordinates.

$$kT_c/\varepsilon = 1.35 \ ; \ V_c \ / \ N_A \sigma^3 = 3.3 \ ; \ P_c V_c/RT_c = 0.34 \tag{18}$$

Phase diagrams of purely repulsive fluids show always a solid-fluid phase transition whatsoever the values of n between 1 and infinite $[n^{-1} = 0$ corresponds to the hard sphere fluid]. Another interesting aspect is the comparison between the solid-fluid transition of a purely repulsive fluid and a Lennard-Jones fluid. For $T^* > 2.5$ i.e $T^* > 2 \ T_c^*$ (L.J), a qualitative agreement is observed between the two fluids but for $T < 2T_c$, the solid-fluid

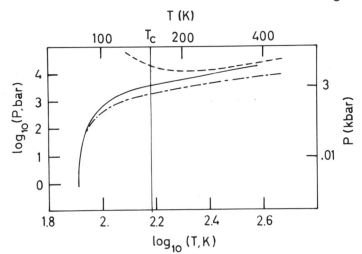

Figure 4. The melting line of argon (— Experimental curve, —·— Longuet Higgins and Widom,—— Rowlinson).

transition of the Lennard-Jones fluids is displaced to higher reduced densities than the one of the repulsive fluid.

Real fluids are little compressible near melting, reflecting the predominance of steric effects of packing. Then repulsive forces must determine this type of behavior. Thus most of the empirical melting laws (equation of Simon, of Kraut and Kennedy or Lindemann criteria) can be exactly formulated from a purely repulsive potential.

The description of the variation of the melting pressure of argon as a function of the temperature by different theories shows the contribution of molecular forces following the temperature range which is considered (fig. 4). Attractive forces are taken into account by a perturbation theory to a first order of a repulsive potential of hard spheres (H.S., $n = \infty$). By adding to H.S. an attractive potential of infinite range and of depth equal to zero (Van der Waals potential), Longuet-Higgins and Widom (21) have given a correct value of the slope of the melting curve up to the triple point, but their prediction of the melting curve is inaccurate above $T_c = 151$ K. Rowlinson (22) has used a H.S. potential perturbated by a softer repulsive potential. Above $2\,T_c$, the melting curve is well described but not below T_c. McQuarrie and Katz (23) have combined a pertubation treatment of attractive and repulsive forces following a Lennard-Jones potential and the description is improved. Melting properties calculated by Zha et al. (8) using variational perturbation theory with a Wigner–Kirkwood quantum correction and an exponential six form potential were in excellent agreement with their experiments.

Consequently above $2\,T_c$, the fluid is little sensitive to the perturbation to a first order of the H.S. potential. Moreover as the Rowlinson treatment reproduces correctly the high temperature behaviour, it follows that only repulsive forces have to be taken into account and that it is essential to describe their stiffness. This will be equivalent to define a repulsive hard core σ which varies with the temperature.

For helium, Loubeyre (5) has performed theoretical calculations using a Lennard-Jones potential where the repulsive part defined the reference system and the attractive part is treated as a perturbation. He has found a theoretical curve lower by 30 % than the experimental one. A better agreement was obtained with the potential of Aziz (24). Other calculations along the melting curve of helium were made by Ross et al. (25) (26).

Several attempts have been made to predict the melting temperatures at high pressure of hydrogen. Exact calculations were carried on by computer simulations for many body systems with simple interparticle potentials derived from the fit of some thermodynamic properties.

It has been shown (27) that the packing fraction $n = \Pi\, d^3\, \rho\, /6$ where d is the hard sphere diameter and ρ the particle density, should have the constant value 0.468 along the melting line. Ross (28), (29) has calculated pressure melting points of H_2 at temperatures for which $n = 0.468$, using intermolecular potentials deduced from shock experiments. However pressures so calculated are much higher than experimental ones (three times at room temperature).

When we consider the Lennard-Jones potential (equation 17), then the reduced temperatures, densities and pressures are defined in function of the parameters ε and σ of the Lennard-Jones potential by the following relations

$$T^* = kT/\varepsilon \qquad\qquad\qquad (19)$$

$$\rho^* = N \rho\sigma^3 = N\sigma^3/V \tag{20}$$

$$P^* = P\sigma^3/\varepsilon \tag{21}$$

We have used for noble gases the parameters of Lennard-Jones potentials given in table 1. A review of parameters of Lennard-Jones potentials was published by Mourits and Rummens (30). The reduced melting pressures were calculated with respect to the reduced melting temperatures (fig. 5). The values for argon, krypton and xenon are almost on a single curve ; but for neon and helium there is a slight deviation. Thus the melting pressure of heavier noble gases can be represented by a single curve. The reduced melting pressure P_m were least squares fitted to a function of reduced melting temperature following a Simon type equation which follows

$$P_m^* = a \, T_m^{*c} + b \tag{22}$$

Substance	ε/k K	σ $(10^{-10}m)$
Helium	10.22	2.556
Neon	33.74	2.756
Argon	119.8	3.405
Krypton	166.7	3.679
Xenon	225.3	4.07
Hydrogen	36.7	2.958
Deuterium	36.7	2.958
Nitrogen	95	3.70
Methane	142.1	3.79

Table 1 : Parameters of the Lennard-Jones potential.

Substance	a	b
Helium	8.71038	-4.40399
Neon	8.51967	-5.47476
Argon	9.57607	-5.74508
Krypton	9.71005	-5.60802
Xenon	9.79402	-5.71871

Table 2 : Coefficients of the Simon equation.

For noble gases the value of the coefficient c was fixed at 1.5 and coefficients a et b are given in Table 2. For neon we use data of Crawford and Daniels (3). For argon we have used only the data of reference (6). In figure (6) a comparison is given between extrapolated values (P_e) and experimental data of Zha et al. (P_z) (8). There is a good agreement with the lower values of Zha which show a large dispersion. For krypton and xenon the data of Lahr and Eversole (31) have been used.

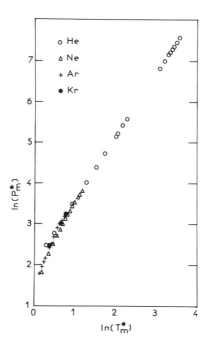

Figure 5. The melting line of noble gases.

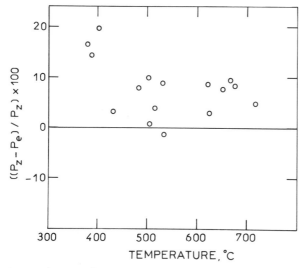

Figure 6. Comparison of extrapolated melting pressure of argon with experimental data of Zha.

Substance	Temperature (K)				
	300	1000	2000	3000	10.000
Helium	11.67	71.2	201	370	2250
Helium	11.67	71.2	201	370	2250
Neon	4.90	30.5	86	159	967
Argon	1.352	9.45	27.2	50.1	306
Krypton	0.831	6.35	18.4	34.1	209
Xenon	0.430	3.96	11.7	21.7	133

Table 3 : Melting pressure (GPa) of noble gases.

As there is some interest to know the phase separation at high temperature for instance in the planet atmosphere, we have extrapolated these curves at higher temperature (Table 3).

In Jupiter which is composed essentially of helium and hydrogen the estimates of the core temperature is 10.000 K and the corresponding pressure is 3500 GPa, this value is much higher than the melting pressures of helium and hydrogen.

If we calculate the melting pressure of noble gases using the same reduced equation and the parameters a and b of argon we get almost the same melting pressure for krypton and xenon but for neon and helium, the value are about 10% higher at 300 K. A general Simon equation which fits all the data of noble gases is

$$P_m^* = 8.71289 \ T_m^{*1.5} - 4.62122 \tag{23}$$

The melting pressure was also calculated for hydrogen, deuterium, nitrogen and methane. The coefficients a and b of the Simon equation are given in Table 4. The c value was fixed at 1.75

Substance	a	b
Hydrogen	6.96848	−0.49061
Deuterium	6.73661	1.1921
Nitrogen	13.453	−6.90054
Methane	11.9331	−5.32234

Table 4 : Coefficients of the Simon equation for molecular gases.

The extrapolated melting pressure at high temperature are reported in table 5. The melting pressure of hydrogen is lower than the melting pressure of helium up to a temperature of about 10000 K. A comparison of the theoretical calculation of the melting pressure (32) for hydrogen with the extrapolated values shows that the theoretical values are 11 % higher at 300 K, 46 % higher at 1000 K and 108 % higher at 2000 K.

As the Simon empirical equation was fitted with succes to a large variety of substances (33), attempts have been made to find a theoretical basis for this equation.

For a purely repulsive potential, it was shown (34) that the exponent c is equal to 5/4 (also if n=12, c=1.25). A similar result was obtained by Hansen (35) who has made Monte Carlo simulation of the melting and has obtained

$$P_m^* = 16 \ T_m^{*5/4}$$ (24)

Substance	Temperatures (K)				
	300	1000	2000	3000	10000
Hydrogen	5.382	44.3	149	303	2493
Deuterium	5.235	42.8	144	293	2410
Nitrogen	2.427	21.3	72	146	1205
Methane	1.398	12.9	43.8	89	735

Table 5 : Melting pressure (GPa) of molecular gases.

4. THE EQUATION OF CLAPEYRON

Along the fluid-solid transition the fundamental equation is the Clapeyron equation

$$(dP_m/dT_m) = (\Delta S_m / \Delta V_m)$$ (25)

where dP_m/dT_m is the slope of the pressure temperature curve, ΔS_m and ΔV_m are the variations of entropy and volume across the transition.
The equation of Clapeyron can be written

$$(T_m dP_m/dT_m) = (\Delta H_m / \Delta V_m) = a + bP_m + CP_m^2$$ (26)

1) If $(\Delta H_m / \Delta V_m) = $ const, the integration of (26) yields an equation similar to the equation of Mukherjee (36)

$$P_m = a\ln (T_m/T_o)$$ (27)

2) If the ratio $(\Delta H_m / \Delta V_m)$ is a linear function of the pressure, the integration of (26) leads to the equation of Simon.

4.1. Entropy change on melting

The entropy change on melting is given by the Clapeyron relation. For helium the result was given as dimensionless quantity (1)

$$\Delta S_m/R = 0.901 \ (P_m + 0.008)^{0.3569} \ x \ (P_m +0.160)^{-0.3569}$$ (28)

For $P_m > 5$ kbar, $\Delta S_m/R$ can be approximated by a constant = 0.901. In this region the fluid-solid entropy difference is almost independent of the melting pressure.
For argon $\Delta S_m/R$ falls continuously with increasing P_m. For argon the melting parameters at 295.7 K are given in table 6 (7).
For hydrogen the melting entropy goes through a minimum at $P_m = 19$ kbar

and thus slowly rises with pressure.

For hydrogen the entropy change on melting is given by (13)

$$\Delta S_m/R = 0.554 \; T_m^{0.724}(T_m - 6.72)^{-0.620} \tag{29}$$

P_m Kbar	V_s (cm^3/mole)	ΔV_m (cm^3/mole)	$\Delta S_m/R$
13.24	20.87 + 0.5	1.11 + 0.02	1.05

Table 6. Melting parameters for argon.

The value is almost constant and at melting temperatures above 300 K, the entropy change can be approximated by

$$\Delta S_m/R = 0.554 \; T_m^{0.1} \tag{30}$$

The entropy difference between fluid and solid increases with T_m (or P_m), this is opposite to N_2.

For nitrogen the entropy change on melting is given by (17)

$$\Delta S_m/R = 0.9943 \; (P_m + 1592)^{0.4412} \; (P_m + 0.831)^{-0.57} \tag{31}$$

At pressures above 15 kbar equation 31 can be approximated by

$$\Delta S_m/R = P_m^{-0.13} \tag{32}$$

The melting entropy can be broken down into two terms, namely

$$\Delta S_m = \Delta S_v + \Delta S_d \tag{33}$$

where ΔS_v is the entropy change caused only by the volume expansion on melting and ΔS_d is the residual entropy change due mainly to disordering of the molecules. Turturro and Bianchi (37) have calculated ΔS_v for various materials from the expression

$$\Delta S_v = \int_{V_s}^{V_f} [dP/dT]_V^{T_m} \; dV \tag{34}$$

in which $(dP/dT)_V^{T_m}$ is the isochoric pressure coefficient evaluated at T_m and is integrated over the fluid-solid volume change at melting ΔV_m

Values of $\Delta S_d/R$ for different fluids are given in table 7.

Substance	$\Delta S_d/R$
Helium	ln 2.1
Argon	ln 1.8
Hydrogen	ln 2.1
Nitrogen	ln 1.9

Table 7 : Values of $\Delta S_d/R$ for different fluids

The value calculated for a soft-sphere molecule is ln 1.9 (38). It was shown by Stishov et al. (39) that the entropy of disorder is obtained by plotting $\Delta V_m/V_s$ against $\Delta S_m/R$. For argon they found that as $(\Delta V_m/V_s) \to 0$, $(\Delta S_m/R) \to$ ln 2.0. In fact for all fluid $\Delta V_m/V_s$ decreases with increasing pressure. But $\Delta S_m/R$ exhibits different behaviour at infinite P_m, decreasing to zero for argon and nitrogen, remaining constant for helium and increasing for hydrogen. Such extrapolation seems to be meaningful.

4.2. Volume change on melting

The ratios $\Delta H_m / \Delta V_m$ for helium, argon, hydrogen, nitrogen and methane are plotted with respect to pressure in figure 7. The variation of $\Delta H_m / \Delta V_m$ is a

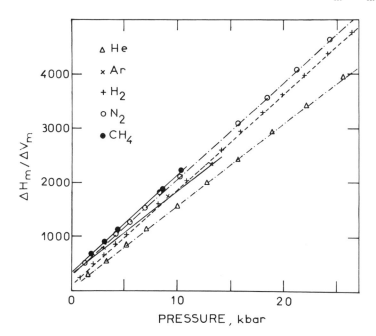

Figure 7. Variation of $(\Delta H_m / \Delta V_m)$ with pressure.

linear function of pressure with the same slope for argon and helium and other slope for hydrogen, nitrogen and methane. Then, the ratio $\Delta S_m/\Delta V_m$ will be a linear function of P_m/T_m

$$(\Delta S_m/\Delta V_m) = A\ (P_m/T_m) \tag{35}$$

As the entropy is almost constant with pressure the volume change on melting will be given by the relation

$$\Delta V_m = B\ (T_m^{(1+x)}/P_m) \tag{36}$$

Coefficients B and x are given in Table 8.

Substance	x	B
Helium	0	0.0488
Hydrogen	0.1	0.0273
Nitrogen	0.13	0.014

Table 8 : Coefficients of equation (36).

5. THERMODYNAMIC PROPERTIES AND EQUATION OF STATE

5.1. Equation of state

The free energy of a Lennard-Jones fluid was determined at high temperature by Hansen (34) ; by computing the contributions of repulsive and attractive parts of the intermolecular potential ; this attractive term was treated as a weak perturbation. Under these conditions, these two terms are function of a single variable $\rho^{*4} T^{*-1}$ (where $\rho^* = N\rho \sigma^3$ and $T^* = kT/\varepsilon$). They have been numerically evaluated using the Percus-Yevick equation for densities lower than 0.7 and by the Monte-Carlo method for higher densities. Then, results were fitted with a polynonial.

$$P^* = \rho^* T^* + \rho^{*2} (B_1 T^{*3/4} - C_1 T^{*1/4})$$

$$+ \rho^{*3} (B_2 T^{*1/2} - 2C_2) + \rho^{*4} (B_3 T^{*1/4} - 3 C_3 T^{*-1/4}) \tag{36}$$

$$+ \rho^{*5} (B_4 - 4C_4 T^{*-1/2}) + \rho^{*6} (- 5C_5 T^{*-3/4}) + \rho^{*11} (B_{10} T^{*-3/2})$$

where

$B_1 = 3.629$, $B_2 = 7.2641$, $B_3 = 10.4924$, $B_4 = 11.459$, $B_{10} = 2.17619$

$C_1 = 5.3692$, $C_2 = 6.5797$, $C_3 = 6.1745$, $C_4 = - 4.2685$, $C_5 = 1.6841$

Molecular dynamics calculations of the pressure and configurational energy of a Lennard-Jones fluid were reported by Nicolas et al. (40) in the density range $0.35 < \rho^* < 1.20$ and the temperature range $0.5 < T^* < 6$. An equation of state for the Lennard-Jones fluid was derived. The equation of state is a modified Benedict-Webb-Rubin (MBWR) equation having 33 constants. This equation fits the data over the density range $0 < \rho^* < 1.2$ and the temperature range $0.5 < T^* < 6.0$.

The MBWR equation is, in dimensionless form :

$$P^* = \rho^* T^* + \rho^{*2} (x_1 T^* + x_2 T^{*1/2} + x_3 + x_4 T^{*-1} + x_5 T^{*-2})$$

$$+ \rho^{*3} (x_6 T^* + x_7 + x_8 T^{*-1} + x_9 T^{*-2})$$

$$+ \rho^{*4}(x_{10}T^* + x_{11} + x_{12}T^{*-1}) + \rho^{*5}(x_{13}) + \rho^{*6}(x_{14}T^{*-1} + x_{15}T^{*-2})$$

$$+ \rho^{*7}(x_{16}T^{*-1}) + \rho^{*8}(x_{17}T^{*-1} + x_{18}T^{*-2}) + \rho^{*9}(x_{19}T^{*-2})$$

$$+ \rho^{*3}(x_{20}T^{*-2} + x_{21}T^{*-3})\exp(-\gamma\rho^{*2})$$

$$+ \rho^{*5}(x_{22}T^{*-2} + x_{23}T^{*-4})\exp(-\gamma\rho^{*2})$$

$$+ \rho^{*7}(x_{24}T^{*-2} + x_{25}T^{*-3})\exp(-\gamma\rho^{*2})$$

$$+ \rho^{*9}(x_{26}T^{*-2} + x_{27}T^{*-4})\exp(-\gamma\rho^{*2})$$

$$+ \rho^{*11}(x_{28}T^{*-2} + x_{29}T^{*-3})\exp(-\gamma\rho^{*2})$$

$$+ \rho^{*13}(x_{30}T^{*-2} + x_{31}T^{*-3} + x_{32}T^{*-4})\exp(-\gamma\rho^{*2}) \tag{37}$$

The values of the parameters are listed in table 9 for $\gamma = 3.0$

i	x_1	i	x_1
1	−0.44480725D − 01	17	0.10591298D + 02
2	0.72738221D + 01	18	0.49770046D + 03
3	−0.14343368D + 02	19	−0.35338542D + 03
4	0.38397096D + 01	20	0.45036093D + 04
5	−0.20057745D + 01	21	0.77805296D + 01
6	0.19084472D + 01	22	0.13567114D + 05
7	−0.57441787D + 01	23	−0.85818023D + 01
8	0.25110073D + 02	24	0.16646578D + 05
9	−0.45232787D + 04	25	−0.14092234D + 02
10	0.89327162D − 02	26	0.19386911D + 05
11	0.98163358D + 01	27	0.38585868D + 02
12	−0.61434572D + 02	28	0.33800371D + 04
13	0.14161454D + 02	29	−0.18567754D + 03
14	0.43353841D + 02	30	0.84874693D + 04
15	0.11078327D + 04	31	0.97508689D + 02
16	−0.35429519D + 02	32	−0.14483060D + 02

Table 9 : coefficients of equation 37

From this equation of state, other thermodynamical quantities can be derived : isothermal compressibility coefficient, heat capacities, sound velocity and so on.

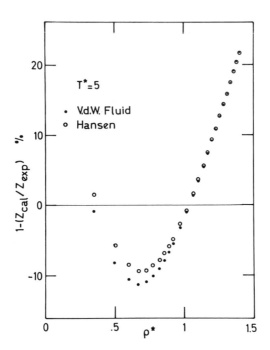

Figure 8. Comparison between compressibility factors deduced from theory and experiment along the reduced isotherm T^*=5.

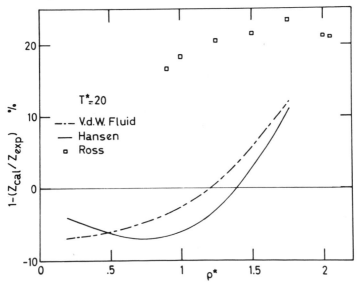

Figure 9. Comparison between compressibility factors deduced from theory and experiment along the reduced isotherm T^*=20.

In figure 8, we have compared the compressibility factor of hydrogen as a function of reduced density for a reduced temperature $T^* = 5$, (an absolute temperature of T = 184 K) to equations of Hansen and Nicolas.

In figure 9, a similar comparison was made with the compressibility factor of helium at a reduced temperature $T^* = 20$ (T = 205 K). These two equations have almost the same behaviour but show a large divergence wich respect to experimental data. However these two equations have their best agreement around $\rho^* = 1.1$.

The same comparison made for argon at a reduced temperature of 2.49 shows a slightly better agreement. The deviation is below 7 % up to the melting line.

We notice that the differences. Z_{exp} and Z_{cal} are always larger than the accuracy of computer calculations (2%) or than the experimental error (0.5 %).

The compressibility factor of helium calculated from (1) is shown in figure 10, as a function of reduced density. Z values were extrapolated up to the melting line which is well above the theoretical one.

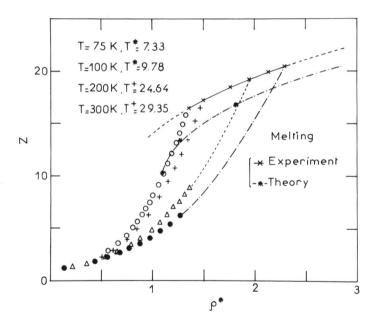

Figure 10. Variation of the compressibility factor of helium with density.

The compressibility factor of hydrogen was calculated with the EOS given by Liebenberg et al. (13) and extrapolated up to the melting line. The compressibility factor is shown in function of reduced density in fig. 11.

The compressibility factor of noble gases at room temperature (41) was calculated as a function of reduced density (fig. 12). The melting line is in good agreement with the melting line of hydrogen but about 15 % higher than the

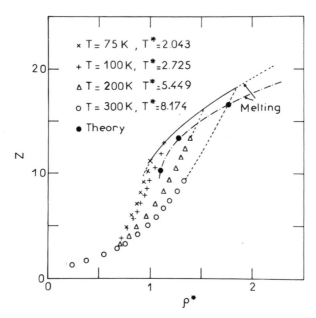

Figure 11. Variation of the compessibility factor of helium with density.

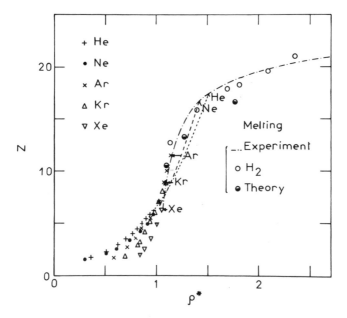

Figure 12. Variation of the compressibility factor of noble gases at room temperature with density.

theoretical one . In summary theoretical predictions lead to lower values of the compressibility factor and the density than the experimental data.

5.2. Velocity of sound

The velocity of sound increases with pressure as shown in fig. 13 where the velocity of sound is plotted for the five noble gases at room temperature (41) and in fig. 14 which represents the velocity of sound of steam at high temperature. At the approach of the fluid solid transition, the velocity of sound increases monotonously, but at the transition there is a drastic change as shown in figure 15. This figure represents the Brillouin frequency shift of fluid and

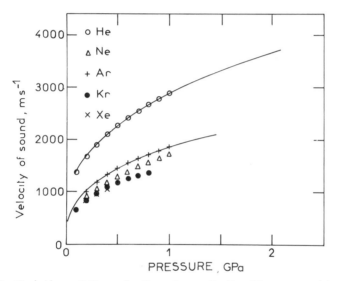

Figure 13. Variation of the velocity of sound of noble gases with pressure.

solid helium at room temperature as a function of pressure. The product of the sound velocity times the refractive index is given on the right side. These measurements were performed by Polian and Grimsditch (42) in a diamond anvil cell. These authors have also calculated the densities along the 297.29 K isotherm. Near melting they get $\rho_m \doteq 0.96$ g/cm^3.

The reduced velocity of sound $c^* = c/c_o$ is given as a function of reduced thermodynamic quantities by

$$c^{*2} = (1/\gamma_o)\,(1/T^*)\,(\partial P^*/\partial \rho^*)_{T*} + (1/C_v^*)\,(1/\rho^*)\,(dP^*/dT^*)^2_{\rho*} \qquad (38)$$

where $c_o = (RT/M)^{1/2}$, $\gamma_o = C_{p.o}/C_{v.o}$, $C_v^* = C_v/R$.

If the second term of eq. 38 is calculated using the analytical equation of Hansen, the deviation with experimental results (43) is about 15 % for xenon, 10 % for krypton and helium and 6 % for neon and argon.

5.3. Heat Capacities

The heat capacity at constant volume is given by the thermodynamic relation

$$C_p = C_{p.o}(T) - T \int_{P_o}^{P} [d^2V/dT^2] \, dP \qquad (39)$$

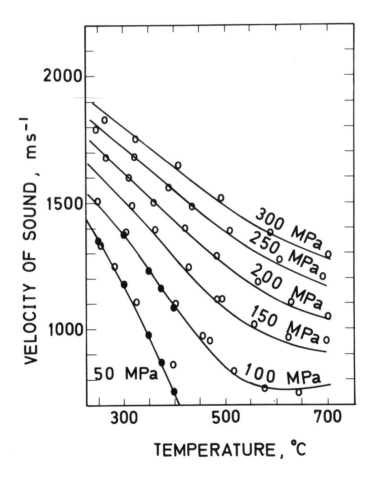

Figure 14. Variation of the velocity of sound of steam with temperature.

In fig. 16, the specific heat at constant pressure is plotted for argon and helium at room temperature as a function of density. The heat capacity at constant pressure is almost constant with density near melting condition, there is no pre-transitional effect at the approach to the melting point.

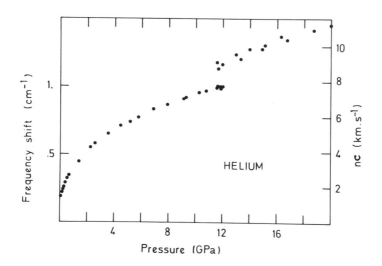

Figure 15. Brillouin frequency shift of fluid and solid helium at room temperature as a function of pressure.

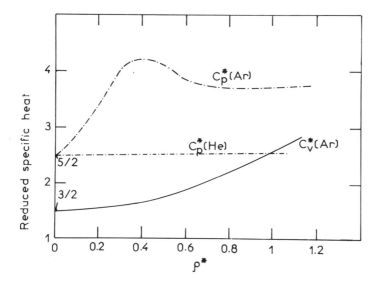

Figure 16. Variation of heat capacities of noble gases with density.

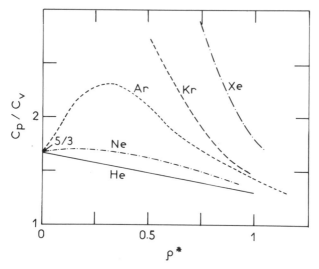

Figure 17. Variation of $\gamma = (C_p/C_v)$ as a function of density for noble gases.

As shown in figure 16 for argon, the heat capacity at constant volume increases slowly with density. The specific heat at constant volume is calculated by equation 5 if   is known or from an EOS by the following relation

$$C_p - C_v = TV \, (\alpha^2/\kappa) \tag{40}$$

where α is the volume expansion.

The ratio of heat capacities which is also the ratio of isothermal and adiabatic compressibilities is plotted as a function of density for noble gases in figure 17.

The theoretical value of γ for an ideal monatomic gas is 5/3. Since C_v increases slowly with increasing density and that C_p is almost constant, γ will decrease along an isotherm to reach an almost constant value near the melting point (for instance $\gamma = 1.08$ for helium).

5.4. Dielectric constant

The Clausius Mossotti function of noble gases has been calculated from dielectric constant measurements (44).

$$CM = (\varepsilon - 1) / (\varepsilon + 2) \, \rho \tag{41}$$

where ρ is the molar density of the fluid.

The Clausius Mossotti function expressed as an expansion in terms of reduced densities for a Lennard-Jones fluid is given by

$$CM(\rho^*) = (\varepsilon-1)/(\varepsilon+2)\rho^* = (4\pi \tilde{\alpha}/3 \, \sigma^3) \, [1 + \Delta(\rho^*, T^*)] \tag{42}$$

The term $\Delta(\rho^* T^*)$ is a positive quantity containing only the classical dipole induced dipole interaction (DID) which has been computed exactly by Alder et al. (45) for several reduced densities and temperatures. Then this DID contribution can be eliminated to obtain the density dependence of the effective polarizability.

$$\alpha = \alpha_0 (1 + B \rho^* + C \rho^* ...)$$ (43)

where α_0 is the polarizability of isolated atoms.

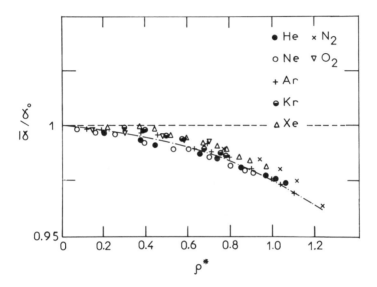

Figure 18. Variation of the effective polarizability versus reduced density for noble gases, nitrogen and oxygen.

$$B = -0.004 \text{ and } C = -0.002$$

This expansion is almost independent of the temperature and the nature of the fluid as shown in fig. 18, where a similar evolution was found for oxygen and nitrogen. On the one hand, the knowledge of the pressure dependence of the dielectric constant combined with relation (43), allows one to calculate the density of the fluid. On the other hand, these dielectric constant measurements can also be used to calculate the density dependence of the refractive index through the Lorentz–Lorenz relation analogous to the Clausius Mossotti function. For instance for the variation of the refractive index of helium with density the following relation was obtained (42)

$$n^2 = 1 + 0.3965 \rho$$ (44)

where ρ is in g/cm^3.

6. TRANSPORT PROPERTIES

In order to make a comparison between experimental values of the viscosity or the thermal conductivity and the result of computer simulation of a hard sphere fluid these properties are reduced to (46) (47).

$$\eta^* = 6.035 \times 10^8 \ V^{2/3}/(MRT)^{1/2} \tag{45}$$

$$\lambda^* = 1.936 \times 10^7 \ V^{2/3}/(RT/M)^{1/2} \tag{46}$$

where all quantities are measured in SI units. Plots of η^* or λ^* versus log V for a given fluid at different temperatures are found to be superimposable by lateral shifts.
   Then

$$\eta^* = F_\eta \ (V/V_o) \tag{47}$$

$$\lambda^* = F_\lambda \ (V/V_o) \tag{48}$$

where V_o is the characteristic close packed volume. The amount by which log V has to be adjusted gives a measure of the effect of temperature changes on V_o. There is apparently no relation between V_o deduced from the analysis of viscosity and V_o deduced from thermal conductivity.
   In terms of the parameters of the Lennard–Jones potential the viscosity and the thermal conductivity are expressed as follows

$$\eta^* = \eta \sigma^2 \ M^{-1/2} \ \varepsilon^{-1/2} \tag{49}$$

$$\lambda^* = \lambda \sigma^2 \ M^{1/2} \ \varepsilon^{-1/2} \ k \tag{50}$$

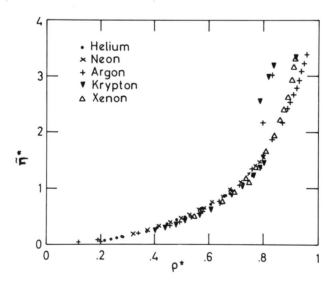

Figure 19. Excess viscosity of noble gases versus density.

It has been found that the reduced fluidity $\Phi^{*} = \eta^{*-1}$ and the reduced thermal resistivity λ^{*-1} of noble gases increase linearly with reduced volume V^{*} in the reduced temperature range $0.72 < T^{*} < 1.30$ (48). Besides it was observed from experiments that the excess viscosity of gases

$$\tilde{\eta} = \eta(\rho,T) - \eta(0,T) \tag{51}$$

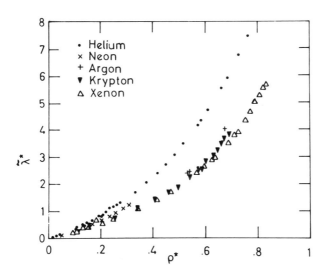

Figure 20. Excess thermal conductivity of noble gases versus density.

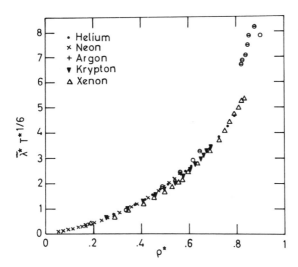

Figure 21. Variation of $\tilde{\lambda}^{*}T^{*1/6}$ versus density.

and the excess thermal conductivity of gases

$$\tilde{\lambda} = \lambda\,(\rho,T) - \lambda(0,T) \tag{52}$$

were temperature independent in a restricted temperature range. By application of the principle of corresponding states to noble gases using relations (49) and (50) we can look to what extend $\tilde{\eta}$ and $\tilde{\lambda}^*$ are temperature independent.

   In figure 19 the reduced excess viscosity is plotted as a function of reduced density and in figure 20 the reduced excess thermal conductivity is plotted as a function of density. The reduced excess viscosity is temperature independent up to a reduced density of 0.8. Near the fluid-solid transition there is a large temperature dependence. On the contrary the reduced excess of thermal conductivity is slightly temperature dependent in the whole density range. The temperature dependence can be approximated by $T^{-1/6}$ as shown in figure 21 where $\tilde{\lambda}^* T^{*1/6}$ is plotted as a function of density.

6.1. Enskog approximation

The first theory for transport properties of dense gases was formulated by Enskog. In his approximation the thermal conductivity is given by :

$$\lambda^* = \lambda^E / \lambda_o = 1/g\,(\sigma) + 4.8\ n + 12.08\ n^2\ g\,(\sigma) \tag{53}$$

and the viscosity by :

$$\eta^* = \eta^E / \eta_o = 1/g\,(\sigma) + 3.2\ n + 12.18\ n^{*2}\ g\,(\sigma) \tag{54}$$

where : $n = \pi N\,\sigma^3\,\rho/6$ for a spherical molecule of diameter σ.

   $g\,(\sigma)$ is the radial distribution at contact which is satisfactory approximated by the relation of Starling and Carnahan.

$$g\,(\sigma) = (1 - 0.5\ n)\,(1 - n)^{-3} \tag{55}$$

   λ_o is the dilute gas approximation for the thermal conductivity and η_o is the dilute gas approximation for the viscosity.
By limiting the expansion of equations (53) and (54) to the fourth power one gets :

$$(\lambda^E - \lambda_o)/\lambda_o = 1.2043\ \rho^* + 3.7915\ \rho^{*2} + 4.3171\ \rho^{*3} + 4.0904\ \rho^{*4} \tag{56}$$

and

$$(\eta^E - \eta_o)/\eta_o = 0.3665\ \rho^* + 3.8196\ \rho^{*2} + 4.3539\ \rho^{*3} + 4.1251\ \rho^{*4} \tag{57}$$

   We have fitted the right hand terms of equations 56 and 57 to the universal curves of thermal conductivity and viscosity. The best fits lead to the following equations :

$$\lambda = \lambda_o + 0.860\,(\frac{\varepsilon^{1/2}k\ T^{*1/6}}{\sigma^2\ m^{1/2}})(1.2043\ \rho^* + 3.7915\ \rho^{*2} + 4.3171\ \rho^{*3}) \tag{58}$$

and

$$\eta = \eta_o + 0.222 \left(\frac{m\varepsilon}{\sigma^2}\right)^{1/2} (0.3665 \, \rho^* + 3.8196 \, \rho^{*2} + 4.3539 \, \rho^{*3}$$
$$+ 4.1251 \, \rho^{*4}) \tag{59}$$

We have compared the calculated values of the thermal conductivity and viscosity coefficients with some experimental values from the literature. The agreement is generally better than 10 % except for the viscosities of xenon and carbon dioxide. The model is too simple to represent the correct behavior of the thermal conductivity of polar substances as NH_3 and H_2O.

6.2. Non equilibrium molecular dynamics approximation

A non equilibrium molecular dynamic method to simulate dense fluid transport has been developed by Ashurt (49), (50). He found that the reduced excess transport coefficients times $T^{1/2+2/n}$ are function of one single variable $x = \rho^* T^{-3/n}$. At high temperature, the repulsive core potential dominates momentum transfer and the high temperature Lennard-Jones shear viscosity must approach the scaling behavior of the twelfth power. The calculated reduced excess shear viscosity $\tilde{\eta}^*$ for $x > 0.5$ is well describe by

$$\tilde{\eta}^* = 7.0 \, (kT/\varepsilon)^{2/3} x^4 \tag{60}$$

which shows a negative temperature derivative

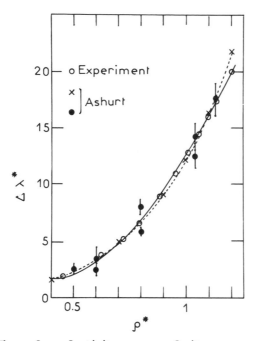

Figure 22. Thermal conductivity excess of nitrogen versus density.

$$(d\ln \eta)/(d\ln T) = - 1/3 \tag{61}$$

A slightly better fit is given by the empirical relation

$$\tilde{\eta}^{*} = 0.0152 \, (\varepsilon/kT)^{-2/3} \, [\, 1 - 0.5 \, (\varepsilon/kT)^{1/2} +$$

$$2.0 \, (\varepsilon/kT) \,]\{\exp \, [7.02 \; (1 - 0.2 \, (\varepsilon/kT)^{1/2}] - 1 \} \tag{62}$$

Experimental excess thermal conductivity data are similar to excess shear viscosity except that the isochoric temperature derivative is positive. The Lennard-Jones thermal conductivity coefficient has been calculated by non equilibrium molecular dynamics simulation of heat flow.

For reduced temperatures greater than one, the results can be represented by a two coefficient exponential function

$$\tilde{\lambda}^{*} = a \, T^{*2/3} \, (\exp \, (bx)-1) \tag{63}$$

where a = 0.36   b = 3.76

We have compared the experimental data of the thermal conductivity of nitrogen with equation 63 in fig. 22. The agreement is surprisingly good.

7. ENERGY TRANSFER IN HIGHLY COMPRESSED FLUIDS

Although an abundance of literature (51) and references therein exist describing condensed phase energy transfer and relaxation phenomenology at ambient pressures and various temperatures, there are only few studies showing behavior as a result of high pressure, of large stress gradients, and of temperatures typical in shock-wave environments. Since understanding molecular energy transfer highly compressed phase is fundamental to understanding shock-induced reactions and detonation, experiments have been initiated to study the effects of pressure up to 3 kbar and temperature (77 K – 400 K range) on highly compressed phase vibrational and electronic energy transfer.

The parameter which is measured in an energy transfer experiment is the vibrational or electronic time τ . In the low density gases τ is connected to the number density, ρ , through the isolated binary collision (IBC) description :

$$\tau = Z P^{*} \tag{64}$$

where Z is the collision frequency and P^{*} the energy transfer probability which does not depend on density. Using the gas kinetic theory, this formula becomes : $\rho\tau=$ constant. An improvement (52) has been made for the liquids, using a simple cell model, which remains an IBC type

$$\tau = A \, (\rho^{-1/3} -\sigma \,) \tag{65}$$

More recently, taking into account the general formulation of IBC model, the radial distribution function g(r) has been introduced (53) : it is proportional to the collision frequency. The problem is to determine where on the potential the g(r) function has to be calculated. This has been partly solved by using a

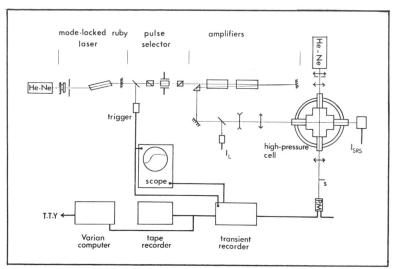

Figure 23. Schematic diagram of the vibrational relaxation time experiment for hydrogen ; the amplified mode-locked ruby laser is the pump. The probe beam intersects perpendicularly the pump beam in the HP cell. I_L and I_{SRS} are the pump laser intensity and the stimulated Raman scattering intensity measurements. The hydrogen molecules are excited on the first vibrational level (v=1) via SRS and the decay of the population monitored by a time resolved Schlieren technique.

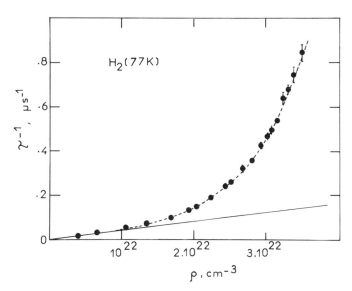

Figure 24. The non-linear dependence of the relaxation rate (τ^{-1}) versus number density (ρ) for hydrogen at 77 K. The straight line represents the IBC model.

hard sphere potential pertubated by an attractive well (54,55).

At L.I.M.H.P., these experiments are based on a pump–probe technique (56,57,58) : the molecules are selectively excited on a well defined molecular energy level via induced infrared absorption or via stimulated Raman scattering during a time shorter than the vibrational (or electronic) relaxation time (fig. 23). The decay of the excited level population is monitored by a time resolved technique which can be or a Schlieren detection, or an infrared (or visible) induced fluorescence. Normal hydrogen and nitrogen have been choosen because their vibrational relaxation time behaviors have been well studied at ambient pressures in the gas and liquid phases. Electronic relaxation time of oxygen hasbeen also investigated using a large pressure gradient pressure up to 2.5 kbar, and temperature (190 K – 300 K) (59).

The well known isolated binary collision model (IBC) could not any longer represent the main features which appear in such experiments, mainly the non-linear pressure dependence of the relaxation rate (inverse relaxation time) : the molecules may not exist individually (fig. 24).

The new environment effect has been introduced in the collision frequency where the radial distribution function for hard sphere potential or softer one has been taken into account. But some discrepancies on the potential parameters remain (60,61) and even so far the problem requires a more sophisticated theoretical approach which has to differ significantly from the frequently used bimolecular collision model (fig. 25).

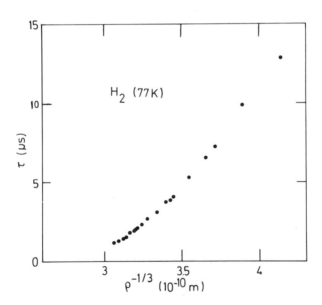

Figure 25. Discrepancy with the simple cell model (see the text) case of H_2 at 77 K.

8. MIXTURES OF DENSE GASES

8.1. Far Infrared Induced Absorption in highly compressed atomic and molecular systems.

In the field of spectroscopic studies, pressure has been currently applied to gaseous molecular samples to observe the density effects on their spectral response. On the other hand, pressure can also be used to simply increase the density and so obtain measurable absorptions from samples with very weak absorption coefficients. This is particularly true for collision induced absorption spectra which originate from the weak dipole moment induced during collisions between two non-polar atomic or molecular systems. The spectroscopic study of

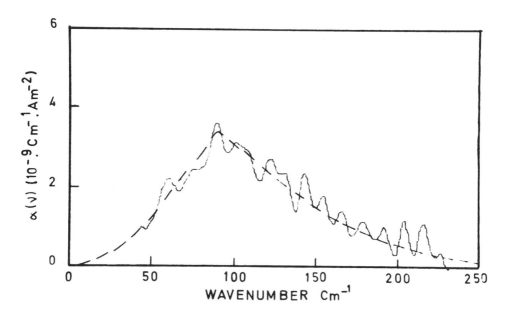

Figure 26. Collision induced translational absorption in an equimolar mixture of He and Ne. $T = 77$ K, $\rho_{He} = \rho_{Ne} = 700$ Am.

such systems, under normal pressure, should require absorption pathlengths of about several thousand meters. A typical example is given by the far infrared collision induced translational absorption in binary mixtures of rare gases. Among ten possible binary mixtures, nine of them have been studied using absorption cells, 1 to 3 m long, at total densities between 60 and 90 Am. However, it is noticeable that twenty five years after the discovery of this absorption process (62,63), no measurable absorption was obtained from the He-Ne mixture for which the induced dipole moment is particularly weak. In fact, Marteau et al. (64) have recently succeeded in recording the absorption spectrum for this pair, using a high pressure cell equiped with diamond windows, 1.25 meter long, at a total pressure of 1500 bar and a temperature of 77 K. Under these experimental conditions the total density of the equimolar mixture was 1400 Am. The same absorption intensity should have been obtained, under normal pressure, through an absorption pathlength of 625 Km. According to the very difficult experimental conditions, the noise could not be eliminated from the recorded spectrum, as shown in fig. 26. Nevertheless it could be concluded to a specific behaviour of the He–Ne pair, compared to all the other rare gas pairs. In particular, the frequency position of the recorded absorption band has been used as a test to confirm some theoretical predictions concerning the polarity of each atom ; as a consequence of the distorsion of the electronic clouds. In the dispersive part of the interaction the following configurations He^- Ar^+, He^- Kr^+, He^- Xe^+ have been established while the He^+ Ne^- configuration might be expected. In fact this prediction seems to be valid in view of the present experimental result.

8.2. Gas–gas equilibria

Only phase equilibria in which one or two of the coexisting phases are in the fluid state will be considered. These systems showing gas–gas equilibrium between fluid phases can persist up to very high pressure. There are two types of gas–gas equilibria. In the first type (fig. 27) the critical line starts at the critical point of the least volatile component and moves continuously to higher temperature when the pressure increases. The second type is shown in fig. 28, the critical line first moves to lower temperature then via a temperature

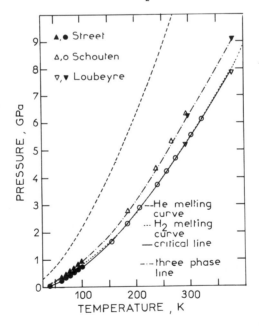

Figure 27. P.T. diagram of He–H$_2$

minimum (double plait point) to higher temperature. In general the critical line ends at the critical endpoint by the appearance of the solid phase. For example in the systeme Ne–Kr the critical curve is intersected by the three phase line at about 0.35 GPa and at 1.5 GPa in the systeme Ne–Xe. The solidification pressure of a binary mixture is given approximately by

$$P_{mix} = P_1 P_2 / (P_1 + x_1 (P_2 - P_1))$$ (66)

where P_1 and P_2 are the melting pressures of pure component 1 and 2 and x_1 is the mole fraction of component 1. The critical line starts at the critical point of the least volatil component (1 for instance). It can be seen from relation (66) that if $x_1 > 0.5$ the value of P_{mix} is closer to the melting pressure P_1 of the least volatile component. The first type of gas–gas equilibria corresponds to the system H_2–He. This effect was first investigated by Streett (65) up to 0.9 GPa. Streett hás made a suggestion about the form of the phase diagram at higher

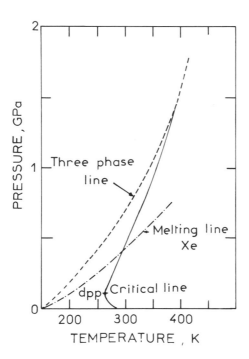

Figure 28. P,T diagram of Ne–Xe

pressure (fig. 29) in the region of pressure induced solidification of the molecular phase. Schouten et al. (66) have reported measurements in a diamond anvil cell of the position of the critical line and the three phase line up to pressure of 5 GPa. These two sets of experiments has shown (fig. 27) that the pressure difference between these two lines increases with temperature. Its follows that the PT range of the fluid region will not be limited by the pressure induced solidification of the less volatile component (H_2).

Recently the H_2–He binary phase diagrams have been studied by Loubeyre et al. (67) (68) up to 373 K and 15 GPa. The concentration of the mixture is

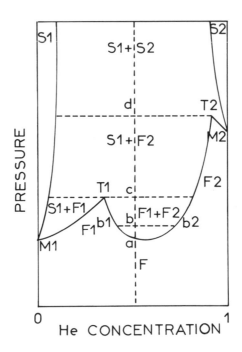

Figure 29. Binary phase diagram of H_2–He.

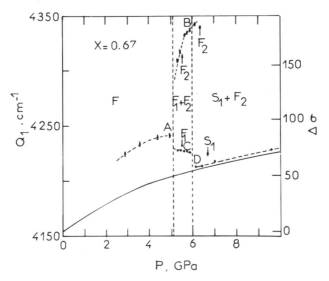

Figure 30. Raman Q_1 vibrational frequencies of an H_2 molecule in the different phases of an initial composition (x=0.67) fluid mixture.

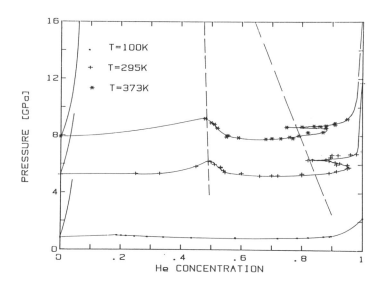

Figure 31. Isothermal binary phase diagrams of H_2-He mixture at three temperatures.

obtained from the measurement of the Raman shift of the vibron Q_1 of the H_2 molecule surrounded by dense H_2-He mixture. Measurements of the Raman shifts of the vibron Q_1 in pure H_2 leads to a melting pressure of 5.23 GPa at 295 K. At 300 K four mixtures have been studied. In fig. 30, the variation of the Q_1 mode versus pressure for an initial concentration of Helium x = 0.67 is shown.

Raman measurements of pure molecular H_2 agree with those of reference (69) (full line) in Fig. 30. Up to 5.1 GPa the fluid is homogeneous and the Q_1 frequency is shift upwards with respect to pure H_2. Between 5.1 and 6 GPa demixing occurs into F_1 (H_2 rich) and F_2 (He rich) fluids. The phase separation is only weakly discontinuous as the concentration is close to critical (x_c = 0.58). Above 6 GPa, F_1 cristallizes into a solid S_1 with an isobaric enrichment in H_2. No discontinuity shift is observed for F_2.

The ratio

$$S(x) = [\sigma(x.P) - \sigma(0,0)] / [\sigma(0.P) - \sigma(0.0)] \qquad (67)$$

were observed to be independent of pressure, where $\sigma(x,P)$ is the Q_1 frequency of the mixture of concentration x in He. $\sigma(0.0)$ and $\sigma(0.P)$ are the frequencies of the Raman active band for pure hydrogen at atmospheric pressure and at pressure P. The experimental phase diagrams presented in figure 31 confirm the evolution of the fluid-fluid domain towards its closing up by an upper critical point (Fig.32) It can be concluded that if the He concentration is less than 0.49 (the Saturn He concentration one is supposed to be 0.35) there will be no fluid separation. If the concentration is larger a layer of demixed fluid fluid phase will take place at a certain depth which can influence the convection modes and the energy flow of the planet Saturn.

The gas-gas equilibria of the second type are of special interest. Their critical line passes through a temperature minimum as a function of pressure (or composition) called double plait point. The critical divergence for a given quantity must be independent of the path of approach to a point of the critical line at constant overall composition except when approaching the critical line tangentially. In the system Ne-Kr showed in fig. 33 the critical exponent for a given quantity in the region where P_c varies smoothly with the temperature, should be the same whether the variable is $T-T_c$ (path 1) or $P-P_c$ (path 2), the composition being held constant. From measurements of the intensity of the scattered light in the hydrodynamic regime Tufeu et al. (70) have obtained the

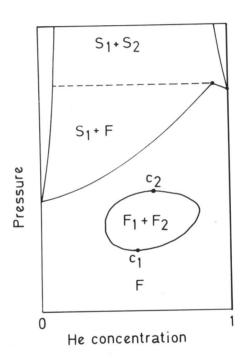

Figure 32. Binary phase diagram of H_2 - He at high temperature.

exponent   governing the divergence of the susceptibility. For path 1 and 2 $\gamma = 1.24$, for path 3 $\gamma = 2.45$. As these results are in good agreement with result obtained for binary liquids, the system Ne-Kr near the double plait point ressembles a liquid-liquid system more than it does a liquid vapor system.

9. CRITICAL ENHANCEMENTS

We have seen that under high compression, at the approach of the fluid solid transition, the compressibility factor is large but the compressibility coefficients are small and there is not an important variation of the thermodynamic and transport properties with the exception of the absorption of sound, the viscosity becomes infinite in the solid. To conclude we would like to show how large variations of the isothermal compressibility coefficient can govern the divergence of other quantities.

At the critical point the compressibility factor of pure substances is of the same order of magnitude. For noble gases and spherical molecules with rough sphere type interactions, Z_c varies from 0.28 to 0.29. When this type of interaction is perturbed the value of Z_c decreases.

The neighborhood of the liquid-gas critical point is characterized by large fluctuations of the order parameter $(\rho - \rho_c)$ where ρ is the actual density and ρ_c the critical density. These fluctuations lead to the divergence of the

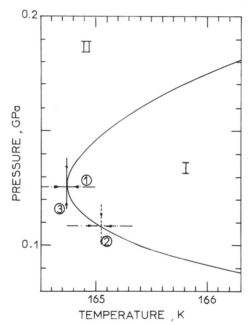

Figure 33. P,T diagram of Ne-Kr.

isothermal compressibility κ.

Density fluctuations are spatially correlated and the spatial extent of these fluctuations is defined by the correlation length ξ. In other words, the range of the order parameter correlation function is represented by the correlation length. Along the critical isochor, the divergence of the isothermal compressibility is expressed by a single power law

$$\kappa = \Gamma \, t^{-\gamma} \tag{68}$$

where Γ is the amplitude of the isothermal compressibility

$$\gamma = 1.24$$

$$t = (T-T_c)/T_c$$

The same expression gives the divergence of the correlation length

$$\xi = \xi_o t^{-\nu} \tag{69}$$

where $\xi_o = 0.63$

Then, the divergence of the correlation length can be expressed as a function of the isothermal compressibility

$$\xi = \xi_o (\kappa/\Gamma)^{\nu/\gamma} \tag{70}$$

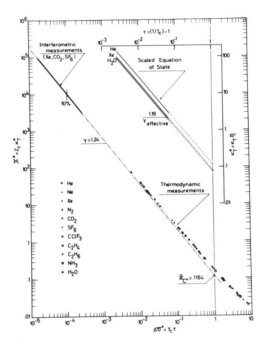

Figure 34. Variation of the scaled isothermal compressibility versus the scaled temperature.

In fig. 34, the variation of the reduced isothermal compressibility of several gases is presented as a function of reduced temperature. The single curve obtained confirms the universal behavior of the divergence.

In fig. 35 , the susceptibility $\rho^2 \kappa$ of ammonia has been plotted as a function of the density. This quantity goes through a maximum at the critical density.

We have seen that the difference of heat capacities is related to the isothermal compressibility coefficients. The divergence of κ will lead to a divergence of C_p as shown in fig. 36 where C_p of ammonia is plotted as the function of the temperature difference along the critical isochore.

The critical thermal conductivity $\Delta\lambda_c$ is given by

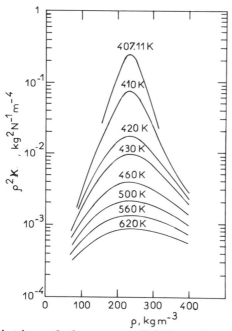

.Figure 35. Variation of the susceptibility of ammonia versus density.

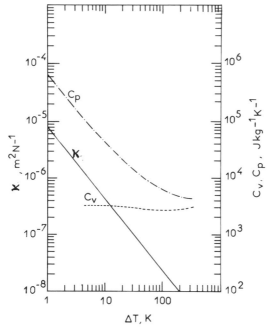

Figure 36. Heat capacities of ammonia versus temperature differences.

$$\Delta\lambda_c = \lambda(\rho,T) - \lambda_B(\rho,T) \tag{71}$$

where $\lambda_B(\rho,T) = \lambda(\rho) + \lambda_o$ is the background thermal conductivity.

This quantity diverges in the critical region as shown for propane in fig. 37.

The mode-mode coupling theory relates the divergence of the critical thermal conductivity to the divergence of the isothermal compressibility by the relation

$$\Delta\lambda_c = \Lambda \frac{k\,T^2}{6\pi\eta\xi} \left(\frac{dP}{dT_\rho}\right)^2 \kappa \tag{72}$$

where ξ can be calculated by eq. 70, the viscosity is η.

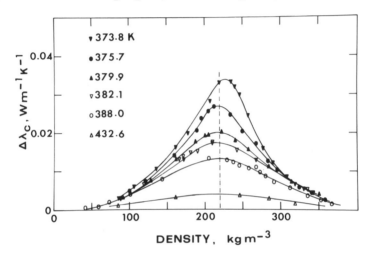

Figure 37. Thermal conductivity excess of propane versus density.

10. REFERENCES

(1) R.L. Mills, D.H. Liebenberg and J.C. Bronson, Phys. Rev. B. **21**, 5137 (1980).

(2) E.R. Grilly and R.L. Mills, Ann. Phys. (N.Y.) **8**, 1 (1959).

(3) R.K. Crawford and W.B. Daniels, J. Chem. Phys. **55**, 5651 (1971)

(4) J.P. Pinceaux and J.M. Besson, in 7th International High Pressure AIRAPT Conference Le Creusot, France edited by Ph. Marteau and B. Vodar, Pergamon, Oxford

(5) P. Loubeyre, Thesis, University of Paris VI (1982)

(6) W.H. Hardy, R.K. Crawford and W.B. Daniels, J. Chem. Phys., **54**, 1005 (1971)

(7) D.H. Liebenberg, R.L. Mills and J.C. Bronson, J. Appl. Phys. **45**, 741 (1974)

(8) C.S. Zha, R. Boehler, D.A. Young and M. Ross, J. Chem. Phys., **85**, 1034 (1986)

(9) V. Diatschenko, C.W. Chu, D.H. Liebenberg, D.A. Young, M. Ross and R.L. Mills, Phys. Rev. B, **32**, 381 (1985).

(10) F. Simon, M. Ruheman and W.A.M. Edwards, Z. Phys. Chem., **B6**, 331 (1930).

(11) R.L. Mills a,d E.R. Grilly, Phys. Rev. 101, 1246 (1956).
(12) V.V. Kechin, A.I. Likhter, Ya M. Pavlyuchenko, L.Z. Ponizovskii and A.N. Utyuzh, Zh Eksp Teor Fiz, 72, 345 (1977).
(13) D.H. Liebenberg, R.L. Mills and J.C. Bronson, Phys. Rev. B. 18, 2663 (1978).
(14) R.L. Mills, D.H. Liebenberg, J.C. Bronson and L.C. Schmidt, Rev. Sci. Instrum., 51, 891 (1980).
(15) H.K. Mao and P.M. Bell, Science 203, 1004 (1979).
(16) H. Shimizu, E.M. Brody, H.K. Mao and P.M. Bell, Phys. Rev. Lett. 47, 128 (1981).
(17) R.L. Mills, D.H. Liebenberg and J.C. Bronson, Journal of Chemical Physics 63, 4026 (1975).
(18) R.L. Mills and E.R. Grilly, Phys. Rev. 99, 480 (1955).
(19) E.R. Grilly and R.L. Mills, Phys. Rev. 105, 1140 (1957).
(20) V.M. Cheng, W.B. Daniels and R.K. Crawford, Physical Review B, 11, 3972 (1975).
(21) H.C. Longuet-Higgins and B. Widom, Mol. Phys. 8, 549 (1964).
(22) J.S. Rowlinson, Mol. Phys. 7, 349 (1964) ; 8, 107 (1964).
(23) D.A.Mc Quarrie and J.L. Katz, J. Chem. Phys., 44, 2393 (1966).
(24) R.A. Aziz et al., J. Chem. Phys., 70, 4330 (1979).
(25) M. Ross and D. Young, Phys. Lett. 78, 463 (1980).
(26) D. Young, A.K. Mac Mahan, M. Ross, Phys. rev. B24, 5119 (1981).
(27) N.W. Ashcroft and J. Lekner, Phys. Rev. 145, 83 (1966).
(28) M. Ross, J. Chem. Phys., 71, 1567 (1979).
(29) M. Ross, J. Chem. Phys. 60, 3634 (1974).
(30) F.M. Mouritz and F.H.A. Rummens, Can. J. Chem. 55, 3007 (1977).
(31) P.H. Lahr and W.G. Eversole, J. Chem. Eng. Data, 7, 42 (1962).
(32) M.Ross, F.H. Ree and D.A. Young, J. Chem. Phys.79,1487,(1977).
(33) M. Ross, F.H. Ree and D.A. Young, J. Chem. Phys. 79, 1487 (1983)
(34) S.E. Babb Jr, Rev. Mod. Phys. 35, 400 (1963).
(35) J.P. Hansen, Phys. Rev. A2, 221 (1970).
(36) K. Mukherjee, Phys. Rev. Letters, 17, 1252 (1966).
(37) A. Turturro and U. Bianchi, J. Chem. Phys., 62, 1668 (1975).
(38) N.G. Hoover, M. Ross, K.W. Johnson, D. Henderson, J.A. Barker and B.C. Brown, J. Chem. Phys. 52, 4931 (1970)
(39) S.M. Stishov and V.I. Fedosimov, J.E.T.P. Lett. 14, 217 (1971).
(40) J.J. Nicolas, K.E. Gubbins, N.B. Street and D.J. Tildesley, Mol. Phys. 37, 1429 (1979).
(41) D. Vidal, R. Tufeu, Y. Garrabos and B. Le Neindre, in "High Pressure Science and Technology" Eds B. Vodar and Ph. Marteau, p. 692, Pergamon Press. (1980).
(42) A. Polian and M. Grimsditch, To be published in Europhys. Lett.
(43) D. Vidal, L. Guengant and M. Lallemand, Physica 96 A, 545 (1970).
(44) M. Lallemand and D. Vidal, J. Chem. Phys. 66, 4776 (1977).
(45) B.I. Alder, H.L. Strauss and J.J. Weiss, J. Chem. Phys., 62, 2328 (1975).
(46) J.H. Dymond, Physica 75, 100 (1974) ; 85A, 175 (1976).
(47) S.F.Y. Li, R.D. Trengove, W.A. Wakeham and M. Zalaf, Int. J. Thermophysics 7, 273, 1986.
(48) J.J. Van Loef, Physica, 124B, 305 (1984).
(49) W.T. Ashurst and W.G. Hoover, Phys. Rev. A, 11, 658 (1975).
(50) W.T. Ashurst, PhD Dissertation University of California (1974).
(51) J. Chesnoy and G.M. Gale, Ann. de Physique, 9 (1984) 893.

(52) W.M. Madigosky and T.A. Litovitz, J. Chem. Phys. $\underline{34}$ (1961) 489.

(53) P.K. Davis and I. Oppenheim, J. Chem. Phys. $\underline{57}$ (1972) 505.

(54) C. Delalande and G.M. Gale, J. Chem. Phys. $\underline{71}$ (1979) 4804.

(55) J. Chesnoy, Chem. Phys. $\underline{83}$ (1984) 283.

(56) M. Châtelet, B. Oksengorn, G. Widenlocher and Ph. Marteau, J. Chem. Phys. $\underline{75}$ (1981) 2374.

(57) M. Châtelet, J. Kieffer and B. Oksengorn, Chem. Phys. $\underline{79}$ (1983) 413.

(58) M. Châtelet and J. Chesnoy, Chem. Phys. Letters, $\underline{122}$ (1985) 550.

(59) M. Châtelet, A. Tardieu, W. Spreitzer and M. Maier, Chem. Phys. $\underline{102}$ (1986) 387.

(60) A. Tardieu, M. Châtelet and B. Oksengorn, Chem. Phys. Letters, $\underline{120}$ (1985) 356.

(61) M. Châtelet, J. Chesnoy, M. Maier, A. Tardieu and B. Oksengorn, Physica, $\underline{139B}$ (1986) 512.

(62) Z.J. Kiss and H.L. Welsh, Phys. Rev. Lett., $\underline{2}$, 166 (1959).

(63) D.R. Bosomworth and H.P. Gush, Can. J. Phys. $\underline{43}$, 751 (1965).

(64) Ph. Marteau, J. Obriot and F. Fondère, Can J. Phys., $\underline{64}$, 822 (1986).

(65) W.B. Street, Astrophys. J. $\underline{186}$, 1107, (1973).

(66) J.A. Schouten, K.C. Van der Bergh and N.T. Trappeniers, J. Chem. Phys. $\underline{114}$, 401, (1985).

(67) P. Loubeyre, R. Le Toulec and J.P. Pinceaux, Submitted to Phys. Rev. B.

(68) P. Loubeyre, R. Le Toulec and J.P. Pinceaux, Am. Phys. Soc. $\underline{32}$, 7611 (1985).

(69) S.K. Sharma, H.K. Mao and P.M. Bell, Phys. Rev. Lett. $\underline{44}$, 886 (1979).

(70) R. Tufeu, P.H. Keyes and W.B. Daniels, Phys. Rev. Letters $\underline{35}$, 1004 (1975).

EXPERIMENTAL STUDIES OF COMPRESSED FLUIDS

E.U. Franck
Institut für Physikalische Chemie
Universität Karlsruhe
Kaiserstraße 12
D7500 Karlsruhe
Federal Republic of Germany

ABSTRACT. With high pressures at elevated and supercritical temperatures
fluids can be studied over a very wide range of density. Fluids with
strongly interacting polar molecules - for example water - are of
particular interest. Results for several thermophysical properties are
presented. PVT-data of sodium, high pressure-high temperature viscosity
of decane and the dielectric constant of supercritical water-benzene
mixtures are discussed. The continuous transition from ionic to
metallic states is shown with the cesium-cesium hydride system to 800 °C.
High pressure far-infrared spectroscopy reveals the dynamics of compressed polar methyl halides. Quantitative determinations of the
solubility of solids in dense supercritical gases are presented. A
number of phase diagrams and critical curves of binary aqueous systems
to 400 °C and 2000 bar have been determined. A new rational equation of
state for such systems is presented. "Hydrothermal" combustion of
methane in supercritical water is possible and a flame, burning at 1000
bar is shown.

1. INTRODUCTION

Remarkable changes of properties and phenomena can be produced in fluid
systems with pressures up to a few thousand bar. This is particularly
obvious at elevated temperatures. Pressures of such magnitude in general
do not cause appreciable changes within molecules, although important
and interesting examples exist (1). Such pressures are sufficient,
however, to produce density variations from gas-like to liquid-values.
Isolated molecules can be brought to conditions where they remain
predominantly or entirely within the range of the intermolecular
potentials of their neighbours. Thus applications of high pressure can
be considered as probes into the region of intermolecular interaction
for scientific reasons as well as means to achieve special states and
transformations for technical purposes.
    Intermolecular interaction is particularly strong between polar
particles. Compounds with high dipole moments usually have higher

R. van Eldik and J. Jonas (eds.), High Pressure Chemistry and Biochemistry, 93–116.
© *1987 by D. Reidel Publishing Company.*

critical temperatures, however, and tend to be corrosive. Since
continuous and wide density variations are often desirable, supercritical temperatures have to be applied. Experimental equipment has to
be designed which permits observations and quantitative, accurate
measurements at conditions of high pressures, high temperatures and
corrosive action. Interesting developments are described in different
contributions of these proceedings. Several selected examples shall be
mentioned here.

At first, methods and results concerning certain thermophysical
properties of fluids will be described. A discussion of two types of
ionic fluids will follow. The section on spectroscopic investigations
at elevated pressure will be limited to far-infrared experiments and
the ultraviolet spectra of gaseous supercritical solutions. Critical
curves and high pressure phase diagrams of aqueous systems and their
calcualtion will be discussed in the next chapter. As a conclusion,
hydrothermal combustion and the generation of"supercritical" flames will
be treated.

2. THERMOPHYSICAL PROPERTIES

Data of the pressure-volume-temperature relation - equation of state
data - for fluids with strongly interacting particles are important as
such, but must also supplement other thermophysical results for improved
interaction. Among the highly polar substances water is certainly the
most extensively investigated. The existing steam tables reach to 1000
bar and nearly 1000 $^{\circ}$C (2). Static density measurements have been
extended to nearly 1000 $^{\circ}$C and 10 kbar (3)(4)(5)(6). Shock wave data
reach into the 100 kbar range (7). Another group of fluids for which
density measurements at high temperatures and pressures are of great
interest are fluid sub- and supercritical metals. So far only mercury
and alkali metals have been investigated to the supercritical region.
Experimental data have been published for mercury (8)(9)(10), for
cesium (11)(12) and rubidium (13). In addition numerous determinations
of the critical points for alkali metals are available (14). Recently
an apparatus was designed and used to determine the density of sodium
to 2600 K and 500 bar (15). Critical temperature, pressure and density
of sodium are assumed to be at 2485 K, 298 bar and 0.30 g cm^{-3}.

The autoclave used for the sodium measurements was filled with
pressurized argon and had an internal heating. A schematic drawing is
shown in Fig. 1. The most important part is a cell of 4.5 cm^3 volume
made of tungsten-rhenium alloy. This cell is brought to the high,
desired temperature with a tungsten resistance heater within the
pressurized argon. The cell is connected by a tungsten-rhenium capillary
with a stainless-steel bellows at a temperature, lower but somewhat
higher than the melting temperature of sodium. The elongation of the
bellows can be precisely measured by a stainless steel with a magnetic
tip within a coil arrangement. Bellows and cell were filled with a
weighed amount of liquid sodium. The pressures of argon and sodium
were equilibrated with the bellows. The expansion of the sodium in the
cell produced by heating could be determined from the bellows extension.

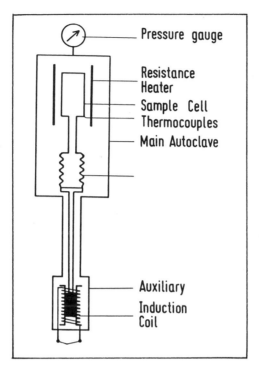

Fig. 1: Schematic drawing of the cell, mounted within an autoclave, to measure PVT-data of sodium to 2600 K and 500 bar (15).

Temperatures were measured by several tungsten-rhenium thermocouples in close contact with the cell. The sodium pressure was obtained from the argon pressure,

Fig. 2 presents a set of experimentally determined density-pressure isotherms (15). Above 1500 K and above 26 bar no other experimental data for sodium in the homogeneous region beyond the coexistence curve were available.

Another important thermophysical property is the viscosity of fluids at high pressure, which permits estmates for diffusion coefficients and ion mobilities in solutions. An extensive literature exists. For the viscosity of dense fluids see for example K. Stephan and K. Lucas (16). Viscosity measurements with water could be extended to 500 °C and 5 kbar (T_c=374 °C, p_c = 221 bar) (17). Determinations with hydrocarbons can usually not be made at such high temperature because of thermal decomposition. An apparatus was built to measure the viscosity of supercritical decane-methane mixtures to 300 °C and 3 kbar (18). The method applied was that of the oscillating disc suspended on a torsion wire within an autoclave. The oscillation time was measured with a laser beam reflected from a small mirror below the disc, using a high pressure sapphire window. The purpose of the operation was to investigate the influence of admixed methane on the momentum transport of decane. In the literature existed so far only

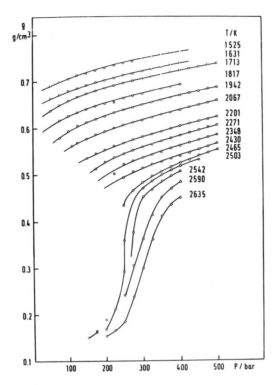

Fig. 2: Experimentally determined isotherms of the density ρ of sodium as functions of pressure (15).

viscosity data of decane to 200 °C and above 75 °C only at pressures to 500 bar. A selected number of new decane data are shown in Fig. 3 in a diagram of experimental viscosity isotherms as functions of pressure, The isobaric temperature dependence is very pronounced. It is less strong if equation of state data are used to calculate the temperature dependence at constant density. Fig. 4 shows it very clearly with a comparison of log η against 1/T curves at constant pressures or densities. Accordingly the activation energies are different: E_p (constant pressure) = 10.5 kJmol^{-1} and E_v (constant volume) is between 3.4 and 0.9 kJ mol^{-1} (for high and small molar volume). If equation of state data are available, extrapolation of viscosities of fluids are easier along isochors. This is true for the thermal conductivity as well, for example for water (19).

The static dielectric constant is of primary importance for electrolytic solvents. For water this property was measured to 550 °C and 5000 bar (20)(21). Fig. 5 shows a three-dimensional diagram based on measurements and on calculations above 500 °C. Although the calculations have a certain degree of uncertainty, it is evident. that at supercritical density and temperature there is an extended region with dielectric constants between 10 and 25, with means, that

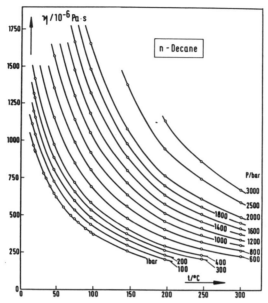

Fig. 3: Experimentally determined isobars of the viscosity η of decane as functions of temperature to 300 °C (18).

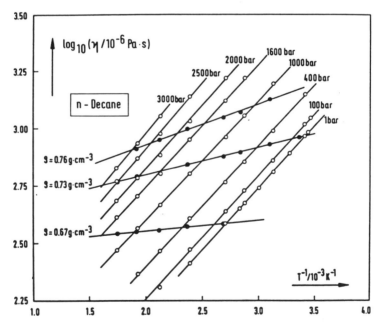

Fig. 4: The logarithms of the viscosity η of decane as functions of the reciprocal temperature at constant densities (isochors) and constant pressures (isobars)(18).

electrolytic dissociation of solutes is possible. Comparatively little
is known of the dielectric constant of mixtures of water and a nonpolar
second component.

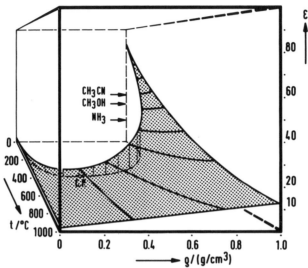

CH₃CN
CH₃OH
NH₃

Fig. 5: The static dielectric constant of water, ε, as a function of
temperature t and density ρ in an approximate presentation (20)(21).

measurements with such mixtures can obviously not normally be made at
room temperature because of insufficient miscibility - with the
exception of some cases, where the nonaqueous molecules, for example,
have high local dipole moments or a suitable quadrupole moment like
dioxane, which is completely miscible with liquid water. Besides a
general interest in dielectric properties of high-pressure supercritical
polar-nonpolar mixtures, such aqueous phases with carbon dioxide,
methane, nitrogen or other partners are important as hydrothermal fluids
in geochemistry.

     The critical temperature of water is 374 °C. At 400 °C water is
completely miscible with benzene even at high density. Measurements of
the static dielectric constant could be made with such fluid binary
mixtures at 400 °C and up to 2000 bar (22). Fig. 6 shows results as
isobars over the whole composition range. At 2000 bar the variation
of the dielectric constant ranges from 2 for pure benzene to 20 for
pure water. The addition of 10 moles percent benzene to water reduces
the value to one half. On the other side it needs 50 moles percent of
water, added to benzene, to increase the dielectric constant only from
2 to 4. Relative small amounts of water seem to reduce the remaining
water "structure" considerably. Theoretical descriptions in the
literature for the dielectric constant of liquid mixtures have so far
not been extended to include wide variations of the total density. It
appears, however, that a relatively uncomplicated function can
represent the supercritical water-benzene mixture if suitably
defined volume fractions are used (23)(24)(25). It is interesting, to

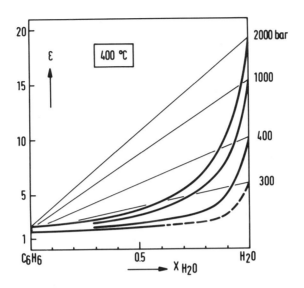

Fig. 6: The static dielectric constant, ε , of supercritical mixtures of water and benzene at 400 °C and four different pressures as a function of the water mole fraction (22).

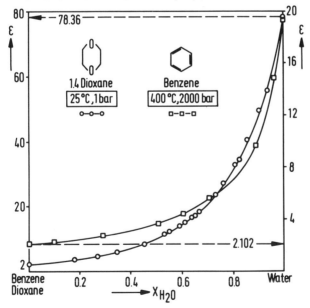

Fig. 7: The static dielectric constant ε as a function of the water mole fraction for water-benzene mixtures at 400 °C and 2000 bar (right-hand scale) and water-dioxane mixtures at 25 °C and 1 bar (left-hand scale).

compare the high temperature-high pressure fluid with liquid water-
dioxane mixtures at ordinary conditions in Fig. 7 (22). Different
scales for the dielectric values have been used. The similarity between
the two curves for such very different conditions is remarkable (26).
Very recent experimental results for methane-methanol mixtures at
high pressures, but below the critical point of pure methanol show a
somewhat different behaviour, however.

3. IONIC FLUIDS

The ionic dissociation of solutes in polar solvents is often increased
by high pressure, because solvation spheres around ions are more
densely packed than the free solvent (1). This applies also to the
self-ionization of pure water. Fig. 8 gives a three-dimensional diagram.
similar to Fig. 5, of the "ion-product" of water as a function of
temperature and density to 1000 $^\circ$C (27). Electrochemical measurements
of various kinds - at high temperatures mainly the analysis of
conductance determinations - lead to a function which describes the
ion product for this region. At 1000 $^\circ$C and a density of 1 g cm^3 the
ion product is calculated to be 10^{-6} (mol·l^{-1})^2. There may be an
uncertainty of one order of magnitude, but there is good reason to
believe, that from room temperature to 1000 $^\circ$C the ion product at
constant density increases by a factor of 10^7 to 10^8. This is supported
by static conductance measurements to 100 kbar (28) and more recently
by conductance data from shock waves to several thousand K and 500 kbar
(29). The shock wave experiments appear to confirm an earlier suggestion
that water at 1000 K and above and at a density of 2 g cm^{-3} and more
should become an ionic fluid like molten sodium hydroxyde. It is
possible, that compressed ammonia undergoes an analogous change. Fully
ionized, high temperature-high density water may occur in the large,
outer planets (30). Its high electric conductivity would be of
consequence for the electric and magnetic properties of these planets.
    Dilute solutions of electrolytes in supercritical water can produce
a molar conductivity, nearly ten times as high as those of normal
liquid solutions (31)(32)(33). The low viscosity in the supercritical
phase causes high ion mobilities. It is a question, whether higher
solute concentrations could lead to very high conductivities. This is
not the case: The concentration of effective charge carriers in the
solutions decreases rapidly with increasing concentration as shown in
Fig. 9 for sodium chloride solutions at 2000 bar (34). On the right
side the molar conductivity of molten sodium chloride is shown together
with extrapolated values for 400 and 500 $^\circ$C, assuming that the liquid
salt could be supercooled. The temperature dependence is not great and
it is also small for the aqueous solutions at these conditions. Thus
the diagram shows, that the conductivity at a mole fraction of 0.1 of
salt approaches already the values for molten salts. For sodium chloride
the whole range of mole fractions could not be covered by experiments.
Sodium hydroxide, however, melts already at 320 $^\circ$C and complete
miscibility exists above this temperature with pressurized water. Very
recent measurements, made with the sodium hydroxide - water system over

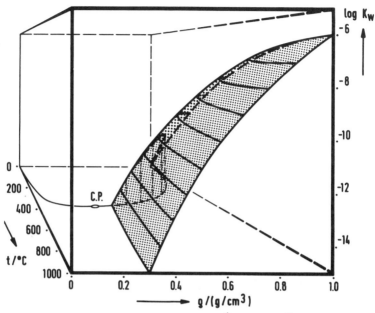

Fig. 8: The ion product of water, log K_w=a(H^+) x a(OH^-) as a function of temperature t and density ρ in an approximate presentation (27).

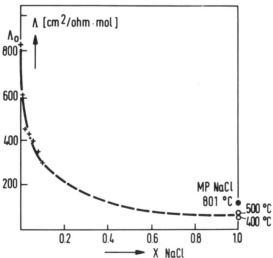

Fig. 9: The molar conductivity Λ of aqueous sodium chloride solutions as a function of the NaCl mole fraction x at 400 and 500 °C. +++: Experimental results. The difference of data at 400 and 500 °C is not relecant in this scale. For the liquid salt: 0 = experimental, 0,0 = calculated.

the whole range of concentrations at 400 °C and 2000 bar, show indeed
the conductance behaviour suggested by Fig. 9.

Conductance measurements with fluids at high pressures and temperatures can also demonstrate the continuous transitions from insulating
to metallic states (35) or from ionic to metallic states. The latter
phenomenon can be observed with liquid mixtures of fused alkali halides
and the corresponding liquid alkalimetals (36). Alkali hydrides are
salt-like compounds similar to the halides. Since the negative hydrogen
ion is the simplest anion, it is attractive to study the ionic-metallic
transition with hydrides. With the exception of lithium hydride,
however, the alkali hydrides at melting temperature and higher are
stable only under elevated hydrogen pressure. Thus, so far only phase
diagrams for lithium hydride (37), sodium hydride (38) and cesium
hydride have been thoroughly investigated (39)(40)(41). Fig. 10 gives
experimentally determined isotherms in   hydrogen pressure produced
cesium-cesium hydride mixtures with increasing hydride content. Below
the critical solution temperature of 665 °C the isotherms pass through
a "plateau pressure" which indicates the region of two coexisting
liquid phases. Above 665 °C the isotherms ascend uninterrupted to high
hydrogen pressures. These measurements were made with cylindrical
autoclaves made from nickel-base super-alloys which could be used with
hydrogen to 800 °C and 1000 bar. Typical dimensions were: 350 mm
length, 80 mm o.d. and 50 mm i.D. Technical difficulties are caused
by the diffusion of hydrogen into container metals and by the corrosive

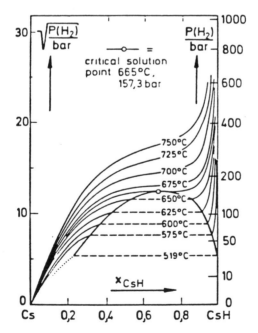

Fig. 10: The system Cs-CsH. Isotherms of the equilibrium hydrogen
pressure p(H$_2$) between 500 and 750 °C according to Szafranski (39)(41).

action of hydrogen and cesium. External heating was used. The interior
of the autoclave was divided into two concentric compartments with
equalised hydrogen pressures. Hydrogen loss from the outer compartment
could be replaced. The liquid cesium and the liquid mixtures were
contained in a cylindrical capsule of very pure iron with thin walls
and a few cm^3 volume. The capsule was mounted within the inner
compartment. Hydrogen could easily diffuse through the iron. The
amount of hydrogen absorbed by cesium at a given temperature and
pressure determined the hydride mole fraction of the mixture.

    After the phase diagram was determined, measurements of the electric
conductance were undertaken in the one-phase region above 665 °C over
the whole range of concentrations. The apparatus was similar to the one
for the phase equilibrium determinations. The conductance cell, also
of thin-walled pure iron, was placed within the inner compartment
instead of the capsule. In the high conductivity (metallic) region
the resistance was measured parallel to the axis of the sample-filled
cell with a four-pole arrangement with DC. Conductance of the cell
walls had to be subtracted. In the low conductivity (ionic) region two
electrodes with alumina isolation with AC were applied.

    Fig. 11 shows a selection of conductance isotherms in a logarithmic
scale as functions of the mole fraction of hydride. The total variation
from liquid metal to liquid salt extends over three and a half orders

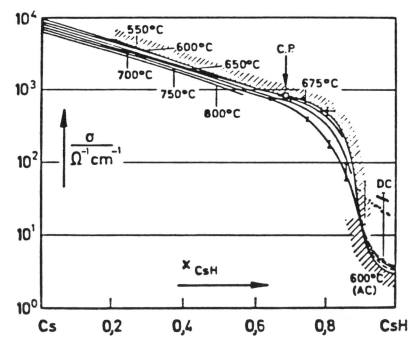

Fig. 11: Experimental conductivity isotherms of liquid Cs-CsH-mixtures
as functions of the mole fraction x(CsH).   :liquid-liquid phase
separation.

of magnitude. From the pure metal to a mole fraction of 0.6 the
logarithm of the conductance decreases moderately and linearely. A very
steep drop of the conductance and a change of the temperature derivative
from negative to positive occurs around a mole fraction of 0.88. In the
range of mole fractions from zero to 0.6 the so-called "nearly free
electron" (NFE) model can be applied. The metallic-ionic transition
does not, as in pure cesium (42)(43), coincide with the critical
concentration (x(CsH) = 0.68, see Fig. 10). To investigate the same
phenomenon with the sodium-sodium hydride system, much high temperatures and hydrogen pressures would be necessary (38).

4. SPECTROSCOPY

Spectroscopic methods of various kinds are particularly well suited to
study molecular interaction in fluids under pressure. Two recent
examples shall be discussed here. The first is concerned with the
dynamics of small polar molecules in dense fluid phases. Far infrared
spectra are used to study rotational motions which are markedly
influenced by density and temperature. Above the critical temperatures
the density varied from low, gas-like, to high, liquid-like, values.
    $CHClF_2$ and CH_3F are examples for a class of small, nearly spherical
polar molecules. Investigations of other properties of the fluid
phases have already been made. Critical data of $CHClF_2$ and CH_2F are:
96,2 $^\circ$C, 50 bar, 0,51 g cm^{-3} (=5,93 mol dm^{-3}) and 44.4 $^\circ$C,
58 bar, 0.28 g cm^{-3} (= 8.23 mol dm^{-3}). The dipole moments are 1.43 D
für $CHClF_2$ and 1.86 D for CH_3F. FIR-spectra have been obtained for
$CHClF_2$, CH_3F and CH_3F-argon mixtures (44)(45)(46). The spectra were
obtained with a fourier transform infrared interferometer, adapted to
meet the technical requirements for high pressure application. A
special optical cell was designed with a sample length between windows
which is variable at operating conditions. A schematic drawing is
shown in Fig. 12. Pressure and temperature can be raised to 1000 bar
and 300 $^\circ$C. The windows are of z-cut crystalline quartz from 10 to
90 cm^{-1} and of pure silicon from 60 to 300 cm^{-1}. The molar absorptivity
of the sample as a function of wave number is always obtained from two
intensity spectra, recorded at different path lengths. Thus influences
of the window-sample interfaces can be eliminated.
    The molar absorptivity curves of $CHClF_2$ and CH_3F at moderate
densities have pronounced maxima at 26 cm^{-1} and at 40 cm^{-1}
respectively. At constant temperature the absorptivity at these wave
numbers is reduced with increasing pressure or density. Instead the
absorptivity around 60 cm^{-1} or at 80 cm^{-1} rises. The interpretation
assumes two types of molecular motion: inertia determined rotation and
also libration - a kind of torsional vibration of the whole molecule.
The decrease near 26 and 40 cm^{-1} indicated the decrease of the inertia
determined rotation while the increase with growing density around
60 and 80 cm^{-1} shows the rise of librational motion. An extensive
discussion is given elsewhere (44).
    Since growing density suppresses the inertia determined rotation,
it is of interest to investigate to what extent this influence depends

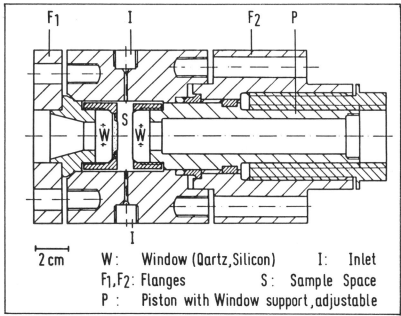

Fig. 12: High pressure - High temperature optical cell for FIR investigations (44)(45).

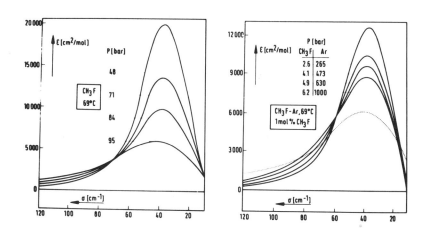

Fig. 13: a) Molar absorptivity $\varepsilon(\sigma)$ of CH_3F between 120 and 10 cm^{-1} at 69 °C at several pressures. b) Molar absorptivity of 1 mol% CH_3F in argon at several total pressures. Dotted line: Pure CH_3F at the same total molar density as at the 2.6 - 265 bar condition (45)(46).

on the polar character of the environment of individual molecules. For this purpose, fluid mixtures were prepared, which contained a small amount of CH_3F (about 1 mol percent) in pressurized argon. It should be mentioned, that the methyl fluoride molecule has nearly the same dipole moment as the isolated water molecule. Fig. 13 shows some of the spectra for a constant temperature of 69 $^{\circ}$C. The left side of Fig. 13 is for pure CH_3F and shows clearly the reduction with increasing pressure. The isobar for 95 is also shown as a dotted curve on the right side. The total molar density is the same as that of the 2.6 - 265 bar mixture (Highest isobar). It is evident that at the same number density argon has a much less pronounced influence on the CH_3F-rotation than a polar environment of methyl fluoride. The influence of further increased argon pressure (to 1000 bar argon) is also shown. The libration range is much less sensitive to the environmental difference. Unlike to normal maxtrix isolation spectrosocpy the dissolution of the methylfluoride in the high pressure argon provides a non-oriented environment with variable interparticle distances.

The second example for a spectroscopic investigation of intermolecular interaction is the determination of solute concentrations in supercritical gases by optical absorption. Dense supercritical gases can be very good solvents for certain solids and high boiling liquids. The solute can easily be precipitated again from the solution by pressure reduction or increase of temperature. This phenomenon is already technically used for various purposes in the chemical and food industry and in the oil and gas industry (47). One of the best known applications is the extraction of caffeine with supercritical carbon dioxide. It can be expected that carbon dioxide with its high quadrupole moment is a better solvent than, for example, methane. Although qualitative investigations with technical objectives are numerous, quantitative determinations of the solubility and its temperature and pressure dependence are limited. A number of such measurements were made with several systems, for which the equilibrium concentrations of the solute were determined by optical absorption in the near ultraviolet region (40)(49).

Special, externally heated high pressure cells were built from a nickel-base superalloy for use to 300 $^{\circ}$C and 2000 bar. They had two windows of cylindrical quartz or sapphire pieces, fitted to Poulter-type, flat surface seals. The interior of the cell contained a support on which the respective solid solute was placed. The pressurized gaseous solvent was introduced through a cappillary. A stainless steel bellows was also placed within the sample space. It was filled from outside with another auxiliary gas and served to increase or decrease the density of the solvent gas without changing its quantity. The whole cell was mounted within the smaple compartment of a UV-spectrometer. Certainly, only such solutes could be investigated which have a suitable absorption band in the near ultra-violet region. Also, the solute should have relatively high melting tmeperature, so that it can be expected to dissolve only very little solvent gas in the solid state. The vapour pressure of the pure solute should be low, so that the high pressure solutions remain always dilute. The advantage is, however, that no solution sample have to be taken and that several solubility data at different temperatures and pressures can be obtained from one sample.

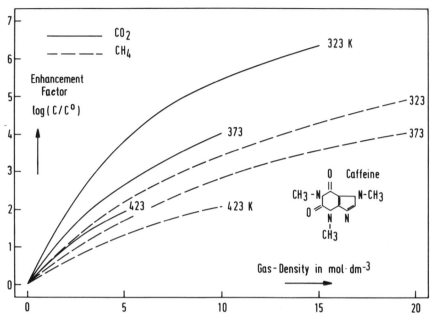

Fig. 14: The logarithms of the enhancement factors log (c/c_o) for the solubility of caffeine in high pressure CO_2 and CH_4 as functions of the molar density ρ(49).

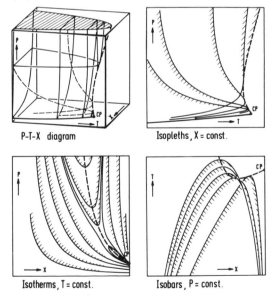

Fig. 15: Schematic phase diagram in the critical region for a binary system water-nonpolar gas. Shaded area: Two-phase region. CP: Critical point of water. x: Mole fraction of water. ---: Critical curve.

Some results for caffeine, dissolved in methane and carbon dioxide are shown in Fig. 14. Here caffeine was used, not because of its practical importance but because of its suitable properties: UV-absorption, vapour, pressure, and partly polar molecule. The solubility is given by "enhancement factors", c/c_o. Here c is the concentration of solute in the high pressure gaseous solvent and c_o the solute concentration at the same temperature caused by the solute vapour pressure in the absence of solvent gas. Fig. 14 shows that, as expected, the enhancement factors increase with growing solvent gas density. The factors reach values between $1o$ and 10^6, which means, that the solubility enhancement of both solvent gases is very high. The c_o-values were determined with the same method. The existing high pressure PVT-data of methane and carbon dioxide made the plots as functions of solvent density possible. The enhancement by carbon dioxide is clearly more pronounced. At constant solvent density the temperature coefficient is always negative. Values for solvent-solute interaction energies can be derived, which tend to reach constant values at very high solvent density.

5. PHASE DIAGRAMS

Binary systems of water and a nonpolar partner can be completely miscible and homogeneous even at high liquid-like density if the temperature is high enough, that is above the critical curve of the particular system. Such critical curves begin at the critical points of the two pure components and extend into the three-dimensional temperature-pressure-composition space (50). For water and nonpolar partners such curves are usually interrupted. The main "upper" branch begins at the critical point of water. The curve   is an high temperature envelope of the region of two coexistent fluid phases. Fig. 15 shows a schematic phase diagram of this type. Qualitatively it describes the water-nitrogen system. In addition to the three-dimensional temperature-pressure-composition diagram the three two-dimensional projections are given. A comprehensive discussion of critical phenomena of high temperature fluids is found elsewhere(51,52).
     A number of aqueous systems have been experimentally investigated to about 400 °C and 2000 bar and more. Among these are H_2O-Ar(53), H_2O-Xe(54), H_2O-He(55), H_2O-H_2(56), H_2O-N_2(57), H_2O-O_2(58), H_2O-CO_2(59), H_2O-CH_4(60), H_2O-benzene (61) and others. With the exception of H_2O-He and possibly H_2O-Ar and H_2O-H_2 all the critical curves pass through a temperature minimum before they proceed to high temperatures and high pressures. - Two of these systems have been investigated only recently. One of these is water-oxygen, for which "isopleths", curves for constant composition, are shown in Fig. 16. The critical curve has a slight minimum. The data were taken, like most of the data for the other systems, with the use of the so-called synthetic method: Known quantities of both partners were introduced into the sample space of known volume. Starting from room temperature isochoric pressure-temperature curves are recorded, which usually show a break-point when the two-phase sample becomes homogeneous. The location of this breakpoint can be confirmed by visual observation through sapphire windows. The handling of water-oxygen mixtures to 400 °C and 2000 bar requires certain precautions. A set of such break-

point measurements gives an isopleth. The two-phase equilibrium ("binodal") surface and the critical curve can be constructed from these isopleths. Although at room temperature the solubility of oxygen in water
is twice as high as that of nitrogen, this difference disappears, when
the critical region is approached. The phase diagrams of water-oxygen
(58) and water-nitrogen (57) at high temperatures and pressures are very
similar. Thus the range of complete miscibility of water with oxygen and
also with air to high densities is known and can be used.

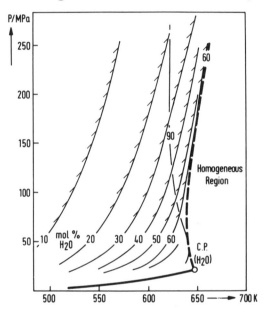

Fig. 16: The phase diagram water-oxygen. Experimentally determined
curves for constant water contents (isopleths) which outline the two-
phase region in the PTx-space. ---: Critical curve (58).

It is a question of basic as well as of practical interest, to what
extent the addition of an electrolyte as a third component will influence
the miscibility and critical curves of binary aqueous systems as discussed above. It is to be expected, that the two-phase region would be
shifted, presumably to higher temperatures, by the addition of a soluble
salt, for example sodium chloride. This is indeed the case, as was first
shown by Takenouchi and Kennedy for water - carbon dioxide - sodium
chloride (62) in the high pressure-high temperature range. A more extensive investigation of this system was performed later (63). The result
is, that a concentration of six weight percent of sodium chloride, relative to water, increases the maximum temperature of the occurance of
two fluid phases by about one hundred degrees at one thousand bar pressure. This effect is of considerable importance in geochemistry.

Recently the water-methane-sodium chloride system was investigated
to 2500 bar and 500 °C (64). At several constant values of water-salt
ratios, curves on the three-dimensional pressure-temperature-composition

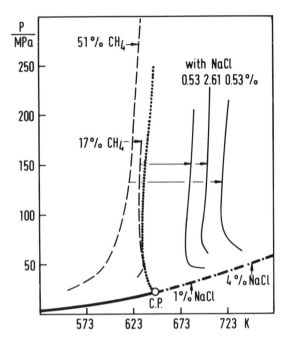

Fig. 17:   The ternary system water-methane-sodium chloride. Two isopleths
(dashed curves) for the binary system water-methane. Two critical curves
for water-methane (...) and water-sodium chloride (-.-.-). Full curves
show the temperature shift caused by the addition of sodium chloride
(64)..

two-phase boundary surface for ten different methane concentrations,
"isopleths", were determined. The experimental method was analogous to
the one mentioned above for the binary systems. Some results are shown
in Fig. 17 for two water-methane ratios. The figure contains also the
critical curve for the binary water-methane combination. Critical curves
for the ternary, salt-containing system cannot be given, because equilibrium tielines across the two-phase region could not yet be determind.
It is evident, however, that small amounts of elctrolyte produce shifts
of the two-phase region of similar magnitude as in the carbon dioxide
containing system. The curve in the lower part of Fig. 17 is part of the
critical curve for the binary H_2O-NaCl-system (65). It appears to be
possible to treat the ternary system with moderate salt contents as quasi
binary, by making use of appropriate points on the H_2O-NaCl curve.
     It is desirable, to have the possibility to calculate the important
features of the binary systems: The fluid-fluid equilibria (binodal surfaces) in the pressure-temperature-mole fraction (PTx) space, the critical curve and excess thermodynamic functions in the supercritical homogeneous region. A considerable number of equations have been proposed
and successfully tested(see for example (66) (67) (68)) for compilation.
The existing equations, however, were often not well suited for mixtures
with highly polar molecules like aqueous systems and for extrapolations

and predictions. Systematic investigations of the influence of the polarity of one of the two partners on the character of the critical curve
were made by Gubbins et al. (69). Experimental data are often difficult and
expensive to obtain.

For the special purpose to predict phase equilibria and critical
curves of binary mixtures which include highly polar partners to high
temperatures and pressures, a new "rational" equation of state was designed which is based on a defined model of intermolecular interaction
(70). This ("CF") equation contains only a limited number of parameters
which can be interpreted on the basis of the model. The equation is of
a modular type to permit later extensions and refinements. Intermolecular repulsion is described by a term of the Carnahan-Starling type with
a slightly temperature dependent sphere diameter. Attraction is taken
into account with a square well potential. Diameter and depth of the
square wells are derived from the critical data of the pure partners.
Two adjustable parameters are used to obtain averaged values of diameters
and depths for the binary combinations: The square root of the product
of the square well depths of the two partners is taken and multiplied by
an "asymmetry factor", ξ , generally smaller than unity. The linear
average of the diameters has also a forefactor, ζ , which, however, often
can be taken as close or equal to unity. The equation is extensively
discussed elsewhere (70) (71).

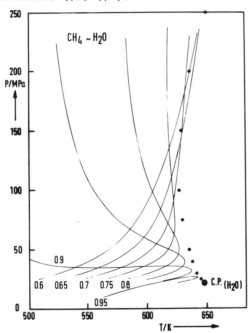

Fig. 18:   Pressure-temperature diagram for the binary system water-
methane. CP: Critical point of pure water.   oooo: Experimental points
on the critical curve (60). Curves: Spinodal isopleths for several
water mole fractions x, calculated with the CF equation (70).

Fig. 18 gives a set of calculated "spinodal isopleths" (curves of constant composition on the suface of mechanical or material stability). The critical curve is an envelope of the set of spinodal isopleths as well as of the binodal (phase stability) isopleths. For the calculation of the water-methane system the same set of coefficients as for the water-nitrogen system was used. Even for the water-xenon system the same values are applicable (54). The agreement with the experimental critical curves is satisfactory. The equation with the unchanged coefficients can be used to calculate the excess volume of the dense homogeneous supercritical region. This is shown again for water-methane in Fig. 19.

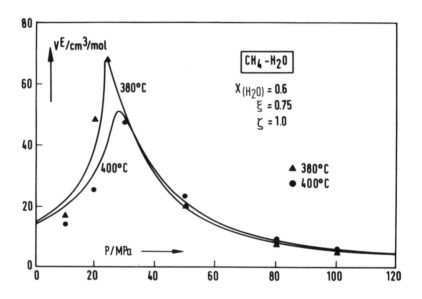

Fig. 19: The binary system water-methane. The molar excess volume $\overline{V}^E$ at the two supercritical temperatures 380 and 400 °C as a function of pressure. ooo: Experimental points. Curves: Calculated with the CF-equation (70).

The excess volume gives the deviation of the molare volume of the mixtures from the values calculated assuming ideal additivity of the molar volumes of the partners. In this case the excess volumes are positive. The agreement with experiments is satisfactory (70). Calculations of other systems, for example water-carbon dioxide - which is of particular geochemical interest, have been made with slightly different coefficients (70).

6. COMBUSTION

It has been shown, that at 400 °C methane and also oxygen are completely miscible with the dense, supercritical water. Very probably also a mixture of both of these gases will form homogeneous phases with water at such conditions. Thus one should expect, that "hydrothermal combustion" could be supported in such phases. To investigate such combustion, an autoclave was designed with sapphire windows to hold and observe water-gas mixtures to 500 °C and 2000 bar. Through a narrow nozzle injection of methane or oxygen into the dense, supercritical phase is possible with flow rates of a few microliters per second. The procedure makes use of two auxiliary autoclaves at room temperature which contain stainless steel bellows filled with the high pressure gases. The space between bellows and autoclave walls is filled with water which is in turn connected with the main reaction autoclave. A slow controlled compression of one of the bellows can produce an oxygen injection into the main vessel. The total pressure can be kept constant by arranging a circular movement of the fluids.

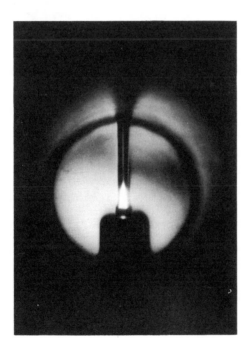

Fig. 20: "Hydrothermal Flame" at 1000 bar. Oxygen is injected at about 3 mm^3 s^{-1} into a supercritical homogeneous fluid of 30 mole percent of methane and 70 mole percent of water at 450 °C. The flame is 3 mm high. Sapphire windows.

It turns out, that a slow injection of oxygen into a homogeneous mixture of water with 30 mole percent methane causes spontaneous ignition and flame formation already at temperatures as low as 430 °C. Steadily burning flames are observed at 1000 bar, about 3 mm high and 0,5 mm wide. A photograph of such a flame is shown in Fig. 20. No phase separation occurs around this "supercritical" flame. The formation of the flame is remarkable, since normally spontaneous ignition temperatures for oxygen-methane mixtures are reported to be around 550 °C (72) (73) (74). The quenching effect of the dense water is not as pronounced as could be expected. The burning velosity of these flames is exceptionally small: of the order of one cm sec^{-1} or even less. High pressure flames described so far in the literature, seem not to exceed about 70 bar or pressure.

7. CONCLUSION

The combination of high pressure with high temperatures in the study of fluids permits the investigation of a very wide range of phenomena. Extensive variation of thermophysical properties from gas-like to liquid-like values can be realized. Unusual miscibilities and high chemical reaction rates are possible. Continuous transitions like those from molecular to ionic and even to metallic states are observed. With the advent of more refined investigation methods and construction material, further extension of the field is to be expected and new technical applications can be foreseen.

REFERENCES

1. See the contribution of W.J. LeNoble in these Proceedings.
2. Properties of Water and Steam in SI-Units, U. Grigull, Ed., Springer, Berlin, 1982.
3. R. Hilbert, K. Tödheide, E.U. Franck, Ber. Bunsenges. Phys. Chemie, 85, 636 (1981).
4. C.W. Burnham, J.R. Holloway, N.F. Davis, Amer. J. Sci. 276 A, 70 (1969).
5. H. Köster, E.U. Franck, Ber. Bunsenges. Phys. Chemie, 73, 116 (1969).
6. S. Maier, E.U.Franck, Ber. Bunsenges. Phys. Chemie, 70, 639 (1969).
7. See the contribution of M. Ross in these Proceedings.
8. F. Hensel, E.U. Franck, Ber. Bunsenges. Phys. Chemie, 70, 1154 (1966).
9. J.K. Kikoin, A.R. Senchenkov, Phys. Metals Metallogr., 24, 843 (1967).
10. G. Schönherr, R.W. Schmutzler, F. Hensel, Phil.Mag., B40, 411 (1979).
11. V.A. Alekseev, V.G. Ovcharenko, F. Ryzhkov, A.P. Senchenkov, J. Exp. Theor. Phys. USSR, 12, 306 (1970).
12. G. Franz, W. Freyland, F. Hensel, J. Physique Colloque C-8, No. 8. (1980).
13. H.P. Pfeiffer, W. Freyland, F. Hensel, Ber. Bunsenges. Phys. Chemie, 80, 716 (1976).
14. R.W. Ohse, Ed., "Handbook of Thermodynamic and Transport Properties of Alkalimetals", Blackwell, London, (1985).
15. H. Binder, Thesis: "PVT-Daten, Kritische Größen und Zustandsgleichun-

gen des Natriums bis 2600K und 500 bar".
Inst. f. Phys. Chem. Univ. Karlsruhe (1984).

16. K. Stephan, K. Luca, "Viscosity of Dense Fluids" Plenum Press, New York, (1979).

17. K.M. Dudziak, E.U.Franck, Ber. Bunsenges.Phys.Chem. $\underline{70}$, 1120 (1966).

18. L.D. Naake, Thesis: "Die Viskosität von n-Dekan und Methan-Dekan Mischungen bei 300 °C und 3000 bar", Inst. für Phys. Chem., Univ. Karlsruhe (1984).

19. F.J. Dietz, J.J. de Groot, E.U. Franck, Ber. Bunsenges. Phys. Chemie $\underline{85}$, 1005 (1981).

20. K. Heger, M Uematsu, E.U. Franck, Ber. Bunsenges. Phys. Chem. $\underline{84}$, 750 (1980).

21. M. Uematsu, E.U. Franck, J. Phys. Chem. Ref. Data, $\underline{9(4)}$, 1291 (1980).

22. R. Deul, "Dielektrizitätskonstante von Wasser-Benzol-Mischungen bis 400 °C und 3000 bar", Thesis, Inst. f. Physikal. Chem. Karlsruhe Universität, (1984).

23. C.J.F. Böttcher, "Theory of Electric Polarization" Vol. 1, Elsevier, Amsterdam (1973).

24. H. Looyenga, Physica, $\underline{31}$, 401 (1965).

25. L. Landau, F. Lifschitz, "Electrodynamics of Continuous Media", Addison-Wesley, N.Y. (1960).

26. D. Büttner, H. Heydtmann, Z. Phys. Chem. N.F., $\underline{63}$, 316 (1969).

27. W.L. Marshall, E.U. Franck, J. Phys. Chem. Ref. Data, $\underline{10}$, 295 (1981).

28. W.B. Holzapfel, E.U. Franck, Ber. Bunsenges. Phys. Chem., $\underline{70}$, 1105 (1966).

29. J.C. Mitchell, W.J. Nellis, J. Chem. Phys. $\underline{76}$, 6273 (1982).

30. D.J. Stevenson, Ann. Rev. Earth and Planetary Sci. $\underline{10}$, 257 (1982).

31. E.U. Franck, Z. Phys. Chem. N.F. $\underline{8}$, 92, 107 (1956).

32. K. Mangold, E.U. Franck, Ber. Bunsenges. Phys. Chemie, $\underline{73}$, 21, (1969).

33. A.S. Quist, W.L. Marshall, J. Phys. Chem. $\underline{72}$, 684 (1968).

34. W. Klostermeier, "Electrical Conductivity of Concentrated Salt Solutions to High Temperatures and Pressures", Thesis, Inst. of Physical. Chem. Karlsruhe Univ. (1973).

35. F. Hensel, E.U. Franck, Ber. Bunsenges. Phys. Chem. $\underline{70}$, 1154 (1966)
F. Hensel, S. Jüngst, B. Knuth, H. Uchtmann, M. Yao, Physica 139 + 140 B, 90 (1986).

36. M.A. Bredig, in M. Blander, Ed.: "Molten Salt Chemistry" Interscience, New York (1964).

37. E. Veleckis, E.M.von Deventer, M.Blander, J.Phys.Chem. $\underline{78}$, 1933 (1974).

38. W. Klostermeier, E.U. Franck, Ber.Bunsenges.Phys.Chem. $\underline{86}$, 612 (1982).

39. A.W. Szafranski, E. Keller, E.U. Franck, to be published in Ber. Bunsenges. Phys. Chem. (1987).

40. E. Keller, "Electrical Conductance and Thermodynamic Properties of Liquid Cs-CsH-Mixtures to 800 °C and 1000 bar. Thesis, Inst. of Phys. Chem. Univ. Karlsruhe (1985).

41. E. Keller, E.U. Franck, Z. Phys. Chemie, N.F. (1987) in press.

42. H. Renkert, F. Hensel, E.U. Franck, Ber. Bunsenges. Phys. Chem. $\underline{75}$, 507 (1971).

43. G. Franz, W. Freyland, F. Hensel, J. Physique, Colloque C $(\underline{8})$, 70 (1980).

44. M.A. Bohn, Thesis: "Intermolecular Interaction and Dynamics of

CHCIF$_2$ to High Temperatures and Pressures". Inst. f. Physikal. Chemie
der Univ. Karlsruhe, 1984.

45. M.A. Bohn, M. Frost, E.U. Franck, Physica, 139 + 140 B, 122 (1986).
46. M. Frost, Thesis, "FIR-Spectra of CH$_3$F-Argon Mixtures at High Pressures", Inst. f. Physikal. Chemie, Univ. Karlsruhe, (1982).
47. "Supercritical Fluid Solvents" Discussion Meeting of the Deutsche Bunsenges., Ber. Bunsenges. Phys. Chem. 88, 784-923 (1984).
48. G.L. Rössling, E.U.Franck,Ber.Bunsenges.Phys.Chem. 87, 882 (1983).
49. H. Ebeling, E.U. Franck, Ber. Bunsenges. Phys. Chem. 88, 862 (1984).
50. J.S. Rowlinson, F.L. Swinton, "Liquids and Liquid Mixtures" 3rd. Ed., Butterwoths, London, (1982).
51. G.M. Schneider, in "Chemical Thermodynamics" Vol. 2, Chap. 4, 105 - 146, McGlashan Ed.: A Specialist Periodical Report, London, 1978.
52. G.M. Schneider, Fluid Phase Equilibria 10, 141 (1983). Ber. Bunsenges. Phys. Chem. 88, 841 (1984).
53. H. Lentz, E.U. Franck, Ber. Bunsenges. Phys. Chem. 73, 28 (1969).
54. E.U. Franck, H. Lentz, H. Welsch, Z. Phys. Chem. N.F. 93, 95 (1974).
55. N.G. Sretenskaja, M.L. Japas, E.U. Franck, to be published in Ber. Bunsenges. Phys. Chem. (1987).
56. T.M. Seward, E.U. Franck, Ber. Bunsenges. Phys. Chem. 85, 2 (1981).
57. M.L. Japas, E.U. Franck, Ber. Bunsenges. Phys. Chem. 89, 793 (1985).
58. M.L. Japas, E.U. Franck, Ber. Bunsenges. Phys. Chem. 89, 1268 (1985).
59. K. Tödheide, E.U. Franck, Z. Phys. Chem. N.F. 37, 388 (1963).
60. H. Welsch, Thesis, "The Systems Xenon-Water and Methane-Water at High Pressures and Temperatures" Inst. Phys. Chem.,Univ. Karlsruhe, 1973.
61. Z. Alwani, G.M. Schneider, Ber. Bunsenges. Phys. Chem. 71, 633 (1967).
62. S. Takenouchi, G.C. Kennedy, Amer. J. Sci., 263, 1055 (1964).
63. M. Gehrig, H. Lentz, E.U. Franck, Ber. Bunsenges. Phys. Chem. 90, (1986).
64. T. Krader, E.U. Franck, Physica, 139 + 140 B, 66, (1986).
65. S. Sourirajan, G.C. Kennedy, Amer. J. Sci., 260, 115 (1962).
66. E.U. Franck, "Fluids at High Pressures and Temperatures" Rossini-Lecture, J. Chem. Thermod. 19, (1987) in press.
67. S.M. Walas, "Phase Equilibria in Chemical Engineering", Butterworth, Boston, (1985).
68. K.C. Chao, R.L. Robinson, Eds. "Equations of State", American Chem. Soc., Washington, 1986.
69. K.E. Gubbins, C.H. Twu, Chem. Eng. Sci. 33, 863, 879 (1978).
70. M. Christoforakos, E.U. Franck,Ber. Bunsenges.Phys.Chem.90,780(1986).
71. M. Christoforakos, Thesis, "Supercritical Systems at High Pressures A new Equation of State",Inst. f.Phys.Chem. Univ.Karlsruhe, (1985).
72. W. Jost, "Explosions- und Verbrennungsvorgänge in Gasen", Springer, Berlin, 1939.
73. A.G. Gaydon, H.G. Wolfhard, "Flames", 4. Ed. (Kapenau and Hall, London, (1979).
74. R.A. Strehlow, "Combustion Fundamentals" McGraw-Hill, New York,(1984).

HIGH TEMPERATURE - HIGH PRESSURE TECHNIQUES FOR THE STUDY OF FLUID
ELECTRONIC CONDUCTORS

F. Hensel
Institute of Physical Chemistry
University of Marburg
Hans-Meerwein-Straße
D-3550 Marburg / FRG

ABSTRACT. The demands of new technologies and the continuing interest
in the metal-insulator transition in expanded fluid metals have resulted in an increased demand for reliable measurements of the properties of fluid electronic conductors up to very high temperatures and
pressures. Equipments and techniques have been developed for high
pressure investigations of the equation of state, neutron diffraction,
electrical transport, optical, and magnetic properties at high temperatures up to more than 2000 K. The purpose of the present paper is to
review recent advances in high temperature - high pressure techniques
with special emphasis on simultaneous measurements of density and electrical transport data, optical reflectivity technique, and neutron
scattering experiments.

1. INTRODUCTION

There is currently widespread interest in both the electronic and thermodynamic properties of electronically conducting fluids at high temperatures and pressures. Part of this interest is motivated by the fact that
a favorable combination of physical properties such as high heat transfer coefficients, high electrical conductivity, and high latent heat of
evaporation makes fluid electronic conductors attractive as heat transfer media in nuclear power cycles and as high temperature working fluids
in turbine power converters. For the selection of a particular fluid and
for the design and operation of processes the knowledge of the thermophysical properties of fluid conductors is essential. The necessity for
safety analysis and risk assessment requires to model the evolution of
hypothetical accidents which can be postulated to occur. Consequently,
one needs to know the properties, in particular equation of state data,
up to conditions far above the proposed operating conditions, preferably
up to and beyond the liquid-vapour critical point temperatures.
    The measurement of such properties is complicated, because fluid
electronic conductors are difficult to experiment with, because, as
table I shows, a combination of high temperature and pressure is required to bring the sample near its critical point. The critical point

R. van Eldik and J. Jonas (eds.), High Pressure Chemistry and Biochemistry, 117–135.

is at low enough temperature and pressure to be studied with conventional static techniques for only a very few metals (Hg, Cs, Rb, K, and
Na) and for the liquid semiconductors S, Se, and Te. The estimated locations of the critical point of other metals are well above that of the
above metals, making it impossible to study them in the critical region
statically. Only transient experiments such as shock waves and exploding
wires /5,6,7/ are reaching temperatures and pressures high enough to
explore e.g. the critical point of molybdenum.

TABLE I   critical data of fluid electronic conductors

Substance	$T_c|K|$	$p_c|bar|$	$\rho_c|g/cm^3|$	reference
Hg	1750	1673	5.8	/1/
Cs	1924	92.5	0.38	/2/
Rb	2017	124.5	0.29	/2/
K	2280	161	0.19	/3/
Na	2485	248	0.30	/4/
Mo	14300	5700	2.9	/5/
S	1313	203	0.58	/8/
Se	1863	380	-	/9/

2.   EXPERIMENTAL PROCEDURES

A major experimental problem encountered in working with electronically
conducting fluids at high temperatures and pressures is their highly
reactive nature which makes their containment in uncontaminated form
very difficult, and which decreases the accuracy with which properties
can be measured. Materials found to be suitable for sample containers
for Hg, Cs, Rb, K and Na are pure tungsten, molybdenum, the alloys
tungsten-rhenium, tungsten-molybdenum and for Hg, Se and S the sintered
nonmetal aluminum oxide. Ideally suitable materials for optical measurements with Hg, Se and S are corundum single crystals up to 2000 K and
quartz up to 1600 K. These materials are chemically inert in contact
with the fluid sample; but cells from these materials cannot be constructed in such a way that they can withstand a high internal pressure
at temperatures above the critical temperatures of the above fluids. An
arrangement is therefore used in which the different cells, together
with the necessary furnaces, are placed in larger autoclaves, which are
filled with Argon or Helium under the same pressure as the fluid metal
inside the cell. This avoids any mechanical stress on the cells. However, at the necessary high pressures of more than 1500 bars argon has
a density larger than 1 g/cm³ and carries heat about by convection
currents that are difficult to eliminate. Severe problems of temperature measurement and control arise. The most conventional way is to
fill the space inside the autoclave as completely as possible with a
solid thermally insulating material. High temperature – high pressure
equipments of this type have been described in the literature for measurements of optical absorption /10/ and reflectivity /11/, for NMR-
experiments /12/ and for measurements of equation of state data and

electrical properties /13,14,15/.

The problem of temperature stability and homogeneity becomes particularly important in studies close to the critical point. These difficulties are directly related to the fact that two thermodynamic deriva-
tives, the isothermal compressibility $\chi_T = 1/\rho(\partial\rho/\partial p)_T$ and the isobaric
expansion coefficient $\alpha_p = 1/\rho(\partial\rho/\partial T)_p$ diverge strongly at the critical
point. The divergence of α_p and χ_T implies that a fluid kept at constant
pressure p or constant temperature T will undergo large variations in
density when T or p fluctuate. Small inaccuracies in T are serious
enough to make it impossible to correlate reliably density data ρ from
one experiment with a property, say, conductivity σ from another experiment because the temperatures T of the two samples are not really the
same and both ρ and σ can change rapidly with T near the critical point
/14/. The most effective way to reduce this difficulty is to measure the
different properties simultaneously in the same experimental set-up and
on the same sample /16,17,18/ and conditions of optimal temperature stability and uniformity.

A comprehensive review of the existing experimental procedures employed for the investigation of the high temperature – high pressure
behaviour of fluid electronic conductors is unnecessary because of the
surveys that exists and in the following I will restrict myself to those
novel developments which have not been adequately reviewed elsewhere,
e.g. simultaneous measurements of density and electrical transport data,
optical reflectivity studies and the analysis of reflectivity data which
are limited to a small photon energy range, and neutron diffraction
measurements.

3. HIGH-PRESSURE – HIGH-TEMPERATURE INSTRUMENTS

3.1. Autoclave for density and electrical transport measurements

Figures 1a and 1b show as one typical example an experimental set-up
which has been used to determine simultaneously the equation of state,
the electrical conductivity and the thermoelectric power of mercury up
to 1600°C and 2500 bar /19/. The high-pressure – high temperature
apparatus consisted of a steel pressure vessel (fig. 1a) with an electric resistance furnace which is thermally insulated from the steel
walls. The furnace consisted of three independently controlled heating
elements made of molybdenum wire. The pressure medium was argon gas.

Fig. 1b shows the principal arrangement of the cell inside the high-
pressure vessel. The mercury samples were contained in a molybdenum cell
with a relatively large volume of about 3 cm^3 which is surrounded by the
furnace. The cell has a capillary tube extending into a cold region of
the high pressure vessel interior, where it was connected with Teflon
bellows, which were kept at about room temperature. The density variation of fluid Hg with pressure and temperature was then determined by
the bellows which are also completely surrounded on their outside by
mercury. The expansion of the bellows was monitored via the corresponding level change of the mercury inside the level indicator. The position
of the mercury level was measured by a capacitive method. In order to

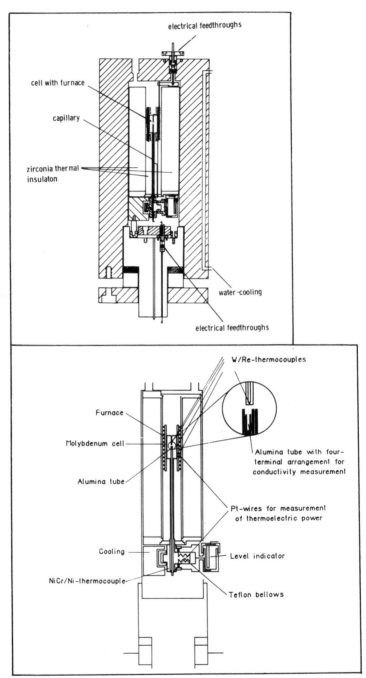

Figure 1a und 1b. Apparatus for simultaneous measurement of density, electrical conductivity and thermoelectric power.

minimize errors arising from incertainties in temperature and temperature homogeneity seven thermocouples were attached at different positions along the sample, in the center of the sample, along the capillary
and at the bellows. For the measurement of the conductivity the molybdenum cell incorporated a 6-bore alumina tube which left the cell via
the capillary into the cold closure. With four molybdenum wires as
electrodes in the holes of the alumina tube the conductivity was measured by a conventional potentiometer four-terminal arrangement. Two
Pt wires at the cold end of one of the bore-holes and at the center of
the cell permit the measurement of the e.m.f. of the Hg-Pt-thermocouple
from which the absolute thermoelectric power of Hg can be derived. In
this version the conductivity and the thermoelectric power samples were
located in the centre of the density sample and therefore the three
quantities were measured on the same or adjacent samples at the same
definite temperature and pressure. In addition we controlled and calibrated the temperature of the different samples by employing the most
accurate vapor pressure data /19/. At a given pressure the vaporization
temperature could easily be found as abrupt changes in conductivity,
thermopower or density when the mercury sample is vaporized. The repro-
ducibility of this procedure was checked over numerous experimental
runs and was found to be better than $\Delta T/T \approx 10^{-3}$. Also the temperature
gradients ΔT within the cell were about $\Delta T/T_A \approx 10^{-3}$, where T_A is the
average temperature in the large cell volume.

These improvements and extensions of temperature control and homogeneity made it possible to measure the equation of state and the
electrical transport properties close to the critical points of Hg /1,
19/ Cs and Rb /2/ with quite high precision. This new experimental information is accurate enough to allow one to determine the exact location of the critical points of Hg, Cs and Rb (table I) and to study the
asymptotic behaviour of the physical properties of metals near the vapor-liquid critical point.

3.2. Autoclave for optical measurements

Useful information about the behaviour of the electronic structure of
electronically conducting liquids can be obtained from studies of the
optical properties, particularly if careful measurements are made of
the dielectric dispersion. However, this approach to probe the electronic structure at high temperatures and pressures has, until recently,
confined to Hg, Se and S. The reasons are clear. The cell, the furnace
and the high pressure vessel have to be provided with windows of a
chemically inert material.

Both the transmission and the reflectivity technique have been
used to determine the optical constants ε_1 and ε_2 of Hg /10,20,21,22/,
Se /23/ and S /11,24/. Figures 2a and 2b show as one example the high-
temperature – high-pressure set-up for reflectivity measurements on
fluid sulphur. The fluid sulphur samples were contained in cylindrical
quartz cells with a polished quartz window at one end. The main part is
a quartz tube with an outer diameter of 15 mm. The inner surface was
polished to a smooth finish. The tube contains at one end a similarly
polished quartz rod of 15 mm diameter and 120 mm length which leads out

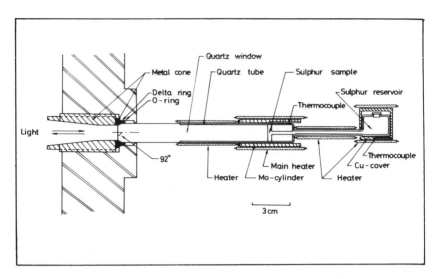

Figure 2a.   Optical high–temperature quartz cuvette for reflectivity measurements.

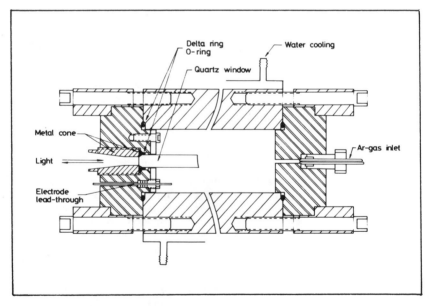

Figure 2b.   Internally heated autoclave for reflectivity measurements at high temperatures.

to the cold end of the cell. The cells with sample thicknesses of about
3 mm were constructed by sealing together the rod and the tube over the
whole length of the tube. The sample was connected via a 2 mm tube with
a sulphur reservoir.

The cell was heated by three independently controlled heating elements made of molybdenum wire. The temperature was measured by three
thermocouples attached at different positions external to the sample in
a molybdenum cylinder in close contact with the part of the cell where
the sulphur sample was located. As mentioned before it was necessary to
subject the fluid sulphur to high pressures at temperatures up to 1100°C.
For this purpose the cell, together with the surrounding heaters, was
mounted inside a high pressure stainless steel autoclave the details of
which are shown in figure 2b. The autoclave was pressurized with pure
argon and the pressure of argon and sulphur were balanced inside the
autoclave within the sulphur reservoir. The space between the heater and
the autoclave wall was filled as completely as possibly with thermally
insulating materials to obtain good thermal insulation and to prevent
convection in the compressed argon. The axis of the autoclave was horizontal. The reflectivity data were obtained for the incident light
nearly normal to the surface between the sulphur sample and quartz.

3.3. Analysis of reflectivity data of high temperature fluids

The investigation of the normal incident reflectivity spectra of fluids
at high temperatures and pressures is limited to relatively small photon
energy ranges because the quartz or sapphire windows needed for the containment of the fluids have a low- and high energy cutoff of transmission. Hence, instead of using the standard Kramers-Kronig inversion
/25/ the optical reflectivity curves have to be analyzed using a classi-
cal-oscillator-fit method to extract the real part $\varepsilon_1(\omega)$ and the imaginary part $\varepsilon_2(\omega)$ of the dielectric constant. In order to demonstrate
that this method compares favorably in convenience and accuracy with the
standard Kramers-Kronig analysis we use it in the following to analyze
the pressure and temperature dependence of the dielectric dispersion of
liquid sulphur close to the polymerization transition.

As is well known liquid sulphur possesses the unusual feature of a
thermodynamic transition between two distinct liquid modifications /26,
27/. Between the melting point (113°C for orthorhombic sulphur) and the
critical polymerization temperature of 159°C it forms a light yellow
liquid of relative low viscosity consisting in the main of S_8 molecules
which have the shape of puckered rings; above 159°C it forms a highly
viscous liquid in which conversion of a significant fraction of S_8-rings
to polymeric chains has occurred. The chains can contain as many as 10^6
atoms and increase their concentration as the temperature increases /28,
29/. At sufficiently high temperatures liquid sulphur behaves like an
electronic semiconductor /30/.

Almost all physical properties are influenced by the temperature
dependent structural changes and changes in the molecular composition:
for instance, the two ends of each chain give rise to unpaired localized
electrons, as static susceptibility /31/ and E.S.R. /32/ measurements
have shown. The colour of the liquid also changes rapidly from yellow to

red upon heating. Its absorption coefficient has been measured /33/ up
to a supercritical temperature of 1100°C and elevated pressures.

The temperature- and pressure dependence of the near-normal-inci-
dence reflectivity R(ω) spectrum is shown in detail in figs. 3a, 3b and
3c for the polymerization range around 159°C. For comparison the vacuum

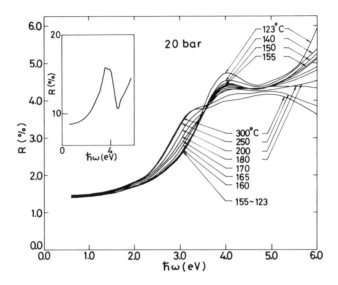

Figure 3a.   Temperature dependence of the reflectivity spectra R(ω) of
liquid sulphur behind a quartz window at a constant pressure of 20 bar.
The inset shows the reflectivity spectrum of crystalline orthorhomic
sulphur at 25°C in vacuum.

reflectivity of crystalline orthorhombic sulphur at 25°C /34/ is given
in the inset of fig. 3a. The structure of the spectrum in the solid
state is very similar to that observed in the liquid below the polymerization transition. The transition shows itself clearly, with a well defined break in the temperature coefficient of R(ω); immediately above
the transition temperature the temperature dependence becomes very large.
With increasing pressure there is a redshift of the R(ω) spectrum
(figs. 3a and 3b); this shift can easily be attributed to an increase in
density and the negative pressure dependence of the critical polymerization temperature /35/.

The method used to extract from the R(ω)-curves the optical con-
stants $\varepsilon_1(\omega)$ and $\varepsilon_2(\omega)$ is based on the use of a simple analytical form
for the dielectric function /36/, which consists of a sum of contribu-
tions from damped oscillators:

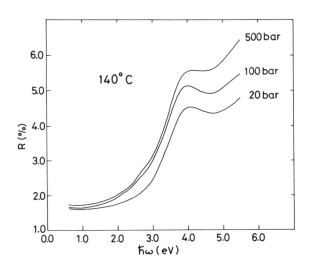

Figure 3b. Pressure dependence of the reflectivity spectrum R(ω) of liquid sulphur at a constant temperature of 140°C.

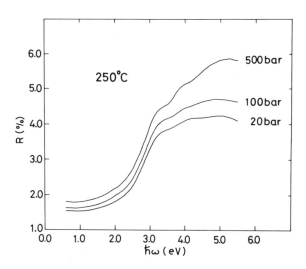

Figure 3c. Pressure dependence of the reflectivity spectrum R(ω) of liquid sulphur at a constant temperature of 250°C.

$$\varepsilon(\omega) = \varepsilon_1(\omega) + i\varepsilon_2(\omega) = \varepsilon_\infty + \sum_{i=1}^{N} \frac{f_i \omega_i^2}{\omega_i^2 - \omega^2 - i\Gamma_i \omega \omega_i} \tag{1}$$

Here f_i, ω_i and Γ_i are the strength, angular frequency, and width of the ith oscillator, respectively, ε_∞ is the contribution of high-frequency excitations to the dielectric function and ω is the light angular frequency. This model dielectric function satisfies the Kramers-Kronig dispersion relation. Verleur /36/ has demonstrated that a limited number of oscillators is sufficient to describe a dielectric function that accounts for the experimental reflectivity curves over a wide frequency range.

The fitting procedure involves an optimization process where the parameters of the oscillators are chosen, reflectivities are calculated and compared with the experimental results. This procedure is repeated to yield good agreement between the calculated and the experimental results. For the analysis of $R(\omega)$ in figs. 3a, b and c we have characterized the dielectric function (1) in terms of 10 to 30 oscillators. At the end of the fitting procedure the relative difference between the calculated and the measured values of the reflectivities was less than 2% over the entire frequency range. However, despite of this very good correspondence between calculated and experimental data, it can not be concluded that the fitting of a small portion of any spectrum will yield the correct results /36/. This is because more than one set of dielectric constants may fit equally well the same portion of a reflectivity spectrum. In order to extract the correct set of dielectric constants we had to use the independently determined absorption data of ref. /33/. Adding absorption coefficients at some particular frequencies as an additional requirement to the automatic fitting procedure gave always the same correct set of dielectric constants regardless of the values of the initial set of paramters.

In figures 4a and 4b and 5a and 5b we show the real and imaginary parts of the complex dielectric function $\varepsilon = \varepsilon_1 + i\varepsilon_2$, derived by the oscillator-fit method of the $R(\omega)$ spectra in figs. 3a, 3b and 3c. Although the reflectivity data extend to 6 eV, the $\varepsilon_1(\omega)$- and $\varepsilon_2(\omega)$-curves are truncated at 5.5 eV because the oscillator-fit results become suspect near the end of the range for which R is known. It is necessary to scrutinize the accuracy of the ε_1- and ε_2-values because the absolute magnitude of the ε-values can depend fairly critically on the general consistency of the oscillator-model-analysis. We show, therefore, in fig. 6 the temperature dependence of the refractive index n derived from ε_1-values (fig. 4a) close to the polymerization transition for a photon energy of 2.1 eV, at which it can be compared directly with the most recent refractive index measurements /37/. The directly measured values of n /37/ compare favorably with those obtained with the oscillator-fit method. The value of n is strongly influenced by the polymerization. Its dependence on the temperature shows a sharply defined cusp at 159°C.

We note that useful information about the electronic structure of melts at high temperatures and pressures can be obtained from studies

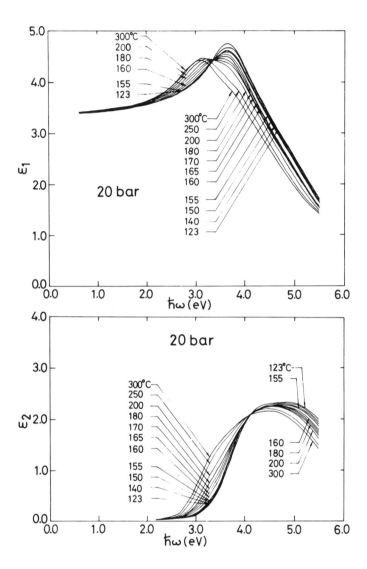

Figure 4a and 4b. Temperature dependence of $\varepsilon_1(\omega)$ and $\varepsilon_2(\omega)$, the real and imaginary parts of the complex dielectric constant of liquid sulphur at a constant pressure of 20 bar.

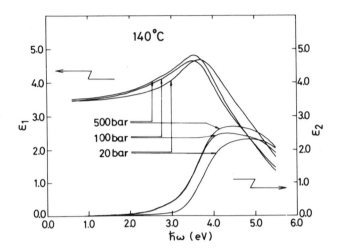

Figure 5a.   Pressure dependence of $\varepsilon_1(\omega)$ and $\varepsilon_2(\omega)$, the real and imaginary parts of the complex dielectric constant of liquid sulphur at a constant temperature of 140°C.

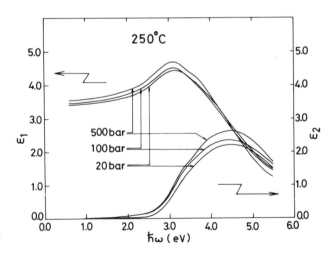

Figure 5b.   Pressure dependence of $\varepsilon_1(\omega)$ and $\varepsilon_2(\omega)$, the real and imaginary parts of the complex dielectric constant of liquid sulphur at a constant temperature of 250°C.

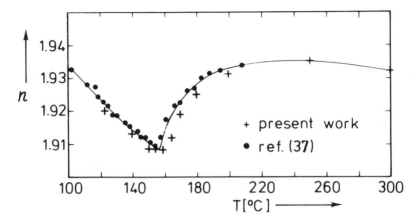

Figure 6. Temperature dependence of the refractive index n at a photon energy of 2.1 eV compared with the data of Donaldson and Caplin /37/. The polymerization transition is at 159°C.

of the reflectivity spectra. This approach to probing the electronic structure of melts appears to have been inhibited by the problems of constructing suitable optical cells and by the inapplicability of the standard Kramers-Kronig inversion. Now that this has been overcome, we anticipate a wide application of this technique.

3.4. Apparatus for neutron diffraction studies on fluids up to 2000 K
    and elevated pressures

The combination of high pressure and high temperature presents a more formidable problem for neutron scattering measurements on fluids than in most other experiments. The sample environment cell, pressure transmitting medium Ar, furnace and pressure vessel make access to the sample difficult. Structure factors have been measured only for expanded Cs /38/, Rb /39/, and Se /40/.
      Fig. 7 shows the experimental set-up which has been used for measurements of the structure factor of liquid Rb and Cs along the coexistence line up to 2000 K. The sample container (1) consists of a thin-walled molybdenum cylinder of 0.2 mm wall-thickness which is closed to the top-end (2) and which is connected with a liquid reservoir (7) at the bottom end. This sample cell is mounted inside the axis of a high pressure vessel (11), made from an aluminium-alloy of high tensile strength. The high temperatures along the measuring compartment (1) are produced by a tungsten-resistance furnace. This is made from two concentric tungsten cylinders. Currents of 100 to 150 Å necessary for high temperatures are led through a high current – high pressure plug (8) at the top end of the autoclave. The temperature along the measuring volume (1) is measured by two W-Re-thermocouples at the top (2) and bottom (3) end of the thin-walled molybdenum cylinder. The temperature profile along the cell is controlled with the aid of a second furnace (5). In

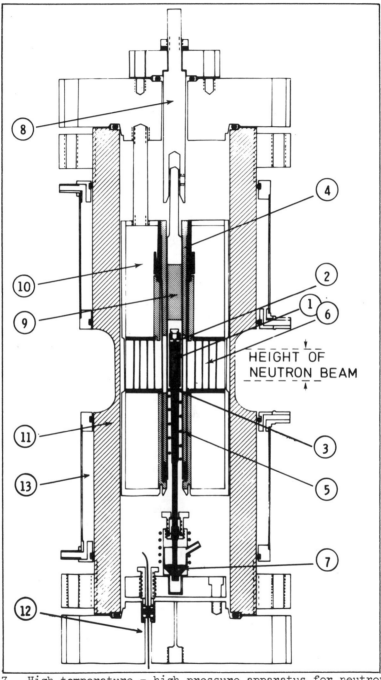

Figure 7. High temperature – high pressure apparatus for neutron diffraction studies on fluids.

the height of the neutron beam heat conduction and radiation between
the axis and the inside wall of the autoclave is reduced by several heat
shields (6) made from 50 μm vanadium foils. With this construction part
of the gas convection around the measuring compartment is reduced, too.
Thermal and electrical insulations above and below the neutron beam are
made by different parts and tubings from boron–nitride (4), (5), (9) and
from zirconia (10). The use of boron–nitride – a high neutron absorber –
has the further advantage, that scattering contributions from above and
below the fluid sample can be completely eliminated. As the high pressure
neutron window must be kept at the minimum possible thickness the temperature at the window must remain low. This is achieved by an effective
water cooling (13), keeping the neutron window at room temperature even
for 1700°C at the axis of the autoclave. All the electrical feed-throuhgs
(12) for thermocouples, additional heaters etc. are led through the
bottom flange of the autoclave.

3.5. Typical data correction for high temperature – high pressure neu-
     tron scattering experiments

The neutron scattering experiments have been carried out at the spectrometer D4B of the high flux reactor at Grenoble. In order to extract the
structure factor S(Q) from the measured scattering intensities, corrections for background, absorption, multiple scattering and inelasticity
have to be considered. In the following only that part of the data
correction will be described, which is specific for high temperature –
high pressure experiments, i.e. the background scattering and absorption
due to the high pressure furnace.
     In an ideal case, when measurements can be performed under vacuum,
the scattering intensity of the liquid sample alone, I_s, is given by –
see e.g. /41/ –

$$I_s = (I^e_{s+c} - A_{c,s+c}/A_{c,c} \cdot I^e_c)/A_{s,s+c} \qquad (2)$$

where I^e_{s+c} and I^e_c denote the experimental intensities of sample plus
container and empty container, respectively. The $A_{i,k}$ are the absorption
coefficients introduced by Paalmann and Pings /41/. In the present experiments, the corresponding intensities, I^e_{s+c+m} and I^e_{c+m} have to be
corrected for the respective background scattering, I^b_{s+c+m} and I^b_{c+m},
originating from the high pressure furnace (m). These background intensities can be obtained from an interpolation between two separate experimental runs: one of a Cadmium-rod, I^e_{Cd+m}, measured instead of the
sample cell inside the high pressure furnace, and another one of the
empty furnace, I^e_m. The first one represents the minimum background, the
second one its maximum. Thus the interpolation yields, written e.g. for
I^b_{s+c+m}:

$$I^b_{s+c+m} = I^e_{Cd+m} + A_{s,s+c} (I^e_m - I^e_{Cd+m}) \qquad (3)$$

The final intensities needed for equation (2) above can be obtained by subtracting the background intensities, e.g.

$$I^e_{s+c} = \beta \, (I^e_{s+c+m} - I^b_{s+c+m}) \tag{4}$$

and similarly for I^e_c. The coefficient β in (4) defines the attenuation by true absorption and scattering due to (m). In normalizing I_s by the corresponding intensity, I_V, of a Vanadium-rod, which has been measured inside the high temperature furnace and corrected in exactly the same way as the sample intensities, the absorption coefficient β cancels in good approximation, within ±2%. This can be checked by comparing the background corrected intensity I_V with that of the scattering of the same Vanadium-rod measured in air. In the final calculation of $S(Q)$ from I_s, the multiple scattering inside the sample and the Placzek correction have to be taken into account in the usual way - see e.g. /42/.

For Q-values >2.5 Å^{-1} some Bragg-peaks from the container-materials appear. Within these regions, which cannot be corrected with sufficient accuracy, the $S(Q)$-curves must be interpolated. For $Q < 0.28 \, \text{Å}^{-1}$ $S(Q)$ can be extrapolated to $S(0) = nk_B T X_T$, where X_T denotes the isothermal compressibility, known from PVT-data.

Fig. 8a displays as an example the static structure factor $S(Q)$ of fluid cesium at different temperatures and densities near the liquid-vapor coexistence curve, the corresponding Fourier transform, the pair correlation function $g(R)$ is shown in fig. 8b (the wiggles at small R-values are due to cut-off errors of the Fourier transform procedure).

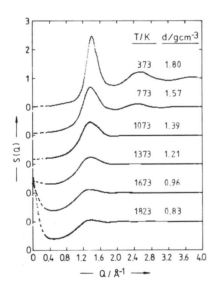

Figure 8a.   Static structure factor $S(Q)$ of expanded liquid cesium.

The following significant changes of $S(Q)$ or $g(R)$ on expansion of the metal are obvious:

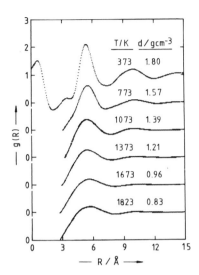

Figure 8b. Pair correlation function g(R) of expanded liquid cesium.

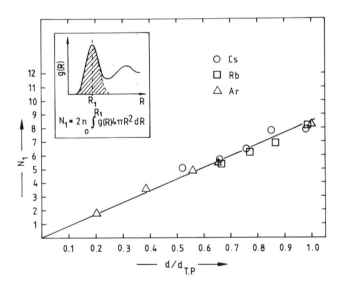

Figure 9. Number of nearest neighbours N_1 as a function of reduced density ($d_{T.P.}$ density at the triple point).

The intensity of the first peak of S(Q) is strongly reduced and broadened, indicating a significant reduction of the number of nearest neighbours N_1 (fig. 9) and a smearing out of the liquid structure. N_1, which has been calculated from the g(R)-curves (see inset in fig. 9) decreases linearly with decreasing density in the region covered by the present experiment, similar to Rb or the simple nonmetallic fluid Ar /48/. This behaviour can be explained as a continously decreasing packing-density during the expansion.

4. CONCLUSION

High pressure – high temperature research on the properties of electronically conducting fluids is a field of vivid interest and widespread activity. The interest is caused by a growing understanding of the electronic transport in disordered systems and by the demands of new technological developments. The activity is a least partly due to several new experimental possibilities. The principal one is certainly the development of optical techniques; this has permitted the study of the dielectric dispersion, which contains valuable information about the electronic structure. Other factors contributing to the improvement of experimental possibilities are improved temperature control and homogeneity which have permitted thermodynamic studies in the critical region.

ACKNOWLEDGEMENT

Financial support by the Deutsche Forschungsgemeinschaft and the Bundesministerium für Forschung und Technologie is gratefully acknowledged.

REFERENCES

/1/   W.Götzlaff, G.Schönherr, and F.Hensel, Proceedings of the 6th International Conference on Liquid and Amorphous Metals, Garmisch-Partenkirchen (1986)

/2/   S.Jüngst, B.Knuth, and F.Hensel, Phys.Rev.Letters, 55, 2160 (1985)

/3/   W.Freyland and F.Hensel, Ber.Bunsenges.Phys.Chem., 76, 347 (1972)

/4/   H.Binder, Doctoral thesis, Karlsruhe 1984

/5/   U.Seydel and W.Fucke, J.Phys.F: Metal Phys., 8, L157 (1978)

/6/   J.W.Shaner, G.R.Gathers, and C.Minichino, High Temperature – High Pressure, 8, 125 (1976)

/7/   M.M.Martyuyuk and O.G.Panteleichuk, High Temp. (USA), 14, 1075 (1976)

/8/   R.Fischer, R.W.Schmutzler, and F.Hensel, Ber.Bunsenges.Phys.Chem., 86, 546 (1982)

/9/   H.Hoshino, R.W.Schmutzler, and F.Hensel, Ber.Bunsenges.Phys.Chem., 80, 27 (1976)

/10/ H.Uchtmann and F.Hensel, Proceedings of the 4th International Conference on High Pressure, Kyoto 1974, p. 591.

/11/ K.Tamura, H.P.Seyer, and F.Hensel, Ber.Bunsenges.Phys.Chem., 90,

581. (1986).

/12/ U.El-Hanany and W.W.Warren, Phys.Rev.Letters, 34, 1276 (1975)
/13/ D.Postill, R.G.Ross, and N.E.Cusack, Phil.Mag., 18, 519 (1968)
/14/ F.Hensel and E.U.Franck, Ber.Bunsenges.Phys.Chem., 70, 1154 (1966)
/15/ I.K.Kikoin and A.R.Sechenkov, Phys.Metals Metallogr., 24, 5 (1967)
/16/ G.Schönherr, R.W.Schmutzler, and F.Hensel, Phil.Mag., 40, 411 (1979)
/17/ F.E.Neale and N.E.Cusack, J.Phys.E, 10, 609 (1977)
/18/ H.P.Pfeiffer, W.Freyland, and F.Hensel, Ber.Bunsenges.Phys.Chem., 80, 716 (1976)
/19/ W.Götzlaff, Doctoral Thesis, University Marburg (1987)
/20/ H.Overhoff, H.Uchtmann, and F.Hensel, J.Phys.F: Metal Phys., 6, 523 (1976)
/21/ W.Hefner and F.Hensel, Phys.Rev.Letters, 48, 1026 (1982)
/22/ U.Brusius, Doctoral Thesis, University of Marburg (1986)
/23/ H.P.Seyer, K.Tamura, H.Hoshino, H.Endo,and F.Hensel, Ber.Bunsenges.Phys.Chem., 90, 587 (1986)
/24/ G.Weser, F.Hensel, and W.W.Warren, Ber.Bunsenges.Phys.Chem., 82, 588 (1978)
/25/ D.L.Greenaway and G.Harbeke, Optical properties and bond structure of semiconductors, Pergamon Press, London (1968)
/26/ A.V.Tobolsky and A.Eisenberg, J.A.Chem.Soc., 81, 780 (1959)
/27/ J.C.Wheeler, S.J.Kennedy, and P.Pfeuty, Phys.Rev.Lett., 45, 1748, (1980)
/28/ A.J.Mäusele and R.Steudel, Z.Anorg.Allg.Chem., 478, 177 (1981)
/29/ J.C.Koh and W.Klement,Jr., J.Phys.Chem., 74, 4280 (1970)
/30/ M.Edeling, R.W.Schmutzler and F.Hensel, Phil.Mag.B, 39, 547 (1979)
/31/ J.A.Pulis, C.H.Massen, and P. van der Leeden, Trans.Faraday Soc., 58, 474 (1962)
/32/ D.C.Koningsberger and T. de Neets, Chem.Phys.Lett., 4, 615 (1970) and ibid 14, 453 (1972)
/33/ G.Weser, F.Hensel, and W.W.Warren, Ber.Bunsenges.Phys.Chem., 82, 588 (1978)
/34/ R.L.Emerald, R.E.Drews, and R.Zallen, Phys.Rev.B, 14, 808 (1976)
/35/ K.Bröllos and G.M.Schneider, Ber.Bunsenges.Phys.Chem., 76, 1106 (1972), ibid 78, 296 (1974)
/36/ H.W.Verleur, J.Opt.Soc.Am., 58, 1356 (1968)
/37/ A.Donaldson and A.D.Caplin, Phil.Mag.B, 52, 185 (1985)
/38/ R.Winter, F.Noll, T.Bodenstein, W.Gläser, and F.Hensel, Proc. of the 6th International Conference on Liquid and Amorphous Metals, Garmisch-Partenkirchen (1986)
/39/ W.Freyland, W.Gläser, and F.Hensel, Ber.Bunsenges.Phys.Chem., 83, 884 (1979)
/40/ M.Edeling ,and W.Freyland, Ber.Bunsenges.Phys.Chem., 85, 1049 (1981)
/41/ H.H.Paalmann and C.J.Pings, J.Appl.Phys., 33, 2635 (1962)
/42/ M.D.North, J.E.Enderby, and P.A.Egelstaff, J.Phys.C (Proc.Roy.Soc.) 1, 784 (1968)
/43/ P.G.Mikolaj and C.Pings, J.Chem.Phys., 46, 1401 (1967)

CRITICAL BEHAVIOUR IN FLUID METALS

F. Hensel
Institute of Physical Chemistry
University of Marburg
Hans-Meerwein-Straße
D-3550 Marburg / FRG

ABSTRACT. Recent experimental results of electrical, optical and thermophysical properties of the fluid metals Cs, Rb and Hg in the immediate vicinity of their critical points made it possible to investigate the interplay between the metal-nonmetal transition and the liquid-vapor critcal point phase transition.

1. INTRODUCTION

Recently, it has been demonstrated that the behaviour of fluid metals near the vapor-liquid critical point is of special interest in the general field of vapor-liquid critical phenomena /1,2,3/. As is well known, under ordinary conditions liquid metals are well described by the nearly-free-electron model /4/, whereas in the dilute vapor the great majority of the electrons are attached to their parent atoms occupying spatially localized atomic orbitals. It follows that the precise nature of the interparticle interaction must change dramatically on going from liquid-like to vapor-like densities, e.g. from metallic cohesion to Van der Waals' type interaction. By contrast, for typical nonmetallic fluids like Ar the character of the intermolecular interaction does not change with density to a good approximation. This contrast has been discussed by Goldstein and Ashcroft /3/ who argued, specifically in relation to the alkali metals Cs and Rb /1/, that the electronic structure change in course of the metal-nonmetal transition may considerably influence the thermodynamic features of fluid metals in the critical region.

The problem of the metal-nonmetal transition in expanded fluid metals has remained a longstanding problem. A number of physical mechanisms for electron localization at the transition have been proposed. The pioneering studies by Mott /5,6/ attributed the transition to Wilson and Hubbard band crossing but modified in an important way by the structural disorder which can cause localization of states by the Anderson process /7/ when the structural disorder is great enough. Other theories of the experimentally observed phenomena have made use of classical percolation ideas /8/, or have postulated, specifically in relation to expanded mercury, the existence of a Frenkel excitonic insulator transition /9/.

R. van Eldik and J. Jonas (eds.), High Pressure Chemistry and Biochemistry, 137–156.

One of the oldest questions, that is still unresolved is, what is the relationship between the metal-nonmetal transition and the liquid-vapor phase transition /10-16/. Is it, for instance, necessary connected with crossing the vapor pressure curve or its prolongation into the critical isochore? Or does it occur wholly in the liquid or wholly in the vapor phase. Questions of this kind were raised as long ago as 1943 by Landau and Zeldovitch /15/ in the context of the liquid-vapor equilibrium in mercury. They suggested that in metallic fluids two critical points may exist, i.e. for the liquid-vapor transition and the metal-dielectric-transition. Furthermore, they found besides the usual triple point for the solid-liquid-vapor coexistence, a triple point for (a) the metallic liquid-metallic gas - dielectric gas coexistence or (b) the metallic liquid - metallic gas - dielectric gas coexistence, respectively. The third possibility is that the liquid-vapor and the metal-dielectric transitions coincide and there is only one critical and one triple point as also proposed for alkali metals by Krumhansl /14/. Another possibility which cannot be excluded theoretically is that plasma transition from a weakly ionized plasma to a highly ionized plasma is completely separated from the liquid-vapor transition, i.e. it forms as isolated line in the p-T diagram without connection with the vapor pressure curve /13/.

2. THE METAL-NONMETAL TRANSITION

From the experimental standpoint it is clear that there is no sharp (first-order) electronic transition except across the liquid-vapor phase change. This is convincingly demonstrated by figures 1 and 2 which present a selection of the most recent and most accurate density ρ and electrical conductivity σ results for fluid monovalent cesium (T_c=1651°C p_c=92.5 bar, ρ_c=0.38 g/cm^3) /17/ and for divalent mercury (T_c=1478°C, p_c=1673 bar, ρ_c=5.8 g/cm^3) /18/ in form of isotherms as a function of pressure at sub- and supercritical temperatures. Apart from the liquid-vapor phase transition no discontinuous changes are indicated in σ and ρ. This behaviour of σ seems to be qualitatively typical for liquid metals and has been observed also for Rb and K and for divalent mercury. Quantitatively certain differences are observed between divalent Hg and the monovalent alkali metals. Near the critical point the conductivity of Hg seems to be more than 2 orders of magnitude smaller than the values observed for the alkali metals.

One criterion often invoked for the onset of metallic behaviour is the minimum metallic conductivity σ_{min} suggested by Mott /19/ to characterize the boundary between delocalized states and states localized by disorder (the Anderson transition /7/). For liquids, σ_{min} typically has values between 200 and 300 ohm^{-1} cm^{-1} which correlates roughly with the value of σ at the critical density ρ_c of Cs, Rb and K. Mott /5/ has argued that the critical points in fluid alkali metals are a consequence of electron-electron interactions which favor localization to reduce double occupancy of individual sites, i.e. the Mott-Hubbard transition. The role of the short-range intraatomic Coulomb interaction in monovalent metals has been discussed by Brinkman and Rice /20/. They showed that the metallic state near the metal-nonmetal transition should be

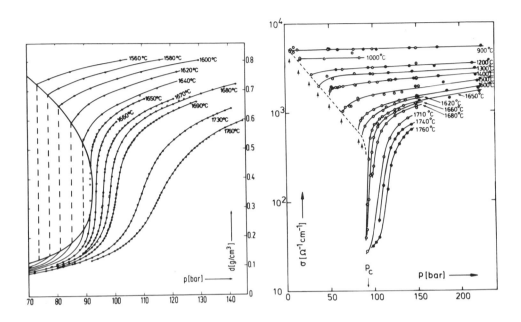

Figure 1. The electrical conductivity and the density of fluid cesium at sub- and supercritical temperatures as a function of pressure.

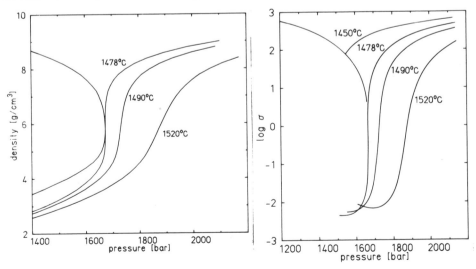

Figure 2. The electrical conductivity and the density of fluid mercury at constant temperatures as a function of pressure.

highly correlated, having a low instantaneous fraction of doubly occupied sites. The correlated metal has an enhanced density of states and,
consequently, enhanced values for the paramagnetic susceptibility. The
presence of large correlation effects in the alkali metals was first
convincingly demonstrated by the observation of a strong enhancement of
the total mass susceptibility for expanded liquid cesium /21/. Similar
susceptibility enhancements have been observed for expanded rubidium
/22/ and sodium /23/. The low density enhancement in Cs has been observed also in nuclear-magnetic resonance experiments /24/.

These experimental observations are entirely consistent with the
proposal that the metal-nonmetal transition in expanded fluid alkali
metals is both of the Mott and Anderson types, in that both electron –
electron interactions and disorder play important roles, and there is
a clear link between the liquid-vapor and the metal-nonmetal transition.

For divalent mercury the situation is more complicated. Here the
metal-nonmetal transition is connected with band-crossing effects. Experimental measurements such as those of the Knight shift /25, 26/ and
optical reflectivity /27,28/ show that as the density is lowered, a gap
in the density of states appears to develop at the relatively high density of 9 g/cm^3. The opening of this gap means that mercury changes
macroscopically to a nonmetallic, effectively "semiconducting" state at
the same density i.e. before the critical point density ρ_c=5.8 g/cm^3
is approached. On the other hand, most theoretical calculations /29,30,
31/ have yielded values for the gap closing density that are very close
to ρ_c. It should be noted, however, that these calculations did not
take into account the situation close to the critical point where critical density fluctuations may strongly affect the electronic properties.
The importance of density fluctuations for fluid mercury densities
smaller than 9 g/cm^3 and larger than 4 g/cm^3 in the temperature region
around T_c is clearly demonstrated by figures 3 and 4 which show the isothermal compressibility χ_T=1/ρ($\partial\rho/\partial p$)$_T$ and the isobaric expansivity
α_p=1/ρ($\partial\rho/\partial T$)$_p$ at constant temperatures or pressures, respectively as a
function of density. χ_T and α_p begin to rise quite rapidly as the density falls below 9 g/cm^3.

3. ELECTRONIC PROPERTIES IN THE CRITICAL REGION

Knowledge of the interplay between these critical point fluctuations
and the electronic properties might be especially important for the
understanding of the metal-nonmetal transition and its relationship to
the liquid-vapor phase transition in fluid mercury. Several attempts
have made to study this interplay experimentally, but the subject has
remained elusive. Part of the difficulty is due to the divergence of
many properties at the critical point. This is shown for α_p and χ_T in
figs. 3 and 4 and for the pressure coefficient 1/σ($\partial\sigma/\partial p$)$_T$ and the temperature coefficient - 1/σ($\partial\sigma/\partial T$)$_p$ of the electrical conductivity σ in
fig. 5 and 6. All quantities become very large in the critical region
so that small pressure or temperature errors cause large density and
conductivity errors. As a consequence, precise temperature and pressure
control and elimination of temperature gradients present demanding

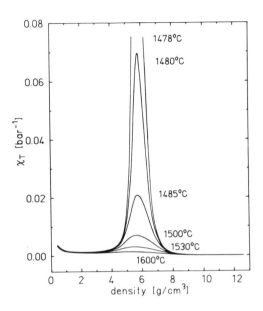

Figure 3.   Isothermal compressibility $\chi_T = 1/\rho(\partial\rho/\partial p)_T$ of fluid mercury at constant temperatures T (T_c=1478°C) as a function of density (ρ_c = 5.8 g/cm^3).

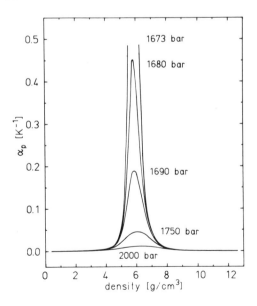

Figure 4.   Isobaric expansivity $\alpha_p = 1/\rho(\partial\rho/\partial T)_p$ of fluid mercury at constant pressure p (p_c=1673 bar) as a function of density (ρ_c=5.8 g/cm^3).

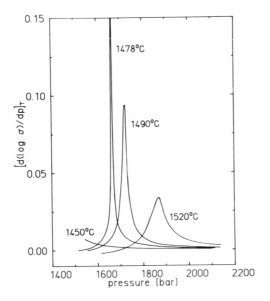

Figure 5. Pressure coefficient of the conductivity σ 1/σ(∂σ/∂p)$_T$ at constant temperatures as a function of pressure (p_c=1673 bar, T_c=1478°C).

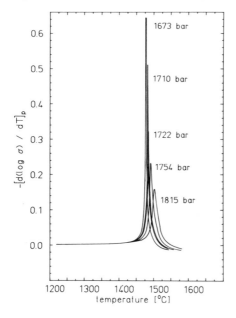

Figure 6. Temperature coefficient of the conductivity 1/σ(∂σ/∂T)$_p$ at constant pressures as a function of temperatures (p_c=1673 bar, T_c = 1478°C).

challenges in these studies. With new developments in high temperature –
high pressure technique (see the preceding lecture) and the possibilities offered by the new electronic devices these challenges can be met.
It was possible to measure the equation of state and electrical transport properties for Cs /1,17/ and Hg /18/ and partly the optical properties for Hg /2,32/ close enough to the critical point $(T-T_c/T_c \approx 10^{-3})$ with
quite high precision and comparatively accurate temperature control and
optimal elimination of temperature gradients. These data seem accurate
enough to allow the determination of the asymptotic behaviour of the
physical properties of Cs and Hg near the liquid-vapor critical point
(see e.g. figures 1 to 6).

The strong interplay between the liquid-vapor critical point and
electrical transport properties of mercury becomes immediately evident
when σ is plotted in the vicinity of the critical point as a function of
density ρ at constant temperature T, as shown in fig. 7 for a number of
selected isotherms. At a density of 9 g/cm^3 the conductivity is about
the "minimum metallic conductivity" (i.e. about 200 ohm^{-1}cm^{-1}). For
densities smaller than 9 g/cm^3, σ falls more rapidly and approaches a
value of about 1 ohm^{-1}cm^{-1} at the critical density ρ_c=5.8 g/cm^3. The
more rapid fall of σ for densities smaller than 9 g/cm^3 has been considered as a strong indication for the onset of the transition to a nonmetallic state. It is clear from the results of figures 7 and 8 that
there is a close correlation between the slope of the σ-ρ-curves and
the critical point. There is no doubt that the steepest fall in the
conductivity of Hg is observed at the critical point. $(\partial \ln\sigma/\partial\rho)_T$, which
is positive and nearly constant for ρ larger than 9 g/cm^3, has a strong
maximum accurately at the critical density. The pattern of the σ-ρ-curve
is especially interesting for densities below the critical density where

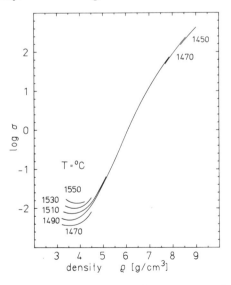

Figure 7.   Electrical conductivity σ of fluid mercury at constant temperatures close to T_c=1478°C as a function of density ρ.

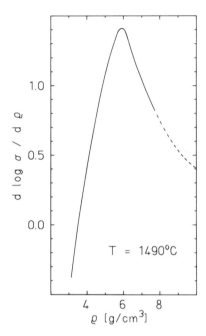

Figure 8. The density coefficient $(\partial \ln \sigma/\partial \rho)_T$ of fluid mercury at constant temperature as a function of density (T_c=1478°C, ρ_c=5.8 g/cm^3).

fluid mercury forms a dense, partially ionized plasma consisting of neutral species, ionic species and electrons. In this case, in addition to
screened Coulomb interaction among the charges, electron-neutral interaction plays an important role for the transport /33,34/. If the density
of neutrals is high enough the electron can interact with many atoms at
the same time: then it can be captured temporarily by density fluctuations or clusters /35,36/. This effect may be expected to be large in
the region of small degree of ionization, i.e. for densities well below
the metal-nonmetal transition range, and for low temperatures. Both the
minimum in the σ-ρ-curves at around 4 g/cm^3 (ρ_c=5.8 g/cm^3) and the
strong positive temperature dependence of σ in this range are completely
consistent with this model.

        As is well known, thermoelectric measurements are especially suited
to study changes in the nature of the electrical transport process /34,
37/. Several experiments have suggested that in Hg at pressures and temperatures near and above critical values the thermoelectric power vanishes /38,39,40/. Neale and Cusack /41/ and more recently Götzlaff /18/
have shown that the thermoelectric powers of Hg at constant pressure p
close to p_c=1673 bar changes sign twice with increasing temperature T
close to T_c=1478°C. Fig. 9 shows as an example the temperature dependence of S at p=1710 bar /18/. From the simultaneous measurements of S
and density ρ Götzlaff /18/ was able to evaluate the density dependence
of the thermoelectric power at a constant temperature near T_c (fig. 10).
It is obvious from the form of the S-ρ curve in fig. 10 that the anoma-

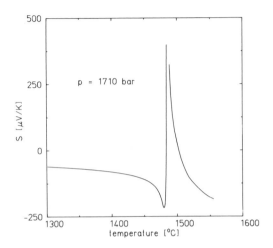

Figure 9. Thermopower S of fluid mercury at a constant pressure of 1700 bar as a function of temperature (p_c=1673 bar, T_c= 1478°C).

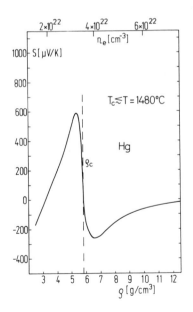

Figure 10. Thermopower S of fluid mercury at a constant temperature as a function of density ρ. (T_c=1478°C, ρ_c=5.8 g/cm^3).

lous temperature dependence of S at constant pressure is mainly caused
by the strong variation of the density close to the critical point. A
remarkably strong increase of S up to large positive values is observed
in the density region where electron-neutral interaction becomes important (cp. also fig. 7). The energy transport by neutrals induced by the
electron current seems to give a large contribution (a drag effect) to
the thermoelectric power in the region of small degree of ionization.
For densities above the metal-nonmetal transition density the thermopower tends to small negative metallic values (fully ionized degenerated
plasma).

From the results in fig. 10 it is clear that the change in the
transport behaviour of expanded Hg is intimately related to the vapor-
liquid critical point.

The interplay between the critical point phase transition and the
rapid changes in the electronic structure, when the nonmetal-metal
transition is traversed, becomes especially evident when the critical
point is approached from the insulating vapor side. This approach has
recently been extensively studied by measurements of the density- and
temperature dependence of the optical properties of the vapor /27,28,
31,32,35/. At very low densities a line-spectrum is observed with the
main absorption lines at 4.89 eV and 6.7 eV corresponding to transitions
between the 6s ground state and the 6p triplet and singlet state of the
Hg-atom. As the density is increased the sharp lines broaden due to
interactions with neighboring atoms resulting in a relatively steep absorption edge which moves rapidly to lower energies with increasing density. Fig. 11 gives a few selected data /32/ for the density dependence
of the edge at a constant temperature 1481°C, i.e. near the critical
temperature. Bhatt and Rice /42/ and Uchtmann et al. /43/ have shown
that a uniform density increase is insufficient to explain a line broadening as large as the observed shift in fig. 11 and that one must take
density fluctuation into account. The absorption edge is then lowered
by the environment of the atom being excited, and the edge is thus explained in terms of absorption by excitonic states of large randomly
distributed clusters. From the large values of the absorption coefficient it can be concluded that the singlet exciton 6^1p_1) with large
oscillator strength broadens faster than the triplet exciton (6^3p_1) with
small oscillator strength. A detailed analysis /43/ of the density dependence of the absorption edge shows that the singlet contribution dominates for densities larger than 1 g/cm^3 whereas for $\rho < 1$ g/cm^3 the
shape of the edge is dominated by the triplet transition.

If one approaches the critical region, i.e. for densities larger
than about 4 g/cm^3 in fig. 11, the optical absorption spectrum qualitatively changes. A third absorption regime appears. At low frequency a
foot extends in the infrared spectral range at least down to 0.5 eV. The
intensity of this absorption foot is strongly density- and temperature
dependent /32/.

The closing of the effective gap in the excitation spectrum (fig.
11) can also be viewed in terms of a nonlinear enhancement of the real
part of the dielectric constant ε_1 with increasing density as demonstrated in fig. 12 which shows results for ε_1 at the constant photon energy
1.27 eV in the form of isotherms plotted versus pressure. As the pressure

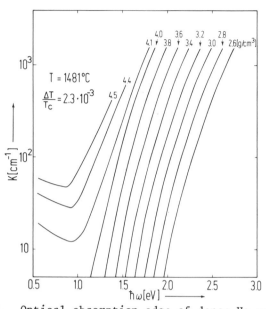

Figure 11. Optical absorption edge of dense Hg vapor at different densities for a constant supercritical temperature of 1481°C; the absorption constant is plotted against photon energy (T_c=1478°C).

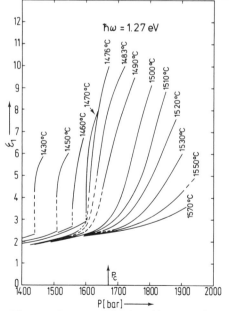

Figure 12. The real part of the dielectric constant ε_1 at the constant photon energy 1.27 eV as a function of pressure p for different sub- and supercritical temperatues (p=1673 bar).

is increased at a constant temperature T larger than about 0.96 T_c (i.e.
1400°C) ε_1 initially follows the Clausius Mosotti model of the polarisi-
bility of induced dipoles before it shows a strong upward deviation from
Clausius-Mosotti behaviour. It is obvious from the pattern of the ε_1-
curves in fig. 12 that the strong dielectric anomaly is inextricably related to the critical point phase transition of mercury.

More physical insight is obtained when ε_1 is plotted in the vicinity of the critical point as a function of density at constant temperature T, as shown in fig. 13 for a number of selected isotherms. Values
of ρ for each set of p-T coordinates were obtained from ref. /18/.

It is obvious that the form of the ε_1-curves is caused by two competing effects. Far away from the critical point (for $T/T_c \gtrsim 1.05$ and
$\rho/\rho_c \lesssim 0.7$) ε_1 is only slightly density dependent at constant temperature
and shows only slight positive deviations from the simple Clausius-Mo-
sotti value. This effect changes close to the phase line where the dielectric transition which precedes the metal-nonmetal transition is
signaled by a steep enhancement of ε_1. The pattern of these curves is
especially interesting close to the usual critical point where ε_1 shows
a strongly temperature dependent part $\Delta\varepsilon_1$, i.e. the different between
the experimental value and the Clausius-Mosotti value. The amplitude of
$\Delta\varepsilon_1$ becomes very large close to the critical point. By contrast, the
anomalous critical point contribution in the dielectric constant of
normal nonpolar fluids is immeasurably small /44/.

As mentioned before the enhancement of ε_1 with density ρ or temperature T can also be viewed in terms of changes in the shape of the

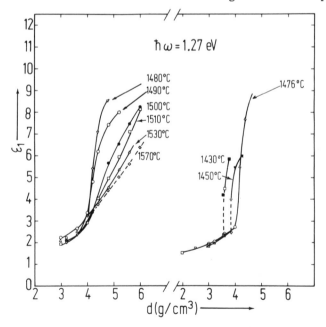

Figure 13.   Real part of the dielectric constant ε_1 of Hg at constant
temperature versus density. (T_c=1478°C, ρ_c=5.8 g/cm³).

absorption spectrum. In fact, ε_1 may be obtained from the standard
Kramers–Kronig integral over the optical absorption coefficient. Therefore, we plotted in fig. 14 the extinction coefficient K of fluid mercury at the constant photon energy 1.27 eV as a function of temperature.
A comparison of figures 13 and 14 shows that the anomalous behaviour of
ε_1 and K close to the critical point of Hg is remarkably similar. This
similarity is strong indication that both features have the same physical origin, i.e. they are consequences of the strong interplay between
the critical point phase transition and the changes in electronic structure as the metal–nonmetal transition is approached.

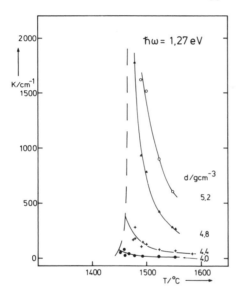

Figure 14.   The extinction coefficient K at constant photon energy
1.27 eV and at constant density ρ as a function of temperature. (T_c
=1478°C, ρ_c=5.8 g/cm^3).

4.   THERMOPHYSICAL PROPERTIES IN THE CRITICAL REGION

For most metals the critical region lies at higher pressures and temperatures than are accessible to conventional experimental techniques.
Consequently, a considerable effort has been carried out to estimate
their critical conditions. The estimation methods are based on different
versions of the relation between the critical data and other thermophysical properties. The most frequently used methods are the principle
of corresponding states and the law of rectilinear diameters which are
quite well obeyed for nonmetallic molecular liquids. However, on theoretical grounds, one should not place too much faith in their validity for
liquid metals. As mentioned before, in the case of metallic liquids the
effective particle interactions in the vapor state are very different
from the metallic binding forces in the liquid state. The magnitude of

this state-dependence distinguishes the particle interactions in metals
from those in insulators.

    Since measurements of high accuracy exist for the equation of state
of fluid rubidium /1/, Cesium /1/ and mercury /18/ we are able to test
experimentally the validity of corresponding state theory and the ex-
actness of the law of rectilinear diameters for fluid metals. For that
purpose we compare in table I the critical compressibility factors Z_c
= $p_c \cdot V_c / RT_c$ for the metals Hg, Cs and Rb and the nonmetal Ar. The data
show that as a group liquid metals do not appear to have very similar

TABLE I

Substance	$Z_c = p_c \cdot V_c / RT_c$
Ar	0.291
Hg	0.385
Cs	0.22
Rb	0.22

values of Z_c. The alkali metals and mercury appear to differ significantly in their values. The limitations of corresponding state theory
are also demonstrated by fig. 15 which gives a plot of the saturated
liquid vapor densities of Cs, Hg and Ar in reduced variables. Poor reduced correlation between Cs and Hg is observed and the metal curves are
clearly distinguishable from that for the group of nonmetallic liquids

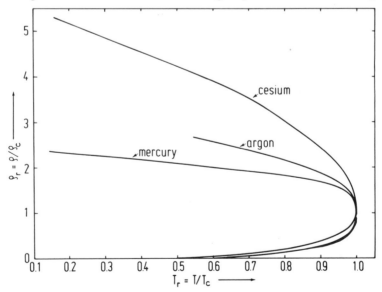

Figure 15. Density variation along the liquid-vapor coexistence curve
for metals in comparison with argon: plotted are reduced densities ρ/ρ_c
versus reduced temperature T/T_c.

(e.g. Ar) which obey a prinicple of corresponding states. Thus the experimental evidence shows that metals and nonmetals cannot be included
together in a group obeying a principle of corresponding states, and
furthermore that reduced correlations are unlikely to hold for liquid
metals as a group. On the other hand, it has been shown that Cs and Rb
which have similar values of Z_c (cp. e.g. table I) can be reduced in
corresponding regions of the phase diagram /2/ so that reduced correlations may be found within groupings of metals.

It is obvious from figures 16 and 17 that the metals cesium and
mercury (the same is true for rubidium /1/) violate the rule of rectilinear diameter over a surprisingly large temperature range. By contrast,
this law is experimentally valid for the coexistence curves of nearly
all simple nonmetallic one-component fluids to within the capacity of
present-day experimentation.

As mentioned before, it has been suggested /3/ that the contrast
between the diameter data of metals, and the apparent experimental
linearity of the diameters of essentially all nonmetallic one-component
fluids arises from many-body effects whose magnitudes distinguish the
particle interactions in metallic fluids from those in nonmetallic
fluids. In particular, it is argues that the strong thermodynamic state
dependence of the effective interactions in a metal, especially as the
metal-nonmetal transition is traversed, corresponds to the mixing of
the thermodynamic fields present in certain solvable lattice models /45,
46,47/. These models, thermodynamic arguments /48,49/ and renormaliza-

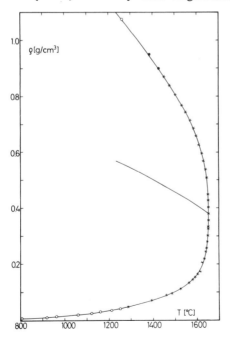

Figure 16. Liquid-vapor coexistence curve of cesium.

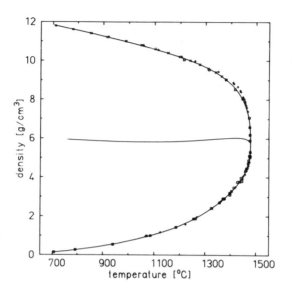

Figure 17. Liquid-vapor coexistence curve of mercury.

tion-group studies /50,51/ predict that the average value of the density, i.e. the diameter, will have the asymptotic form

$$\rho_d = (\rho_L + \rho_V)/2 = \rho_c + D(\tau)^{1-\alpha} \qquad (1)$$

where $\tau = |T_c - T|/T_c$ and the exponent α is the same as that of the divergence of the constant volume specific heat for a pure fluid. Thus the densities $\rho_{L,V}$ in the two branches of the coexistence curve are expected to behave like

$$\rho_{L,V} = \rho_c \pm B(\tau)^{\beta} + A(\tau)^{1-\alpha} + \ldots \qquad (2)$$

It is an empirical fact that the higher order terms in eq. (2) cancel to a large extent when the difference $\rho_L - \rho_V$ is formed. For this difference a power law with an exponent β is found to hold, and the asymptotic range of validity is normally quite large in τ. The latter fact is of great help in the analysis of coexistence curves. In fig. 18 we have plotted $\log(\rho_L - \rho_V/2\rho_c)$ versus $\log|\tau|$ for Cs as and example. Fitting to the equation

$$\frac{(\rho_L - \rho_V)}{2\rho_c} = B|\tau|^{\beta} \qquad (3)$$

we find $\beta = 0.355$. Similar values are found for Rb /1/ and Hg /18/. These experimentally determined β-values are very close to those observed for normal nonmetallic fluids which belong to the same static universality class as an uniaxial ferromagnet represented by the three-dimensional Ising model or the Landau-Ginzburg-Wilson model with an one-component

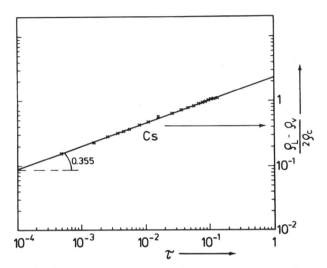

Figure 18. Single power law analysis of the coexisting liquid and vapor densities of Cs.

order parameter. The main difference between the coexistence curves of nonmetallic and metallic fluids seems to be the different magnitude of the $A|\tau|^{1-\alpha}$ term in eq. (2) which manifests itself in large amplitudes of the singularities in the coexistence curve diameters.

The predicted singularity in the diameter (see eq. (1)) is difficult to verify experimentally for several reasons. Firstly, it has been shown /52/ that if one particular function, e.g. ρ, has a $|\tau|^{1-\alpha}$ singularity, then any less symmetric function ρ', where ρ' is an analytic function of ρ (e.g. $\rho = V^{-1}$), behaves as $|\tau|^{2\beta}$. The mass density has long been known empirically to give a more symmetric coexistence curve than volume /53/. However, this is only a strong, but not a conclusive argument for supposing that ρ is the appropriate function for the order parameter. Secondly, studies of different fluid properties have revealed that corrections to asymptotic scaling must be applied when the range accessible to experimentation exceeds the range of asymptotic validity of the scaling laws. For instance the expansion for the diameter has the form

$$\rho_d = \frac{\rho_L + \rho_V}{2\rho_c} = 1 + D_0|\tau|^{1-\alpha} + D_1|\tau| + \dots \qquad (4)$$

Since $(1-\alpha)$ is not very different from unity, the true singularity is difficult to separate out from the analytic temperature term of eq.(4). The coefficient D_1 does not even have to be much larger than D_0 for the analytic term to dominate over the entire range accessible to experimentation. The latter certainly causes the invisibility of the $|\tau|^{1-\alpha}$ anomaly in most nonmetallic fluids.

Up to now the only convincing experimental evidence for the existence of a $(1-\alpha)$ term for one-component system stems from the analysis of the diameters of the metals Cs, Rb /1/ and Hg /18/. An example is given for the analysis of the diameter of Cs in fig. 19. A single power law

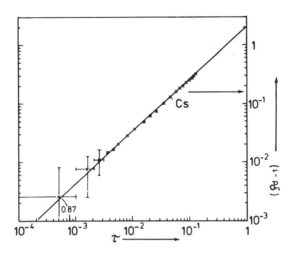

Figure 19.   Single power law analysis of the diameter of Cs.

applies over a range $10^{-3} < |\tau| < 10^{-1}$ and the apparent experimentally determined $(1-\alpha)$ value are very close to the value 0.89 predicted by the renormalization-group theory. This finding strongly supports the suggestion /3/ that the strong state-dependence of the effective interparticle interactions, and especially the changes in such forces in course of the metal-nonmetal transition, lead to very large amplitudes of the $(1-\alpha)$ anomaly in the diameters of liquid-vapor coexistence curves of metals.

ACKNOWLEDGEMENT

Financial support by the Deutsche Forschungsgemeinschaft and the Bundesministerium für Forschung und Technologie is gratefully acknowledged.

REFERENCES

/1/ S.Jüngst, B.Knuth, and F.Hensel, Phys.Rev.Letters, 55, 2160 (1985)
/2/ F.Hensel, S.Jüngst, B.Knuth, H.Uchtmann, and M.Yao, Physica, 139B,
      90 (1986)
/3/ R.E.Goldstein and N.W.Ashcroft, Phys.Rev.Letters, 55 2164 (1985)
/4/ M.Shimoji, Liquid Metals, Academic Press, New York 1977
/5/ N.F.Mott, Metal-Insulator Transition, Taylor and Francis, London
      1974
/6/ Metal Non-Metal Transitions in Disordered Systems, Edit. L.R.Friedman
      and D.P. Tunstall. Scottish Universities, Summer School in Physics,
      Edinburgh, 1978
/7/ P.W.Anderson, Phys.Rev., 109, 1492 (1958)

/8/ M.H.Cohen and J.Jortner, Phys.Rev.Letters, 30, 699 (1973)
/9/ L.A.Turkevich and M.H.Cohen, Ber.Bunsenges.Phys.Chem., 88, 297
      (1984) and Phys.Rev.Letters, 53, 2323 (1984)
/10/ N.F.Mott, Phil.Mag., 37, 377 (1978)
/11/ S. Nara, T.Ogawa, and T.Matsubara, Prog.Theo.Phys., 57, 1474 (1977)
/12/ W.Ebeling and R.Sändig, Ann.Phys. (Leipzig), 28, 289 (1973)
/13/ W.Ebeling, W.D.Kraeft and D.Kremp, Theory of Bound States in Plas-
      mas and Solids, Akademie Verlag, Berlin, 1976
/14/ J.A.Krumhansl, in: Physics of Solids at High Pressures, C.T.Tomi-
      zuka and R.M.Emrick (eds.), Academic Press, New York 1965
/15/ L.Landau and G.Zeldovitch, Acta Phys.Chem.USSR, 18, 194 (1943)
/16/ F.Yonezawa and T.Ogawa, Prog.Theo.Phys. (Japan), Suppl. 72, 1 (1982)
/17/ F.Hensel, S.Jüngst, F.Noll, and R.Winter, in: Localization and
      Metal-Insulator Transition, D.Adler and M.Fritzsche (eds.), Plenum
      Press, New York, 1985, p. 109
/18/ W.Götzlaff, Doctoral Thesis, University of Marburg (1987)
/19/ N.F.Mott, Phil.Mag., 19,835 (1969)
/20/ W.F.Brinkmann and T.M.Rice, Phys.Rev., B2, 4302 (1970)
/21/ W.Freyland, Phys.Rev., B20, 5104 (1979)
/22/ W.Freyland, J.Phys.(Paris)Colloq. 41 C8-74 (1980)
/23/ L.Bottyan, R.Dupree, and W.Freyland, J.Phys.F 13, L173 (1983)
/24/ U.El-Hanany, G.F.Brennert, and W.W.Warren, Phys.Rev.Letters 50,
      540 (1983)
/25/ U.El-Hanany and W.W.Warren, Phys.Rev.Letters, 34, 1276 (1975)
/26/ W.W.Warren and F.Hensel, Phys.Rev., B26, 5980 (1982)
/27/ H.Ikezi, K.Schwarzenegger, A.L.Passner, and S.L.McCall, Phys.Rev.,
      18, (1978)
/28/ W.Hefner, R.W.Schmutzler, and F.Hensel, J.Phys. (Paris), Colloq.
      41, C8-62 (1980)
/29/ L.R.Matheiss and W.W.Warren, Phys.Rev., B16, 624 (1977)
/30/ F.Yonezawa and T.Ogawa, Supp.Prog.Theor.Phys., 72 1 (1982)
/31/ H.Overhof, H.Uchtmann, and F.Hensel, J.Phys.F 6, 523 (1976)
/32/ U.Brusius, Doctoral Thesis, University of Marburg (1986)
/33/ J.P.Hernandez, Phys.Rev., 53 2320 (1984)
/34/ F.E.Höhne, R.Render, G.Röpke, and H.Wegner, Physica, 128A, 643
      (1984)
/35/ W.Hefner and F.Hensel, Phys.Rev.Letters, 48, 1026 (1982)
/36/ J.P.Hernandez, Phys.Rev.Letters, 48, 1682 (1982)
/37/ T.C.Harman and J.M.Honig, Thermoelectric and Thermomagnetic Effects
      and Applications, McGraw-Hill Book Company, New York (1967)
/38/ L.J.Duckers and R.G.Ross, Phys.Letters, A38, 291 (1972)
/39/ V.A.A.Alexeev, A.A.Vedenov, V.G.Ovcharenkov, and Yu F. Ryzhkov,
      Sov. Phys.-JETP Lett., 16 49 (1972)
/40/ M.Yao and H.Endo, J.Phys.Soc.Japan, 51, 1504 (1982)
/41/ F.E.Neale and N.E.Cusack, J.Phys.F.Metal Phys., 9, 85 (1978)
/42/ R.N.Bhatt and T.M.Rice, Phys.Rev., B20, 466 (1979)
/43/ H.Uchtmann, J.Popielawski, and F.Hensel, Ber.Bunsenges.Phys.Chem.,
      85, 555 (1981)
/44/ J.V.Sengers, D.Bedeaux, P.Mazur, and S.C.Greer, Physica, 104A, 573
      (1980)
/45/ J.S.Rowlinson, Adv.Chem.Phys., 41, 1 (1980)

/46/ B.Widom and J.S.Rowlinson, J.Chem.Phys., 52, 1670 (1970)

/47/ N.D.Mermin, Phys.Rev.Letters, 26, 957 (1971)

/48/ M.S.Green, M.J.Cooper, and J.M.H.Levelt-Sengers, Phys. Rev.Letters, 26, 492 (1971)

/49/ J.J.Rehr and N.D.Mermin, Phys.Rev.A, 8, 472 (1973)

/50/ M.Ley-Koo and M.S.Green, Phys.Rev.A, 16, 2483 (1977)

/51/ J.F.Nicoll and P.C.Albright, Proc. 8th Symp. Therophysical Properties, J.V.Senergs (ed.), ASME, New York, 1982, P.377

/52/ M.J.Buckingham, in: Phase Transitions and Critical Phenomena, C.Domb and M.S.Green (eds.), Academic Press, London, 1972, Vol. 2

/53/ J.M.H.Levelt-Sengers, Physica, 73, 73 (1974)

FLUID PHASE SEPARATIONS IN POLYMERIZING SYSTEMS AT ELEVATED PRESSURES

L.A. Kleintjens
DSM Research & Patents
PO Box 18
6160 MD Geleen
Netherlands

R. Van der Haegen and R. Koningsveld
Chemistry Department,
University of Antwerp,
B-2610 Wilrijk, Belgium

ABSTRACT. Phase separations that tend to accompany polymerization reactions are reviewed from a classic thermodynamic standpoint. Both lower and upper critical miscibility may occur and are details of the same phenomenon, partial miscibility in its dependence on pressure and temperature. Since pressure plays a marked role, we discuss fluid phase equilibria in terms of a mean-field lattice gas equation of state capable of adequate descriptions of observed phenomena and of predictions of the influence of pressure, molar mass and polymer chain structure characteristics such as chain branching.

INTRODUCTION

Polymer synthesis in industrial practice usually takes place in bulk, suspension, emulsion or solution. The starting situation is represented by a monomer, either by itself or dissolved in a more or less volatile solvent. After the initiation, polymer chains begin to grow and the system becomes more and more complex. The four polymerization procedures have in common that, at least in an early stage, the growing chains are dissolved, either in their monomer or in an inert solvent. The proceeding polymerization is accompanied by a decrease of the entropy of mixing of the system, taken per unit volume. As a consequence, polymerization reactor contents have a growing tendency to split into two or more fluid phases when the polymer chains get longer.

The latter phenomenon has been known and understood for a long time. In 1941, Staverman and Van Santen [1,2], Huggins [3] and Flory [4] came up independently with a calculation of the entropy of mixing molecules differing in molecular volume, the emphasis being on long chain molecules in a small-molecule solvent. Adding a Van Laar term to account for the heat of mixing, Huggins [5] and Flory [6] were able to

R. van Eldik and J. Jonas (eds.), High Pressure Chemistry and Biochemistry, 157–191.

explain the asymmetry of miscibility gaps, known to occur in polymer solutions. The longer the polymer chains, the more asymmetric the miscibility gap in the sense that it is shifted into the solvent-rich concentration region. The explanation was not new, Van der Waals demonstrated many years before that his equation of state predicts such an asymmetry in binary systems if widely differing values for the two b-parameters (molecular volumes) are used [7].

The statistical-mechanical considerations by Staverman, Huggins and Flory were based on the lattice model of the liquid state [8]. One obvious disadvantage of the model for use in the present paper is that it involves the condition

$$\Delta V^e = 0 \tag{1}$$

where ΔV^e is the excess volume of mixing. Condition (1) implies that thermodynamic properties of mixtures are, within the model, independent of pressure. To remove this limitation models have been developed on the basis of Prigogine's cell theory [9-12] which avoid the rigid lattice site by using a cell as the basic volume unit (BVU) of the system, the volume and occupancy of which may depend on temperature and pressure.

An alternative route is supplied by lattice-gas models in which the BVU of the rigid lattice is retained, and empty sites are introduced at random to allow for variations in volume without a change of the amount of matter [13-20], a strategy we shall adopt in this paper.

To illustrate the general theoretical considerations we describe applications of a simple lattice-gas model to fluid phase relations in polymerizing systems under pressure. We draw on experimental data i.a. on the systems n-alkane/polyethylene and ethylene/polyethylene and discuss a mean-field lattice-gas (MFLG) treatment adequate for the purpose. The polymerization reaction itself is left out of consideration here, we discuss systems in which the reaction has given rise to the occurrence of polymer chains of a certain (average) length and length distribution.

It is always useful to consider the problem from a phenomenological point of view. Classic thermodynamics may not supply relations between measurable and molecular properties, but it does give an ordering principle in that it provides an inevitable scheme of relationships between experimental and thermodynamic properties that every model will have to obey, if it is to be taken seriously. Therefore, we begin with a brief classic treatment of phase relationships as they may be expected to occur in the systems of present interest.

FLUID PHASE RELATIONS

Principles of phase equilibria after Gibbs [21] and Van der Waals [7] supply the ordering procedure mentioned above. The application of these principles has been greatly stimulated by the work of Bakhuis Roozeboom [22], and was recently worked out for polymeric systems [23]. First we consider binary fluid mixtures showing partial miscibility at ambient pressure.

Liquid mixtures of molecules differing little in molecular size show miscibility gaps that are not seldom nearly symmetrical around the 50/50 composition on a mass fraction basis. Figure 1 gives an example and also shows how the miscibility gap distorts if one of the

Figure 1
Miscibility gap in a system with molecules of comparable size (a) and with molecules differing by a factor 10^3 in molar mass (b).
System a: n-hexane/anilin; system b: diphenylether/linear polyethylene (M_w = 155 kg/mole).

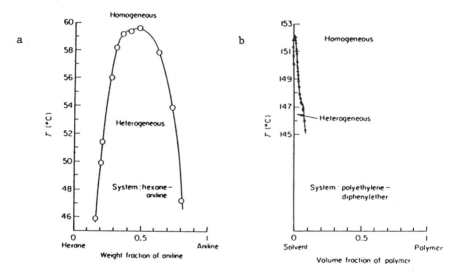

constituents has a molecular size exceeding that of the other by orders of magnitude. As mentioned in the introduction, both lattice model and equation of state treatments explain the extreme asymmetry of the two-phase region in terms of the difference in molecular volume between the constituents. Figure 1 is an arbitrarily selected example of a very familiar phenomenon. There is a maximum (critical) temperature above which the system is completely miscible in all proportions (upper critical miscibility (UCM) behavior). The opposite also occurs, partial miscibility setting in when the temperature is raised. Then, there is a minimum (critical) temperature below which the system is completely miscible (lower critical miscibility (LCM) behavior).

The question now arises what happens when such two-phase equilibria are subjected to a rise in pressure. To understand this we must first ask what happens to the phase diagram at higher temperatures when, eventually, the system will reach its boiling point. We know also that a decrease of the pressure p at constant T will likewise cause the mixture to boil. Thus, there are three variables to deal with, pressure p, temperature T and composition, here characterised by the volume fraction ϕ. A three-dimensional representation $p(T,\phi)$ as suggested by Bakhuis Roozeboom offers a convenient way to visualise the phase relations relevant for the present problem.

Figure 2 shows an example of that part of the binary Bakhuis
Roozeboom diagram we shall need here. It is usual to have p as ordi-
nate and T and ϕ on the other two axes. We note the vapor/liquid
equilibrium curves for the two components 1 and 2, each ending in a
vapor/liquid critical point (C_1 and C_2). Mixtures of 1 and 2 are
characterized by an extra degree of freedom and, at constant p and T,
liquid and vapor equilibrium phases can, and usually do, differ in
composition. $T(\phi)$ and $p(\phi)$ phase diagrams (isobaric and isothermal
sections in the Bakhuis Roozeboom diagram, respectively) will have the
well-known lense shape illustrated in figure 3. If we increase p and T
beyond the critical point of the more volatile component 1, the lense
changes into a loop which shrinks and vanishes in C_2.

Figure 2
Bakhuis Roozeboom $p(T,\phi)$ diagram for vapor (V)-liquid (L) equilibrium
in a binary system with complete miscibility in the liquid phase.

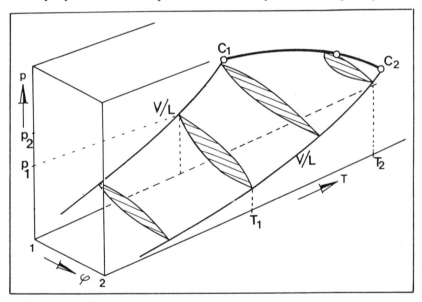

Figures 2 and 3 refer to one-phase liquids (above the curved surfaces representing the liquid and vapor compositions in $p(T,\phi)$ space).
Remembering figure 1 we can now imagine how liquid-liquid phase
separations will appear in a Bakhuis Roozeboom diagram. Depending on
the sign of ΔV^e miscibility increases or decreases with rising
pressure (negative or positive ΔV^e, respectively). Usually $\Delta V^e < 0$ and
we may expect the Bakhuis Roozeboom diagram to have a shape like that
in figure 4. Within the dome shaped three-dimensional miscibility gap
we have two liquid phases in equilibrium and miscibility increases
upon an increase of temperature and/or pressure. The various possible
types of diagram have been comprehensively reviewed by Schneider [24].

Figure 3

T(ϕ) and p(ϕ) phase diagrams (isobaric and isothermal sections from
fig. 2 respectively).

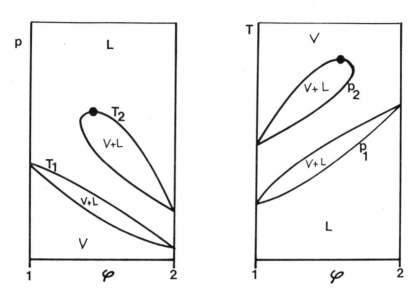

Figure 4

Bakhuis Roozeboom p(T, ϕ) diagram for vapor (V)-liquid (L) and liquid
(L_1)-liquid (L_2) equilibrium in a binary system.

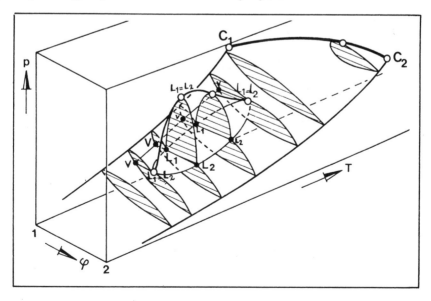

What isothermal and isobaric sections will look like can easily be inferred from figure 4. Examples are shown in figures 5 and 6.

Figure 5   Figure 6

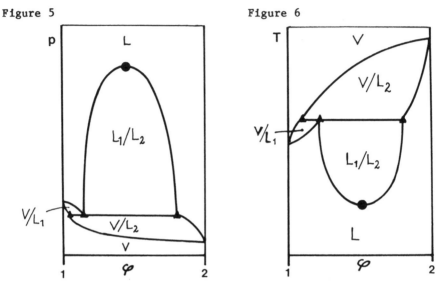

Figure 5
Isothermal section (p(φ) phase diagram) from figure 4.
Three phase equilibrium $V/L_1/L_2$:▲▲▲; L_1/L_2 critical point: ●.

Figure 6
Isobaric section (T(φ) phase diagram) from figure 4.
Three-phase equilibrium $V/L_1/L_2$:▲▲▲; L_1/L_2 critical point: ●.

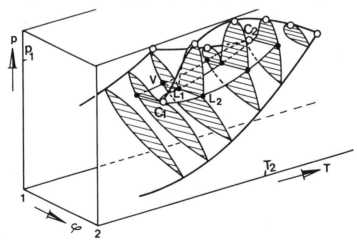

Figure 7
Bakhuis Roozeboom p(T,φ) diagram for vapor (V)-liquid (L) and liquid (L_1)-liquid (L_2) equilibrium in a binary system with interference of L_1/L_2 and V/L critical curves.

It is seen that, at constant T, a lowering of the pressure will give rise to the appearance of a vapor phase. Then, there are three phases in equilibrium and, since T is fixed, the system has become nonvariant i.e. there is just one pressure at which such a three-phase equilibrium is possible at that T (phase rule). Three-phase equilibria in binary systems are usually indicated by horizontal lines with the three equilibrium phase compositions clearly indicated (see figures 5 and 6). On one side of a three-phase equilibrium line there are two two-phase regions and there is one two-phase region on the other side. The disparity between the constituent's vapor/liquid critical points has not been exaggerated in figures 5 and 6 so as to keep essentials distinguishable. Usually, the critical points are far apart and if the second constituent is a polymer, its critical point is a physically meaningless quantity. In considerations like the present it may be thought to lie at a very high temperature. For sake of clarity we ignore the extreme asymmetry brought about by the difference in molecular volume in such cases and illustrate phase relations for a system with widely diverging critical points and a miscibility gap in the liquid phase. The Bakhuis Roozeboom diagram is shown in figure 7 and demonstrates the interference of the liquid-liquid equilibrium with the binary critical curve. Equilibria between a liquid (fluid) phase and a liquid-vapor critical phase will occur. The liquid/liquid critical curve changes character and becomes a vapor/liquid critical curve upon an increase of temperature. Supercritical phenomena are related to this situation [24].

The isothermal sections indicated in figure 7 explain the shape pressure/composition phase diagrams will assume, figure 8 shows some examples. The more familiar 'phase diagrams', temperature/composition graphs, can also be derived from figure 7 [23]. Some are shown in figure 9. We have slightly distorted these diagrams to remind the reader of the subject matter, polymer solutions.

Usually, phase diagrams in the literature show only a part of diagrams like those in figures 5, 6, 8 and 9, figure 10 is an example. The system ethylene/polyethylene is a case in point. Figures 11 and 12, taken from work by Ehrlich [25] and by De Loos and Diepen [26], show the relation between actual data and results of the old theory. A more familiar (T(ϕ)) phase diagram is presented in figure 10 and refers to the systems n-hexane/polyethylene, an example of lower-critical miscibility behavior. The reigning pressures are much lower than in the previous example. Finally, we mention work by MacHugh et al. [27] who studied solutions of an ethylene/propylene copolymer in mixed solvents and reported phase relations similar to those in figure 7 (see figure 11).

If the polymerization takes place under conditions above those of the critical point of the solvent or monomer, we may expect to be faced with the supercritical phenomena described in figures 7-9. Because the second constituent is a polymer we must also expect the phase diagrams to be extremely distorted. Thus, the phase separations observed in the system ethylene/polyethylene may give rise to misunderstanding. It would seem obvious that they represent the well-known feature of extreme asymmetry displayed by miscibility gaps in polymer

Figure 8
Isothermal sections (p(φ) phase diagrams) from figure 7.

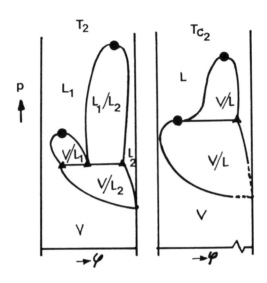

Figure 9
Isobaric sections (T(φ) phase diagrams) from figure 7.

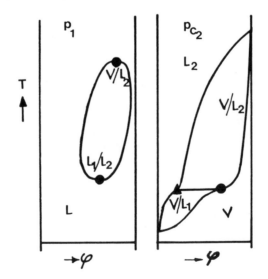

Figure 10
Phase boundary in a solution of linear polyethylene (M_w/M_n = 27; M_z/M_w = 7) in n-hexane. The dashed curve indicates the estimated position of the spinodal. Critical point: 0.

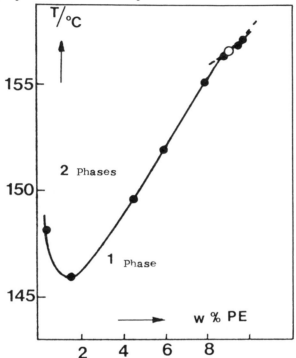

Figure 11
Schematic $p(T,\phi)$ diagram for ethylene/polyethylene (from ref. 25).

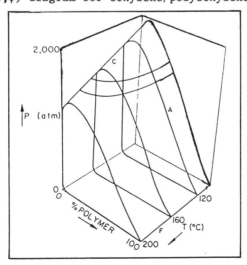

Figure 12
p(φ) phase diagram for ethylene/linear polyethylene.
W_p = overall weight fraction of polymer.
Top: overall schematic diagram; bottom: inset enlarged, data from ref.
26.

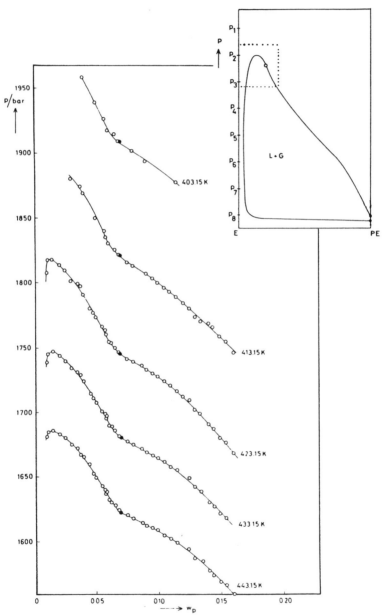

solutions (see figure 1) but a closer scrutiny of figure 12 shows that they really are the upper parts of vapor–liquid loops, shifted to the monomer–rich side of the diagram. The ethylene polymerization in n-alkane solvents is often carried out under conditions below the critical point of the solvent and we have an example of lower critical miscibility. Upon extension to higher T, however, the phase boundary curves will gradually change character and eventually form the upper part of the vapor–liquid loop.

Since we shall use the concept of continuity between vapor and liquid states, we do not need to be concerned by the features discussed above, the same equation of state should be capable of covering the complete range of p and T, vapor, dense gas or liquid alike. We note the obvious technical relevance of studies of fluid phase separations at elevated pressures and proceed now with the development of a molecular model useful for the ordering of observed phenomena and, possibly, for predictive calculations.

MOLECULAR MODELS

Rigid Lattice Model

We use the lattice model in two versions, the rigid lattice serving to explain the simple principles, and the lattice-gas treatment for pressure dependent properties. Rigid lattice representations of mixtures are given schematically in figure 13. Provided mixing is random and without change in internal energy, the small–molecule mixture is characterised by a number of possible arrangements of the N_1 molecules 1 and N_2 species 2, given by

$$\Omega_0 = N!/N_1!N_2! \tag{2}$$

where $N = N_1 + N_2$. We follow Silberberg's procedure [28] to calculate the number of arrangements Ω_1 for a system in which the molecules in component 2 have partially been connected by covalent bonds. If the ν_2 polymer chains so obtained have m_2 units each, then the number of sites 2, $N_2 = \nu_2 m_2$ and

$$\Omega_1 = (N!/N_1!\nu_2!)(V^*/V)^{\nu_2(m_2 - 1)} \tag{3}$$

where V is the total volume of the system and V^* the space available for any of the next units in the polymer chain after placement of the first one somewhere in V.

Assuming V^* not to depend on the composition of the mixture and applying standard procedures one obtains the so-called combinatorial entropy of mixing ΔS

$$\Delta S/NR = -\phi_1\ln\phi_1 - (\phi_2/m_2)\ln\phi_2 \tag{4}$$

where ϕ_1 and ϕ_2 are the site fractions of components 1 and 2 ($\phi_1 = N_1/N$; $\phi_2 = \nu_2 m_2/N$). Within the rigid lattice model site frac-

tions can be identified with volume fractions; they properly reduce to
mole fractions for $m_2 = 1$.

The Gibbs free energy ΔG follows from its definition, $\Delta H - T\Delta S$,
where the enthalpy of mixing ΔH can be set equal to zero within the
assumptions of this model. For condensed systems one can set $\Delta H \simeq \Delta U$,
the change in internal energy upon mixing. Relaxing the condition
$\Delta U = 0$ to allow for differences in internal energy upon mixing that do
not markedly affect the combinatorial entropy of mixing, we have
(strictly-regular mixture [29]).

$$\Delta G/NRT = \phi_1 \ln\phi_1 + (\phi_2/m_2)\ln\phi_2 + \chi\phi_1\phi_2 \tag{5}$$

where the interaction parameter χ is defined as $z\Delta w_{12}/RT$ (z = the
number of nearest neighbors of molecules 1 and chain units (sites),
Δw_{12} = the change in internal energy per site upon forming a 1-2 contact out of 1-1 and 2-2 contacts; mean-field treatment). In practice
it is often found useful to write [30]

$$\chi = \chi_s + \chi_h/T \tag{6}$$

and to treat χ_s and χ_h as adaptable parameters. Eqs (5) and (6) represent the Flory-Huggins-Staverman (FHS) expression [1-6].
<u>Figure 13</u>
Schematical representation of a small molecule mixture (a) and a
polymer solution (b) on a lattice.

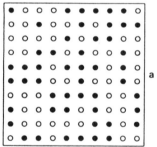

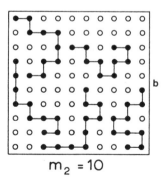

$m_2 = 10$

Figure 14

Left: Gibbs free energy of mixing ΔG in a binary liquid system as a
function of ϕ_2, the volume fraction of the second component, showing
the influence of temperature.
Right: Phase diagram with binodal (——), spinodal (---), critical
point (o) and tie line (●■-■●).

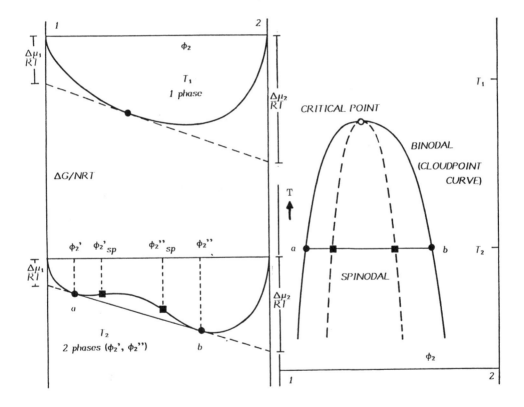

Figure 15
a: $\Delta G/NRT$ as a function of ϕ, for indicated values of χ, calculated
with eq. (5) for $m_2 = 10$. Spinodal points indicated by x.
b: Phase boundaries (binodals in binary systems) calculated with eqs
(5) and (6) for indicated values of m_2 ($\chi_s = 0$, $\chi_h = 200$ K).
Critical point: 0, drawn curves: binodals, dashed curves: spinodals.

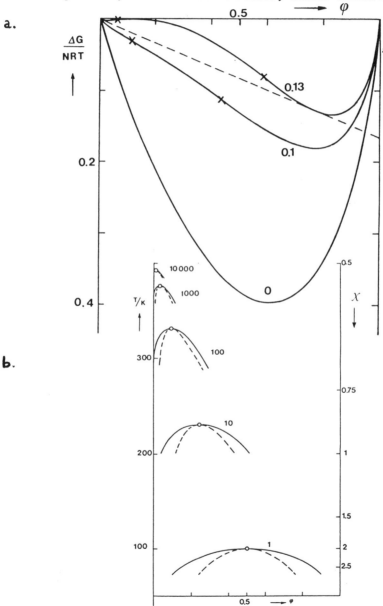

We are now in a position to discuss partial miscibility on the basis of a molecular model. Two-liquid phase equilibria are characterised by a $\Delta G(\phi_2)$ curve showing a plait and, therefore, two points of inflexion and two points where a double tangent touches the curve [31]. The loci of the inflexion points and double tangent points in $T(\phi_2)$ space (a phase diagram) have been called spinodal and binodal [7]. We see in figure 14 that the binodal connects coexisting phase compositions and relates to the chemical potential difference $\Delta\mu_i$ being identical in the two phases. We have

$$\Delta\mu_i/RT = [\partial(\Delta G/RT)/\partial N_i]_{p,T,N_j} = \Delta G/NRT + \phi_i[\partial(\Delta G/NRT)/\partial\phi_i]_{p,T} \quad (7)$$

The temperature plays an important role in that it changes the shape of the $\Delta G(\phi_2)$ curve. It is seen in figure 14 that a one-phase mixture is characterized by $\Delta G(\phi_2)$ having a positive curvature in the full composition range. A change in temperature may bring the double-tangent points and the point of inflexion closer together and between the one-phase and two-phase states there is one temperature at which the four points coincide. Such a state is called a critical or consolute state.

Figure 15 shows curves similar to those in figure 14, but calculated with eqs (5-7). The spinodal curves and critical points have been calculated with expressions derived from eq (5) with the aid of their definitions [7, 21, 23]:

$$\text{Spinodal: } [\partial^2(\Delta G/NRT)/\partial\phi_2^2]_{p,T} = 0: \quad 1/\phi_1 + 1/m_2\phi_2 - 2\chi = 0 \quad (8)$$

$$\text{Critical point: } [\partial^3(\Delta G/NRT)/\partial\phi_2^3]_{p,T} = 0: \quad 1/\phi_1^2 - 1/m_2\phi_2^2 = 0 \quad (9)$$

We note that the simple rigid-lattice model reproduces the experimental findings albeit in a qualitative fashion only [31-33]. Further improvement is possible, up to near-quantitative descriptions for which the reader is referred to the literature [33, 34]. Since the main point of interest here is the influence of pressure we turn to an appropriate extension of the model indicated in the introduction, viz. the mean-field lattice gas.

2. Mean-Field Lattice-Gas Model

Introduction of randomly distributed empty sites on the lattice makes the volume of the system variable without need to change the amount of matter. Figure 16 gives some schematical examples. It is seen that a single component substance can now be treated as if it were a binary mixture of occupied and vacant lattice sites. As a consequence, the free energy can be written in a similar form as eq. (5). We drop the condition $\Delta H \simeq \Delta U$ since we are now dealing with systems in which ΔV^e cannot be neglected. Hence, it is the Helmholtz free energy, ΔA, of mixing N_0 vacant and N_1 occupied sites that is relevant:

$$\Delta A/NRT = \psi_0\ln\psi_0 + (\psi_1/m_1)\ln\psi_1 + \Gamma\psi_0\Psi_1 \quad (10)$$

Figure 16
Schematical representation of a small-molecule and a macromolecular
substance on a lattice with vacancies (lattice-gas model).

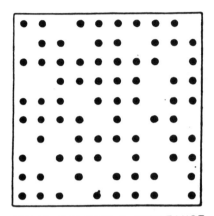

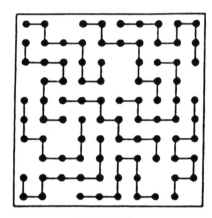

SMALL-MOLECULE SUBSTANCE   POLYMER

$$\psi_0 = 0.25 \qquad\qquad \psi_0 = 0.1$$

where ψ_0 and ψ_1 are the site fractions of vacant and occupied sites,
respectively, and m_1 is the number of sites occupied by a molecule of
the single species 1. The site occupancy ψ_1 is related to the molar
volume V_1 by

$$\psi_1 = v_0 m_1/V_1 \tag{11}$$

where v_0 is the BVU, to be identified here with the molar volume of
empty sites. In numerous applications of the MFLG eq. (10) we have
found that the interaction function is suitably specified as [15,
18–20].

$$\Gamma = \alpha_1 + (\beta_{01} + \beta_{11}/T)(1 - \gamma_1)/(1 - \gamma_1\psi_1) \tag{12}$$

where $\beta_{11} = -z_1 w_{11}/2R$, z_1 = the number of nearest-neighbor contacts
per occupied site, w_{11} = the internal energy contribution per 1-1 contact, $\gamma_1 = 1 - \sigma_1/\sigma_0$, σ_1 = Bondi [35] molecular surface area of species 1; α_1 and β_{01} are empirical constants.
    Eq. (12) makes use of very early findings by Staverman [36] that
the change in internal energy upon mixing, related to the numbers of
contacts broken and formed, must be calculated with reference to the
physical fact that virtually no molecules exist that are identical in
size and shape. Hence, they differ in numbers of nearest-neighbors
and a constant average coordination number for the entire lattice cannot do justice to the actual situation. Staverman suggested setting

the number of nearest—neighbor contacts of a site to be proportional
to the molecular surface area of its occupant. Following this line of
reasoning we arrive at eq. (12), though without the two empirical
constants that are entropic in nature. Molecular surface areas can be
estimated with Bondi's group contribution method [35].

Standard procedures lead to expressions for the mean-field lattice equation of state and spinodal and critial conditions for a
single component, both of the latter being valid at the vapor-liquid
critical point. We have [18-20]

$$- pv_0/RT = \ln(1 - \psi_1) + (1 - 1/m_1)\psi_1 + \left\{\alpha_1 + (\beta_{01} + \beta_{11}/T)(1 - \gamma_1)^2/ \right.$$

$$\left. (1 - \gamma_1\psi_1)^2\right\}\psi_1^2 \tag{13}$$

which equation properly reduces to Boyle's law for $V_1 \rightarrow \infty$ (see eq.
(11)). Further,

$$1/\psi_0 + 1/m_1\psi_1 = 2\alpha_1 + 2(\beta_{01} + \beta_{11}/T)(1 - \gamma_1)^2/Q^3 \text{ (spinodal)} \tag{14}$$

$$1/\psi_0^2 - 1/m_1\psi_1^2 = 6(\beta_{01} + \beta_{11}/T) \gamma_1(1 - \gamma_1)^2/Q^4 \text{ (critical point)} \tag{15}$$

where $Q = 1 - \gamma_1\psi$.

The three equations valid at a critical point, i.e. eq. (13) for
$p = p_c$ (p_c = critical pressure), eqs (14) and (15), allow determination of three of the six parameters from one set of critical data
(p_c, T_c, V_{1_c}). Some further data are then needed to fix the other
three parameters. We have found the quality of fits to data to not
depend a great deal on the value of v_0, the number obtained usually
being within the range 10-40 cm^3/mole. In view of Frenkel's estimation
of an equilibrium hole volume to ca. 20 cm^3/mole [13] we do not only
give preference to a number around that value, but also prefer to make
it a constant value for all substances to be considered, including
mixtures [18-20]. The latter manoeuvre avoids the difficult problem of
defining a combination rule for v_0 in mixed systems.

Having established the parameters for the two components in a
binary mixture from pure component data, we need some binary data to
do the same for the mixture. The present MFLG treatment has the advantage of supplying expressions for binary mixtures that, albeit
complex, do not contain an excessive amount of binary parameters. The
Helmholtz free energy of mixing occupied sites of two kinds 1 and 2
with vacant sites 0 reads [18-20]

$$\Delta A/NRT = \psi_0\ln\psi_0 + (\psi_1/m_1)\ln\psi_1 + (\psi_2/m_2)\ln\psi_2 +$$

$$\left\{\alpha_1 + (\beta_{01} + \beta_{11}/T)(1 - \gamma_1)/Q^*\right\} \psi_0\psi_1 + \left\{\alpha_2 + (\beta_{02} + \beta_{12}/T)(1 - \gamma_2)/Q^*\right\} \cdot$$

$$\cdot\psi_0\psi_2 + \left\{\alpha_m + (\beta_{0_m} + \beta_{1_m}/T)(1 - \gamma_2)/Q^*\right\} \psi_1\psi_2$$

where $N = N_0 + \nu_1 m_1 + \nu_2 m_2$ and $Q^* = 1 - \gamma_1\psi_1 - \gamma_2\psi_2$ \hfill (16)

We observe that the binary expression for ΔA is a superposition of two single-component ΔA equations extended by a third, quite similar, binary interaction term. The first, combinatorial term ($\psi_0 \ln \psi_0$) is the same as before, the concentration of vacant sites in the mixture, ψ_0, being determined by equilibrium conditions. There are three binary parameters, α_m, β_{0_m} and β_{1_m}, but they do not occur in the term Q^*, which arises from the differences in nearest-neighbor contact numbers, already known from the single-component fits. In the MFLG treatment a binary system, with volume fractions ϕ_1 and ϕ_2, is represented by a pseudo-ternary with site fractions ψ_0, ψ_1 and ψ_2. The former can be calculated from the latter with

$$\phi_1 = \psi_1/(1 - \psi_0); \quad \phi_2 = \psi_2/(1 - \psi_0) \tag{17}$$

We refer to the Appendix for details of the determination of binary parameter values with the relevant pseudo-ternary expressions for equation of state, spinodal and critical point and only present an application of eq. (16) to the system ethylene/naphthalene [37]. Part of the binary vapor-liquid critical curve of this system has been measured by Van Welie and Diepen [38], eight data points being available for different (critical) compositions of the mixture (including pure napthalene). Using five of these to fix the three binary parameters in eq. (16), we found the rest of the critical curve to be well predicted (see figure 17). The excess volume ΔV^e also came out quite reasonable within a factor of 2 [37]. Other examples on small-molecule binary mixtures can be found in the literature [18-20, 39].

Figure 17
Vapor-liquid critical curve in the system ethylene/napthalene, projected onto the p(T) and T(x) planes ($X_{C_{10}H_8}$ = mole fraction napthalene). Critical points used in the parameter estimation:✦, predicted course of critical curve:------. Critical points: ●.

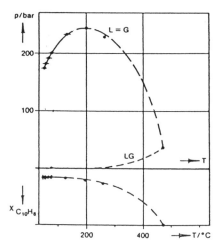

POLYMERS AND MOLAR MASS DISTRIBUTIONS

If one of the constituents is a polymer we have two extra problems. First, there is no vapor-liquid critical point to calibrate eqs (13-15) with and we must use pVT isotherms (eq. (13)). This can be done as is witnessed by figure 18 for PE.

Figure 18

Specific volume of liquid linear polyethylene (M_w = 160 kg/mole) as a function of temperature and pressure. Curves fitted to the points with eq (13).
Parameter values: α_1 = -10.131; β_{01} = 5.51; β_{11} = 2.035 x 10^3 K; γ_1 = -1.114; v_0 = 45 cc/mole.

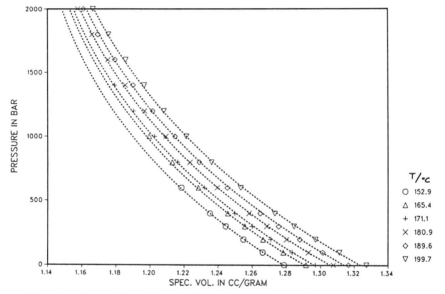

Secondly, in thermodynamic studies like the present, synthetic polymers can never be treated as single-component substances. They are at best multicomponent mixtures of homologues sharing an identical chemical structure in the repeat units. The theory and practice of polymer synthesis predicts and assesses that chain-length or molar-mass distributions must be expected to arise from a polymerization reaction. These distributions do not need to impede thermodynamic analyses though they do complicate them [40].

Theory can deal with chain-length distributions simply by writing as many combinatorial terms in ΔA as there are components in polymeric constituents [33, 40, 41]. Assuming constituent 1 to represent the polymer we only have to replace the term $(\psi_1/m_1)\ln\psi_1$ in eq. (10) by $\Sigma(\psi_{1_i}/m_{1_i})\ln\psi_{1_i}$, where ψ_{1_i} and m_{1_i} are the site fraction of and number of sites occupied by component i in constituent 1. Eq. (10) remains unaltered in every other respect because the approximation used assumes the interaction term not to depend on i and, consequently, only to contain the total site fraction of 1, $\psi_1 = \Sigma\psi_{1_i}$.

The consequence of these seemingly slight alterations are noticeable in eqs (13-15). In eq. (13) m_1 must be replaced by the number-average chain length m_{1_n}, in eq. (14) by the mass-average m_{1_w} and in eq. (15) by $m_{1_z}/m_{1_w}^2$, where m_{1_z} stands for the z-average chain length. These quantities are defined by

$$\psi_1/m_{1_n} = \Sigma\psi_{1_i}/m_{1_i}; \quad \psi_1 m_{1_w} = \Sigma\psi_{1_i}m_{1_i}; \quad \psi_1 m_{1_w}m_{1_z} = \Sigma\psi_{1_i}m_{1_i}^2 \tag{18}$$

Analogous changes have to be made in eq. (16) and the equation of state, the spinodal and critical conditions issuing from it (Appendix).

Spinodals and critical points are firmly connected with the binodals they belong to and are therefore useful tools in scanning phase relations. The equations determining binodals are much less direct since, like the equation of state, they contain logarithmic terms. The objective being semi-quantitative comparisons of theory and experiment here, we refer the reader to the literature for more comprehensive treatments [18, 40, 42, 43], and restrict the discussion mainly to spinodal curves and their relation to measured phase boundaries.

n-ALKANE/POLYETHYLENE

Industrial processes exist in which solution polymerization of ethylene is carried out in mixtures of hydrocarbons, on the average roughly resembling n-hexane. In the ΔA expression relevant for n-alkane/polyethylene we have an interaction term for the repeat units within the alkane (solvent, component 1; parameters α_1, β_{01} and β_{11}) and within the polyethylene (multicomponent constituent 2; α_2, β_{02}, β_{12}). We assume the similarity between n-alkane and (linear) polyethylene to allow considering the interaction terms to be identical (apart from a dependence on chain length [19]). As a consequence, the interchange energy between the units in the solvent and those in the polyethylene can be set equal to zero ($\alpha_m = 0$; $\beta_{0_m} = 0$; $\beta_{1_m} = 0$, see eq. (16)). Hence, quasi-binary phase diagrams for n-alkane/polyethylene systems should be predictable with the parameters for the single constituents [19].

Various authors have reported phase boundaries for the system, using different n-alkanes [44-46]. Such systems exhibit lower-critical miscibility behavior. The authors carried out the experiments in closed tubes and the pressure must have varied along the phase boundary, since it was that of the vapor in equilibrium with the mixture.

We neglect this small change in pressure and calculate spinodals to compare with the phase boundaries, the course of which they must essentially follow. The authors specified the mass-average molar masses of their samples so that we can perform the comparison without doubts about the chain-length distributions that are usually very broad in polyethylene. It is seen in figure 19 that the predicted spinodals come out quite consistently with the measured phase boundaries.

Figure 19
Comparison of experimental phase boundaries for linear polyethylene
(M_W = 177 kg/mole) in n-hexane (▼), n-heptane (●) and n-octane (o),
with spinodals (-–--) predicted with the MFLG model for constant indicated pressure (in bar).
Parameter values in table la.

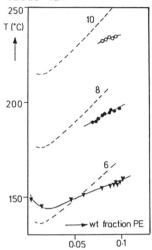

Figure 20
Comparison of experimental phase boundaries for linear polyethylene in
n-heptane (ref. 46). M_W = 49 kg/mole (●), 83 kg/mole (o) and 136
kg/mole (□).
Spinodals predicted with the MFLG model:————. (Constant pressure of
8 bar).
Parameter values in table la.

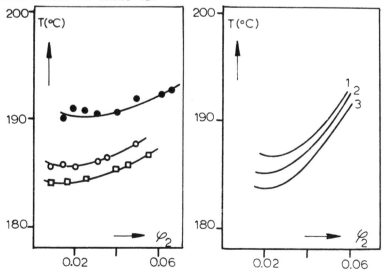

As a further test of the MFLG model we compare the dependence of the phase boundaries in n-heptane on the mass-average molar mass of the polyethylene. In figure 20 we see that the predicted spinodals have the chain-length dependence right within a few degrees C. The pressure dependence of the phase boundary is likewise predicted in the correct order of magnitude by the model (figure 21) [19]..

ETHYLENE/POLYETHYLENE

If we want to treat the systems obtained by bulk polymerization of super-critical ethylene with any thermodynamic model, we run into another problem quite typical for polymers. The high-pressure ethylene polymerization process does not lead to linear chains, but to long-chain branched structures carrying short side chains [47].
The short branches, in particular, have a marked influence on liquid-liquid phase behavior in normal liquids at ambient pressure [43] and should therefore also be accounted for in MFLG treatments. Before describing a simple molecular model for short-chain branched polymers we first discuss an application of the method outlined above to linear polyethylene in supercritical ethylene.

Figure 21
Pressure dependence of the demixing temperature of polyethylene in n-hexane for indicated polymer concentrations in wt %. MFLG prediction :
------   Parameter values in table 1a.

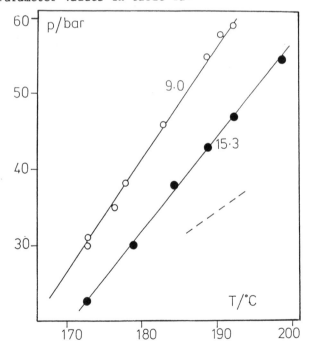

Table 1

a. MFLG parameters for 'pure' n-alkanes (i = 1) and polyethylene
(i = 2).

	α_i	β_{oi}	β_{1i} (K)	γ_i	v_0 (cc/mole)
n-hexane	0.8898	−0.2515	172	−3.1704	20
n-heptane	0.9239	−0.2060	144	−3.6493	22
n-octane	0.9299	−0.1860	130	−4.2200	23
linear poly-	1	0	1100	0.9	20, 22, 23
ethylene					but
					$v_0 m_2 = 1.2245\ M_w$

b. MFLG parameters for the system ethylene (i = 1) - polyethylene
(i = 2).

	α_i	β_{oi}	β_{1i} (K)	γ_i	v_0 (cc/mole)
ethylene	1.2508	−1.4100	530	−1.5488	45
linear	$\Big\{$ 1	0	1100	0.9 $\Big\}$	45
polyethylene	1.65	0	1030	0.95	

1. Ethylene/linear polyethylene

It is in the spirit of the model that the MFLG parameters for
polyethylene used in the preceding section should also be suitable for
other systems in which polyethylene is one of the constituents, like
ethylene/linear polyethylene. We further find the MFLG parameters for
ethylene with the procedure for small molecules, described above [39].
Table 1 lists the parameter values so obtained. The simplying assumption we could make about the binary parameters in n-alkane/poly-
ethylene systems is not allowed here. We use (quasi-)binary data by
Lusthof [48] on a linear polyethylene sample with a relatively narrow
distribution (m_w/m_n = 1.4; mass average molar mass, M_w = 8.6 kg/mole)
to make a rough estimation of α_m, β_{0_m} and β_{1_m}. To this end we compare
spinodals calculated with a trial set of binary parameters with
experimental phase boundaries, remembering that the two are not the
same but must be located in the same $p(T,\phi_2)$ range. Figure 22 shows
such a comparison; the spinodal curves were calculated with the binary
parameter values stated in Table 1. The agreement is not very good,
but we accept the situation since we are in this stage only interested
in qualitative comparisons.

Figure 22
Left: Comparison of experimental phase boundaries (o; ref. 48) for the
system ethylene/linear polyethylene with spinodals calculated with
α_m = 0.7; β_{om} = 1760; β_{1m} = -82.10^4 K and γ_2 = 0.9 for v_0 = 45.
Temperatures indicated in °C.
Right: Influence of the value of γ_2. MFLG spinodales calculated with
γ_2 = 0.9 and γ_2 = 0.95.
Parameter values in table 1b.

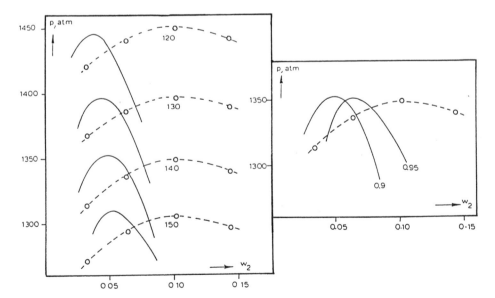

Figure 23
Experimental phase boundaries for ethylene/polyethylene showing the
molar mass dependence at 150 °C.
Ref. 48: $\square$, M_w = 8.6 kg/mole; ref. 49: O , M_w = 315 kg/mole, ref. 50:
$\bullet$, $M_w \approx 10^3$ kg/mole.
Dashed lines are calculated spinodals with M-values indicated; parameters from table 1b and γ_2 = 0.95 were used, α_m and β_m parameters from
fig. 22.

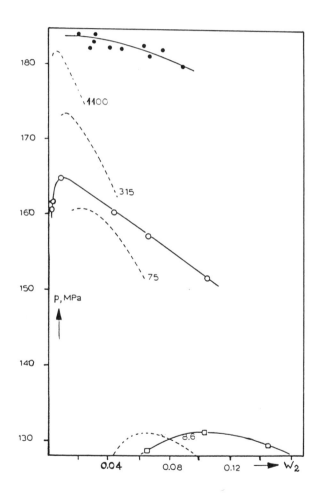

Miscibility gaps for ethylene/polyethylene, taken from different literature sources [48-50], are shown in figure 23. We use these data to test the ability of the set of binary parameters estimated above to predict the location of the phase boundaries for other samples of polyethylene in ethylene. Again, we employ spinodals and plot them in figure 23 as dashed curves. We see that the high molar mass sample is as well reproduced by the MFLG treatment as one may expect in view of the meagre basis on which the prediction rests. The intermediate sample, however, escapes adequate description, at least if the actual mass average molar mass of 315 kg/mole is used in the calculation. Lowering of this value to ca. 75 kg/mole is needed to bring the calculated spinodal into the range of the miscibility gap. We shall see that the discrepancy is caused by chain branching, the calculation showing that a linear polyethylene of M_w = 315 kg/mole is less miscible than a branched sample with the same M_w.

2. Ethylene/branched polyethylene

If chain branching has such an obvious effect on miscibility we must extend the molecular model so as to accommodate branched structures. We know from structure determinations that low-density polyethylene chains are composed of end-, linear middle-, and branched middle units (figure 24). We have chains i, differing in number of units m_i. Let a fraction δ of the m_i units of chain i be branched middle units, then the number of these units per chain is $\nu_i = \delta m_i$. One or two side chains may originate from a branched middle unit; if there are two, they would generally differ in length.

Figure 24 illustrates that the chain geometry may differ widely, also at constant δ and m_i. Depending on the length of the side chain, its end units will more or less be influenced by the main chain. The shorter the side branch, the less pronounced the specificity of the interaction of the end unit will be. We introduce a parameter ε to express the average 'effectivity' of the end units in a number. For very short side branches, the effectivity ε will be much smaller than unity ($0 < \varepsilon < 1$, figure 24a). With growing side-chain length, ε will rapidly go to unity and then remain constant (figure 24b). The total effective number of end units per chain i is thus: $2 + \varepsilon \nu_i$. If two side branches originate from a branched middle unit the relevant ε value will be between 0 and 2.

Applying the usual procedures (strictly regular mixing rules [8, 29]) to the system we obtain the same expression for the interaction function Γ as before (eq. (12)), except for the definition of γ_2 which is now [18, 43]

$$\gamma_2 = 1 - \sigma_2/\sigma_0 - (\sigma_3/\sigma_0 - \sigma_2/\sigma_0)\,\delta - (\sigma_1/\sigma_0 - \sigma_2/\sigma_0)\,(\varepsilon\delta + 2/m_n) \quad (19)$$

where σ_1, σ_2 and σ_3 are the surface areas of end, linear middle and branched middle units. The assumption has been made that the various types of surface are identical in their contribution to the internal energy per contact.

Figure 24
Schematic illustration f various types of branched polymer chains, equal in number of units (m_i) and number of branch units (δm_i) but differing in length of side chains (ε, see text) and functionality of the branch units. Chain a, $0 < \varepsilon \ll 1$; chain b, $\varepsilon = 1$; chain c, $\varepsilon = 2$.

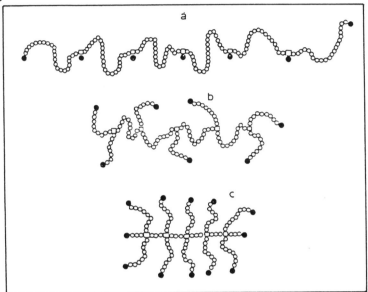

Figure 25
Left: experimental phase boundaries for ethylene/polyethylene (ref. 26) showing the effect of chain branching at roughly identical molar mass distribution, o: linear; ●: branched (5.18 end groups per 100 C atoms). Temperature: 150 °C, M_w = 54 kg/mole.
Right: MFLG spinodals for indicated values of γ_2. (α_2 and β_{02} from tabel 1b for $\gamma_2 = 0{,}95$).

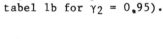
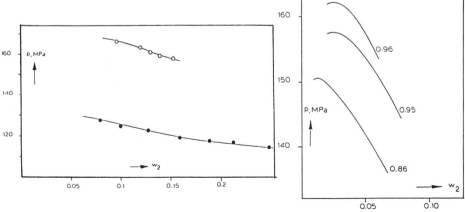

An increase of the concentration of branched middle units in the chains decreases the value of γ_2, at constant values of the σ_1. Hence, we can compare experimental phase boundaries for two polyethylene samples in ethylene, roughly identical in molar mass distribution but differing in chain geometry. Figure 25 shows such data, for a linear and a branched polyethylene, and also presents predicted spinodal (dashed) curves indicating that introduction of eq. (19) indeed brings the required effect about.

CONCLUSION

We have seen that polymerizations at elevated pressures lead to systems that may not a priori be assumed to be simple homogeneous liquid mixtures. Phase separations may occur easily that give rise to the appearance of two fluid phases, one of which is extremely dilute whereas the other, though relatively concentrated, still contains more solvent than polymer. Such phase separations have been known for a long time under milder conditons than considered here, and been interpreted in terms of the extreme disparity in molecular size between the components, viz., monomer or solvent vs polymer. Classic thermodynamic rules apply here too and can be used to understand phase relationships.

The mean-field lattice equation of state as presented here appears to be well capable not only to describe observed phenomena, including the influence of pressure, but also to predict locations of miscibility gaps as a function of average molar mass and chain geometry.

REFERENCES

1. Staverman, A.J.; Van Santen, J.H., Rec. Trav. Chim. 60 (1941) 76.
2. Staverman, A.J., Rec. Trav. Chim. 60 (1941) 640.
3. Huggins, M.L., J. Chem. Phys. 9 (1941) 440.
4. Flory, P.J., J. Chem. Phys. 9 (1941) 660.
5. Huggins, M.L., Ann. N.Y. Acad. Sci. 43 (1942) 1.
6. Flory, P.J., J. Chem. Phys. 10 (1942) 51.
7. Van der Waals, J.D.; Kohnstamm, Ph., Lehrbuch der Thermodynamik, Barth, Leipzig, 1912, Vol. II.
8. See: Guggenheim, E.A., Mixtures, Oxford, Clarendon, 1952.
9. Prigogine, I, The Molecular Theory of Solutions, Amsterdam, North-Holland, 1957.
10. Flory, P.J., J. Am. Chem. Soc. 87 (1965) 1833.
11. Patterson, D., Pure Appl. Chem. 31 (1972) 133.
12. Simha, R., Macromolecules 10 (1977) 1025.
13. Frenkel, J., Kinetic Theory of Liquids, Dover reprint, New York, 1955.
14. Cernuschi, F.; Eyring, H., J. Chem. Phys. 7 (1939) 547.
15. Kanig, G., Kolloid Z. & Z. Polym., 190 (1963) 1.
16. Trappeniers, N.J.; Schouten, J.A.; Ten Seldam, C.A., Chem. Phys. Lett. 5 (1970) 541; Physica 73 (1974) 556.
17. Sanchez, J.C.; Lacombe, R.H., J. Phys. Chem. 80 (1976) 2352, 2568.
18. Kleintjens, L.A., Ph.D. Thesis, Essex UK, 1979.
19. Kleintjens, L.A.; Koningsveld, R., Coll. Polym. Sci. 258 (1980) 711.
20. Kleintjens, L.A.; Fluid Phase Eq. 10 (1982) 183.
21. Gibbs, J.W., The Scientific Papers, Dover Reprint, New York, 1961, Vol. I.
22. Bakhuis Roozeboom, H.W., Die heterogenen Gleichgewichte vom Standpunkte der Phasenlehre, Braunschweig, Vieweg, 1913.
23. Koningsveld, R.; Stockmayer, W.H., Equilibrium Thermodynamics of Polymer Systems, Vol. I, Polymer Phase Diagrams (with Nies, E.), Oxford University Press, to appear in 1987.
24. Schneider, G.M., Chem. Thermod., Vol. II, Specialist Period. Repts, The Chem. Soc., London, 1978, 105.
25. Ehrlich, P., J. Polym. Sci. Part A, 3 (1965) 131.
26. De Loos, Th.; Poot W.; Diepen, G.A.M., Macromolecules 16 (1) (1983) 111.
27. McClellan, A.K.; MacHugh, M.A., Pol. Eng. Sci. 25 (1985) 1088.
28. Silberberg, A., J. Chem. Phys. 48 (1968) 2835.
29. Hildebrand, J.H.; Scott, R.L., The solubility of Non-Electrolytes, New York, Dover, 1964.
30. Rehage, G., Kunststoffe 53 (1963) 605.
31. See e.g. Tompa, H., Polymer Solutions, London, Butterworth, 1956.
32. Flory, P.J., Principles of Polymer Chemistry, Cornell University Press, 1953.
33. Nies, E.; Koningsveld, R.; Kleintjens, L.A., Progr. Coll. Polym. Sci. 71 (1985) 2.
34. Nies, E., Ph.D. Thesis, Antwerp, 1983.
35. Bondi, A., J. Phys. Chem. 68 (1964) 441.

36. Staverman, A.J., Rec. Trav. Chim. $\underline{69}$ (1950) 163.
37. Leblans-Vinck, A.M.; Koningsveld, $\overline{R}$.; Kleintjens, L.A; Diepen, G.A.M., Fluid Phase Eq. $\underline{20}$ (1985) 347.
38. Van Welie, G.S.A.; Diepen, G.A.M., Rec. Trav. Chim. $\underline{80}$ (1961) 659, 666, 673.
39. Nies, E.; Kleintjens, L.A.; Koningsveld, R.; Simha, R.; Jain, R.K., Fluid Phase Eq. $\underline{12}$ (1983) 11.
40. See: Koningsveld, R.; Disc. Farad. Soc. No. 49 (1970) 144.
41. Flory, P.J., J. Chem. Phys. $\underline{12}$ (1944) 425.
42. Koningsveld, R.; Stockmayer, W.H.; Kennedy, J.W.; Kleintjens, L.A., Macromolecules 7 (1974) 73.
43. Kleintjens, L.A.; Koningsveld, R.; Gordon, M., Macromolecules $\underline{13}$ (1980) 303.
44. Freeman, P.I. Rowlinson, J.S., Polymer $\underline{1}$ (1960) 20.
45. Nakajima, A.; Hamada, F., Kolloid Z. & $\overline{Z}$. Polym. $\underline{205}$ (1965) 55.
46. Kodama, Y.; Swinton, F.L., Brit. Polym. J. $\underline{10}$ (1978) 191.
47. Bovey, F.A.; Schilling, F.C.; McCrackin, F.$\overline{L}$.; Wagner, M.L., Macromol. $\underline{19}$ (1976) 76.
48. Lusthof, M., unpublished data.
49. Swelheim, T.; De Swaan Arons, J.; Diepen, G.A.M., Rec. Trav. Chim. $\underline{84}$ (1965) 261.
50. Nees, F., Ph.D. Thesis, Karlsruhe, 1978.

APPENDIX

Lattice gas expressions for the spinodal and for the critical point

For a binary mixture, the lattice gas expression for the spinodal is given by:

$$J_{sp} = A_{\psi_1 \psi_1} \cdot A_{\psi_2 \psi_2} - A_{\psi_1 \psi_2}^2 = 0$$

and the critical point by

$$J_{cr} = \frac{\partial J_{sp}}{\partial \psi_1} \cdot A_{\psi_2 \psi_2} - \frac{\partial J_{sp}}{\partial \psi_2} \cdot A_{\psi_1 \psi_2} = 0$$

where $A = \dfrac{\Delta F}{N_\psi RT} = \psi_0 \ln \psi_0 + \dfrac{\psi_1}{m_1} \ln \psi_1 + \dfrac{\psi_2}{m_2} \ln \psi_2 + \psi_0 \psi_1 g_1 + \psi_0 \psi_2 g_2 + \psi_1 \psi_2 g_{12}$

and $g_1 = \alpha_1 + \dfrac{g_{11} (1 - \gamma_1)}{q}$

$g_2 = \alpha_2 + \dfrac{g_{22} (1 - \gamma_2)}{q}$ (Note that: $g_{ii} = \beta_{0i} + \beta_{1i}/T$)

$g_{12} = \alpha_{12} + \dfrac{g_m (1 - \gamma_2)}{q}$

with $q = 1 - \gamma_1 \psi_1 - \gamma_2 \psi_2$

and $A_{\psi_1 \psi_1} = \partial^2 A / \partial \psi_1^2$ etc.

$$\psi_0 = \frac{n_0}{N_\psi} = \frac{n_0}{n_0 + n_1 m_1 + n_2 m_2} \quad \text{and} \quad \psi_1 = \frac{n_1 m_1}{N_\psi}$$

The involved derivatives of A are:

$$A_{\psi_1 \psi_1} = \frac{1}{\psi_0} + \frac{1}{m_1 \psi_1} - 2 g_1 + 2(\psi_0 - \psi_1) g_{1_{\psi_1}} + \psi_0 \psi_1 g_{1_{\psi_1 \psi_1}} - 2 \psi_2 g_{2_{\psi_1}}$$

$$+ \psi_0 \psi_2 g_{2_{\psi_1 \psi_1}} + 2 \psi_2 g_{12_{\psi_1}} + \psi_1 \psi_2 g_{12_{\psi_1 \psi_1}}$$

$$A_{\psi_2\psi_2} = \frac{1}{\psi_0} + \frac{1}{m_2\psi_2} - 2\,\psi_1\,g_{1_{\psi_2}} + \psi_0\psi_1\,g_{1_{\psi_2\psi_2}} - 2\,g_2 + 2(\psi_0 - \psi_2)\,g_{2_{\psi_2}}$$

$$+ \psi_0\psi_2\,g_{2_{\psi_2\psi_2}} + 2\,\psi_1\,g_{12_{\psi_2}} + \psi_1\psi_2\,g_{12_{\psi_2\psi_2}}$$

and

$$A_{\psi_1\psi_2} = \frac{1}{\psi_0} - g_1 - g_2 + (\psi_0 - \psi_1)\,g_{1_{\psi_2}} - \psi_1\,g_{1_{\psi_1}} + \psi_0\psi_1\,g_{1_{\psi_1\psi_2}} - \psi_2\,g_{2_{\psi_2}}$$

$$+ \psi_0\psi_2\,g_{2_{\psi_1\psi_2}} + (\psi_0 - \psi_2)\,g_{2_{\psi_1}} + g_{12} + \psi_2\,g_{12_{\psi_2}} + \psi_1\,g_{12_{\psi_1}} + \psi_1\psi_2\,g_{12_{\psi_1\psi_2}}$$

where $g_{1_{\psi_1}} = \partial g_1/\partial\psi_1$, $g_{2_{\psi_2}} = \partial\psi_2/\partial\psi_1$ etc. leading to:

$$g_{1_{\psi_1}} = g_{11}\,(1 - \gamma_1)\,\gamma_1\,q^{-2}$$

$$g_{1_{\psi_2}} = g_{11}\,(1 - \gamma_1)\,\gamma_2\,q^{-2}$$

$$g_{1_{\psi_1\psi_1}} = 2\,g_{11}\,(1 - \gamma_1)\,\gamma_1^2\,q^{-3}$$

$$g_{1_{\psi_2\psi_2}} = 2\,g_{11}\,(1 - \gamma_1)\,\gamma_2^2\,q^{-3}$$

$$g_{1_{\psi_1\psi_2}} = 2\,g_{11}\,(1 - \gamma_1)\,\gamma_1\gamma_2\,q^{-3}$$

$$g_{2_{\psi_1}} = g_{22}\,(1 - \gamma_2)\,\gamma_1\,q^{-2}$$

$$g_{2_{\psi_2}} = g_{22}\,(1 - \gamma_2)\,\gamma_2\,q^{-2}$$

$$g_{2_{\psi_1\psi_1}} = 2\,g_{22}\,(1 - \gamma_2)\,\gamma_1^2\,q^{-3}$$

$$g_{2_{\psi_2\psi_2}} = 2\ g_{22}\ (1 - \gamma_2)\ \gamma_2^{\,2}\ q^{-3}$$

$$g_{2_{\psi_1\psi_2}} = 2\ g_{22}\ (1 - \gamma_2)\ \gamma_1\gamma_2\ q^{-3}$$

$$g_{12_{\psi_1}} = g_m\ (1 - \gamma_2)\ \gamma_1\ q^{-2}$$

$$g_{12_{\psi_2}} = g_m\ (1 - \gamma_2)\ \gamma_2\ q^{-2}$$

$$g_{12_{\psi_1\psi_1}} = 2\ g_m\ (1 - \gamma_2)\ \gamma_1^{\,2}\ q^{-3}$$

$$g_{12_{\psi_2\psi_2}} = 2\ g_m\ (1 - \gamma_2)\ \gamma_2^{\,2}\ q^{-3}$$

$$g_{12_{\psi_1\psi_2}} = 2\ g_m\ (1 - \gamma_2)\ \gamma_1\gamma_2\ q^{-3}$$

For the critical condition we further need:

$$\partial J_{sp}/\partial\psi_1 = A_{\psi_1\psi_1\psi_1}\ A_{\psi_2\psi_2} + A_{\psi_1\psi_2\psi_2}\ A_{\psi_1\psi_1} - 2\ A_{\psi_1\psi_1\psi_2}\ A_{\psi_1\psi_2}$$

and

$$\partial J_{sp}/\partial\psi_2 = A_{\psi_1\psi_1\psi_2}\ A_{\psi_2\psi_2} + A_{\psi_2\psi_2\psi_2}\ A_{\psi_1\psi_1} - 2\ A_{\psi_1\psi_2\psi_2}\ A_{\psi_1\psi_2}$$

where

$$A_{\psi_1\psi_1\psi_1} = \frac{1}{\psi_0^{\,2}} - \frac{1}{m_1\psi_1^{\,2}} - 6\ g_{1_{\psi_1}} + 3\ (\psi_0 - \psi_1)\ g_{1_{\psi_1\psi_1}} + \psi_0\psi_1\ g_{1_{\psi_1\psi_1\psi_1}}$$
$$- 3\ \psi_2\ g_{2_{\psi_1\psi_1}} + \psi_0\psi_2\ g_{2_{\psi_1\psi_1\psi_1}} + 3\ \psi_2\ g_{12_{\psi_1\psi_1}} + \psi_1\psi_2\ g_{12_{\psi_1\psi_1\psi_1}}$$

$$A_{\psi_1\psi_1\psi_2} = \frac{1}{\psi_0^2} - 2\, g_{1_{\psi_2}} - 2\, g_{1_{\psi_1}} + 2\,(\psi_0 - \psi_1)\, g_{1_{\psi_1\psi_2}} - \psi_1\, g_{1_{\psi_1\psi_1}}$$

$$+ \psi_0\psi_1\, g_{1_{\psi_1\psi_1\psi_2}}$$

$$+ (\psi_0 - \psi_2)\, g_{2_{\psi_1\psi_1}} - 2\, g_{2_{\psi_1}} - 2\,\psi_2\, g_{2_{\psi_1\psi_2}} + \psi_0\psi_2\, g_{2_{\psi_1\psi_1\psi_2}}$$

$$+ 2\, g_{12_{\psi_1}} + 2\,\psi_2\, g_{12_{\psi_1\psi_2}} + \psi_1\, g_{12_{\psi_1\psi_1}} + \psi_1\psi_2\, g_{12_{\psi_1\psi_1\psi_2}}$$

$$A_{\psi_1\psi_2\psi_2} = \frac{1}{\psi_0^2} - 2\, g_{1_{\psi_2}} - 2\,\psi_1\, g_{1_{\psi_1\psi_2}} + (\psi_0 - \psi_1)\, g_{1_{\psi_2\psi_2}} + \psi_0\psi_1\, g_{1_{\psi_1\psi_2\psi_2}}$$

$$- 2\, g_{2_{\psi_1}} - 2\, g_{2_{\psi_2}} + 2\,(\psi_0 - \psi_2)\, g_{2_{\psi_1\psi_2}} - \psi_2\, g_{2_{\psi_2\psi_2}} + \psi_0\psi_2\, g_{2_{\psi_1\psi_2\psi_2}}$$

$$+ 2\, g_{12_{\psi_2}} + 2\,\psi_1\, g_{12_{\psi_1\psi_2}} + \psi_2\, g_{12_{\psi_2\psi_2}} + \psi_1\psi_2\, g_{12_{\psi_1\psi_2\psi_2}}$$

$$A_{\psi_2\psi_2\psi_2} = \frac{1}{\psi_0^2} - \frac{1}{m_2\psi_2^2} - 3\,\psi_1\, g_{1_{\psi_2\psi_2}} + \psi_0\psi_1\, g_{1_{\psi_2\psi_2\psi_2}} - 6\, g_{2_{\psi_2}}$$

$$+ 3\,(\psi_0 - \psi_2)\, g_{2_{\psi_2\psi_2}} + \psi_0\psi_2\, g_{2_{\psi_2\psi_2\psi_2}} + 3\,\psi_1\, g_{12_{\psi_2\psi_2}} + \psi_1\psi_2\, g_{12_{\psi_2\psi_2\psi_2}}$$

and further:

$$g_{1_{\psi_1\psi_1\psi_1}} = 6\, g_{11}\, (1 - \gamma_1)\, \gamma_1^3\, q^{-4}$$

$$g_{1_{\psi_1\psi_1\psi_2}} = 6\, g_{11}\, (1 - \gamma_1)\, \gamma_1^2\gamma_2\, q^{-4}$$

$$g_{1_{\psi_1\psi_2\psi_2}} \quad 6\, g_{11}\, (1 - \gamma_1)\, \gamma_1\gamma_2^2\, q^{-4}$$

$$g_{1_{\psi_2\psi_2\psi_2}} = 6\, g_{11}\, (1 - \gamma_1)\, \gamma_2^3\, q^{-4}$$

$$g_{2_{\psi_1\psi_1\psi_1}} = 6\ g_{22}\ (1 - \gamma_2)\ \gamma_1^{\ 3}\ q^{-4}$$

$$g_{2_{\psi_1\psi_1\psi_2}} = 6\ g_{22}\ (1 - \gamma_2)\ \gamma_1^{\ 2}\gamma_2\ q^{-4}$$

$$g_{2_{\psi_1\psi_2\psi_2}} = 6\ g_{22}\ (1 - \gamma_2)\ \gamma_1\gamma_2^{\ 2}\ q^{-4}$$

$$g_{2_{\psi_2\psi_2\psi_2}} = 6\ g_{22}\ (1 - \gamma_2)\ \gamma_2^{\ 3}\ q^{-4}$$

$$g_{12_{\psi_1\psi_1\psi_1}} = 6\ g_m\ (1 - \gamma_2)\ \gamma_1^{\ 3}\ q^{-4}$$

$$g_{12_{\psi_1\psi_1\psi_2}} = 6\ g_m\ (1 - \gamma_2)\ \gamma_1^{\ 2}\gamma_2\ q^{-4}$$

$$g_{12_{\psi_1\psi_2\psi_2}} = 6\ g_m\ (1 - \gamma_2)\ \gamma_1\gamma_2^{\ 2}\ q^{-4}$$

$$g_{12_{\psi_2\psi_2\psi_2}} = 6\ g_m\ (1 - \gamma_2)\ \gamma_2^{\ 3}\ q^{-4}$$

NUCLEAR MAGNETIC RESONANCE AND LASER SCATTERING TECHNIQUES AT HIGH
PRESSURE

Jiri Jonas
Department of Chemistry
School of Chemical Sciences
University of Illinois
Urbana, Illinois 61801
U.S.A.

ABSTRACT. This review covers various aspects of nuclear magnetic
resonance spectroscopy and laser Raman and Rayleigh spectroscopy at
high pressure. The presentation is organized into the following main
sections: 1. Introduction; 2. Experimental High Pressure NMR Techniques; 3. Experimental High Pressure Laser Scattering Techniques; 4.
Applications of NMR at High Pressure; 5. Applications of Laser Raman
Scattering at High Pressure. The main emphasis is on studies aimed
towards improving our fundamental understanding of the dynamic structure of fluids but several examples dealing with disordered solids are
also included.

1. Introduction

Considering the fact that NMR spectroscopy at ambient conditions has
been applied to a wide spectrum of problems in chemistry and physics,
it is not surprising to find that high-pressure NMR (1-4) techniques
have also had many applications. In this lecture, the results of
several specific NMR measurements on liquids and disordered solids at
high pressure will be discussed to illustrate the range of problems
that can be studied by this approach.

　　Laser Rayleigh and Raman scattering experiments on fluids at high
pressure have recently provided important and unique information about
dynamic processes in liquids. In this lecture a discussion of recent
applications of Raman and Rayleigh measurements at high pressure to
study the dynamic structure of liquids will provide a convincing
evidence about the power of these experimental techniques to improve
our understanding of molecular motions and interactions in the liquid
state.

　　Since liquids and disordered solids are relatively compressible,
the maximum pressure usually used is 10 kbar; such pressures only
change intermolecular distances. In contrast to change the molecular
electronic structure of a specific material (5), pressures in excess
of 30 kbar have to be used.

R. van Eldik and J. Jonas (eds.), High Pressure Chemistry and Biochemistry, 193–235.

1.1. Information Content of NMR and Laser Scattering Experiments

The large majority of NMR applications in chemistry deal with liquids
in which the NMR lines are narrowed by motional averaging to a natural
linewidth of the order of 0.1 to 1 hertz. High-resolution NMR spectra of complex molecules in the liquid phase usually exhibit a great
deal of structure and yield a wealth of information about the molecule. A discussion of some of the applications of multinuclear
high-resolution Fourier transform (FT) NMR spectroscopy at high pressure is included because this recently developed technique shows a
great promise in the study of chemical processes.

It is well known that an NMR signal from magnetic nuclei can
provide detailed information about the nature and the rate of molecular motions. During the NMR relaxation experiment, after the spin
magnetization is changed from its equilibrium value by a specific
radio-frequency pulse, one determines the time constant of its return
to equilibrium. This time constant, the spin-lattice relaxation time,
T_1, is the time needed to reach thermal equilibrium between the spins
and the lattice (6). The lattice is the collection of atoms of molecules that constitute the sample. Depending on the specific nucleus
and the specific system, nuclei can relax by different mechanisms.

The dipolar, quadrupolar, and spin-rotation interaction mechanisms (6) represent the main relaxation modes observed in studies of
the dynamics of liquids. The relaxation process occurs through fluctuating magnetic and electric fields (for quadrupolar nuclei). Since
these fluctuating fields have their origin in motions of the molecules
in the liquid, the measured relaxation times provide information about
molecular motions. Using well-known theoretical expressions, one can
analyze the experimental spin-lattice relaxation times and determine
the correlation times for the appropriate motion. Physically, a
correlation time represents the average time the molecule needs to
lose memory of its initial position, orientation, or angular momentum.
The information content of the NMR relaxation experiments is very
high. For example, by measuring the relaxation of nuclei at different
parts of a molecule, one can learn how fast the molecule rotates about
its different axes. In addition, one can learn from the NMR spin-echo
experiments (6) about diffusion - a fundamental transport property.

As outlined above the NMR relaxation experiment enables one to
obtain the correlation time for a specific motion. One has to realize
that a correlation time is the integral of the appropriate correlation
function. The great advantage of Raman experiments lies in the fact
that the analysis of Raman lineshape provides information about the
detailed nature of the correlation function - one obtains the time
dependence of the correlation function itself and not only an
integral. It is only natural that the Raman experiment has also some
limitations, namely, only Raman lineshapes of relatively simple
molecules can be analyzed to yield unambigous results. The general
theory of light scattering is well established and at this point a few
pertinent comments about the Raman experiments are appropriate.

From the experimental polarized and depolarized Raman bandshapes
one can obtain the isotropic scattering intensity $I_{iso}(\omega)$ and the

anisotropic scattering intensity $I_{aniso}(\omega)$. Only vibrational (nonorientational) processes contribute to $I_{iso}(\omega)$ whereas both reorientational and vibrational processes contribute to $I_{aniso}(\omega)$. This provides the means of separating reorientational processes from vibrational processes and of calculating reorientational and vibrational correlation functions. Assuming vibrational relaxation to be the major nonreorientational broadening mechanism, one can show (7) that

$$I_{iso}(\omega) = [I_{VV}(\omega) - 4/3\ I_{VH}(\omega)]/\int [I_{VV}(\omega) - \tag{1}$$

$$- 4/3\ I_{VH}(\omega)]d\omega$$

and

$$C_v(t) = \langle Q^v(0)\ Q^v(t)\rangle = \int I_{iso}(\omega)\ \exp\ (-i\omega t)\ d\omega, \tag{2}$$

where $C_v(t)$ is the vibrational correlation function. Similarly, one may write

$$I_{aniso}(\omega) = I_{VH}(\omega)/\int I_{VH}(\omega)\ d\omega, \tag{3}$$

and

$$C_R(t) = \langle Tr\beta^v(0)\ \beta^v(t)\rangle = \frac{\int I_{aniso}(\omega)\ \exp\ (-i\omega t)\ d\omega}{\int I_{iso}(\omega)\ \exp\ (-i\omega t)\ d\omega} \tag{4}$$

where $C_R(t)$ is the reorientational correlation function.

Since the theoretical basis for the use of depolarized Rayleigh scattering for the study of collision induced spectra has recently been discussed by a number of authors (8,9), a discussion of this topic is not necessary at this point.

1.2. Reason for Measurements at High Pressure

The use of pressure as an experimental variable in any experiment leads to some added complexity in instrumentation; however, the information to be gained from this approach fully justifies its use. There are several fundamental reasons for performing experiments on liquids at high pressure.

In most NMR and laser scattering studies of liquids only the temperature has been varied in spite of the fact that the use of pressure provides another dimension over which to investigate the liquid system. Only by using both pressure and temperature as vari-

ables in an NMR experiment can one separate the effects of density and
temperature on molecular motions. Due to the close packing of molecules in a liquid, even a small change in density can produce a considerable change in the molecular dynamics of the system; therefore,
in order to test rigorously a theoretical model of a liquid, or a
model of a specific dynamic process in a liquid, one must perform
isochoric, isothermal, and isobaric experiments. As an illustration
of the importance of separating the effects of density and temperature
on molecular motions, Fig. 1 shows the temperature dependence of the
self-diffusion coefficient, D, in liquid tetramethylsilane (TMS) at
constant density and at constant pressure. The dramatic difference in
the temperature dependence of D at constant pressure or at constant
density is readily apparent.

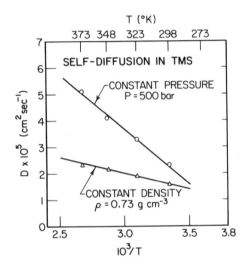

Figure 1. Temperature dependence of self-diffusion in liquid tetramethylsilane at (o) constant pressure and (Δ) constant density.

Recent studies of intermolecular interactions and Fermi resonance
(10,11) may serve as yet another example of the important role of high
pressure experiments. The Fermi resonance between the ν_1 and the
first overtone of ν_4 have been studied in liquid ND_3 as a function of
density and temperature (11). Fig. 2 shows the changes in relative
intensities of these Fermi resonance coupled bands for the extreme
density range of our measurements. Since Fermi resonance parameters
are sensitive to intermolecular potential, we can change them by
varying temperature and pressure. The transition dipole moments of the
ν_1 + $2\nu_4$ bands are found to vary and the Fermi resonance treatment
enables us to estimate the changes in their relative magnitude. The
high pressure experiments provided the critical spectroscopic information needed for the theoretical analysis of intermolecular interactions in ND_3.

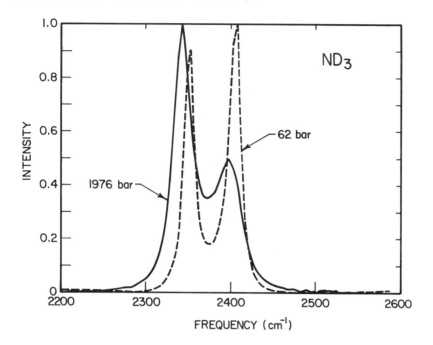

Figure 2. Density effects on the relative intensity of the isotropic $\nu_1 + 2\nu_4$ Fermi resonance lines in ND_3. The full line denotes density $= 0.730$ g cm^{-3} (T=0°C; P = 1967 bar), and the dashed line denotes density = 0.457 g cm^{-3} (T = 100°C; P = 62 bar).

In addition, the use of pressure enables one to extend the measurement range on liquids well above the normal boiling point and allows the study of supercritical dense fluids. As an example, Fig. 3 shows the NMR proton T_1 in water (12) measured both in the low-temperature (T < 30°C) anomalous region and in the high temperature (T > 400°C) supercritical region.

2. Experimental High Pressure NMR Techniques

High resolution NMR spectroscopy at high pressure represents one of the promising new areas of research at high pressure. Recent advances in magnet technology have resulted in the development of magnets capable of attaining a high homogeneity of the magnetic field over the sample volume so that even without sample spinning, one can achieve a very high resolution. At the same time, the Fourier transform techniques make all these high resolution experiments much easier to be performed at high pressure than it was the case with classical CW techniques.

Therefore, our overview of the experimental high pressure techniques focuses on high resolution, multinuclear NMR techniques for work at high pressure. In view of the great promise for high resolu-

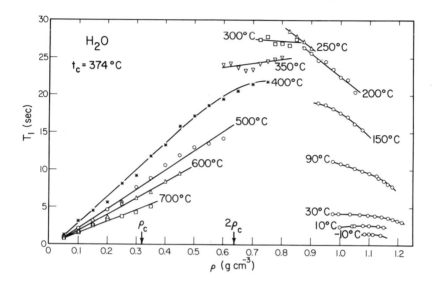

Figure 3. Proton spin-lattice relaxation time, T_1, in water as a
function of density over the temperature range -10°C to 700°C.

tion, high pressure NMR techniques in the field of homogeneous
catalysis a high resolution NMR probe equipped with stirring which was
specially designed for the study of homogeneous catalytic reactions
will be described.

2.1. High Resolution, High Pressure NMR Techniques

In view of the fact that the general design principles for construction of NMR probes have been discussed to a considerable detail in a
number of publications, including the report (1) at the last NATO ASI
dealing with research at high pressure, only Fig. 4 is presented which
gives the schematic drawing of a high pressure NMR probe used in our
laboratory.
        The considerable research activity in the field of high resolu-
tion NMR at high pressure in recent years is illustrated in Table 1
which summarizes the main performance characteristics of individual
high pressure NMR techniques. The first high resolution FT NMR experiments at high pressure have been reported by Wilbur and Jonas (13)
who indicated the possible applications of these experiments. A
wide-gap (9.5 cm) Varian electromagnet model V-3800-1 was used in
these experiments and the details of the design have been described in
detail elsewhere (14). Yamada (15,16) has described a glass capillary
technique which allows measurements up to 2000 bar using a standard
probe in a commercial NMR spectrometer. Oldenziel and Trappeniers
(17) have developed a high pressure vessel in which the sample can be
rotated under high pressure. A gas was used as a pressurizing medium

TABLE 1

Characteristics of high-resolution probes for NMR measurements at high pressure.

Technique[a]	Experimental Range P(bar)	T(°C)	Sample diameter (mm)	Resolution	Proton frequency [MHz]	Ref.
HPV-NR	1-9000	-70 - 300	8	7×10^{-8}	60	14
C-NR	1-2000	-25 - 100	1	1.7×10^{-8}	60	15,16
C-R[b]	1-140	-100 - 100	3.4	3×10^{-9}	360	21
HPV-R[c]	1-2500	29	4	1×10^{-8}	60	17
C-R	1-1000	25[d]	2	4×10^{-8}	60	18
HPV-NR	1-3700	-10 - 80	1.1	1×10^{-8}	60	20
HPV-NR	1-5000	-70 - 200	6	5×10^{-9}	180[e]	22
HPV-NR	1-1000	-20 - 55	6	3×10^{-7}	38.3	23
HPV-NR[f]	1-2000	-50 - 150	8	1×10^{-8}	180[e]	24

a HPV-high pressure vessel; NR- non-rotating sample; R - rotating sample; C - capillary.
b Sapphire tube.
c Gas pressurizing system.
d Temperature range depends on commercial spectrometer used.
e Superconducting magnet.
f Catalysis probe equipped with stirring.

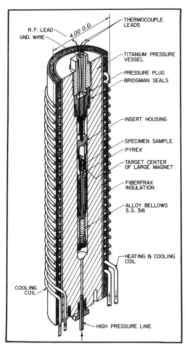

Figure 4.   Schematic drawing of NMR probe for experiments at high
pressure.

and a small DC motor located inside the high pressure vessel was
employed for spinning the sample.   The authors mentioned that deterioration of the motor occurs rather rapidly under the influence of the
pressurized gas and the motor had to be frequently replaced.   Merbach
et al. (18) and Ludemann et al. (19) have described different modifications of the capillary technique (16) which is suitable for work
with a commercial NMR spectrometer.   However, the highest pressure
achievable by using the capillary technique is naturally limited (2000
bar) and the sample volume is necessarily small (i.d.   1-2 mm).
Merbach et al. (18) have discussed the relative merits of the capillary technique and the design using a high pressure vessel.   Their
specific design (20) using a high pressure vessel enables them to work
with samples of 1.5 mm diameter and the resolution is 1 - 3 x 10^{-8}.
Roe (21) has recently described a high resolution technique using
sapphire tubes which allow measurements via a commercial NMR spectrometer to pressure up to 140 bar.
      The schematic drawing of the high-resolution, high-pressure probe
(22) is shown in Fig. 5.   This probe is located inside a titanium alloy
IMI-680 vessel, which can be heated or cooled using thermostating
jackets.   In our earlier design (14) of the high-pressure RF
feedthroughs, we used the conventional electrical feedthrough design
using SS-316 cones soldered on copper wire and the plastic (SP-1
Vespel polyimide, Du Pont de Nemours & Co., Wilmington, DE) cone was

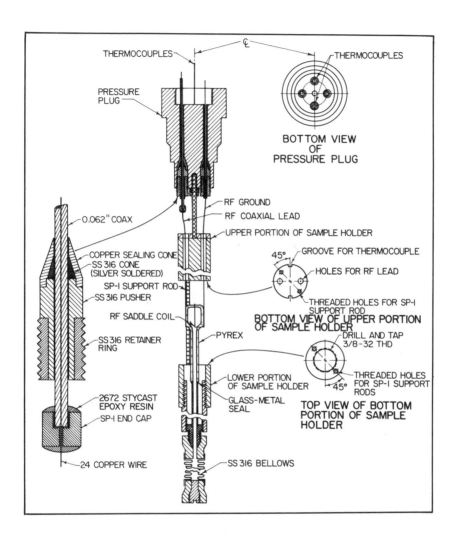

Figure 5. Assembly drawing for components of the high resolution NMR probe for work at 180 MHz.

used as a sealing cone to provide insulation from the plug body.
However, the stray capacitance of such a RF feedthrough is   20pF is
too high for a saddle coil (L = 0.1 μH) and tapped-parallel-tuned
circuit at 180 MHz.

Our feedthrough (22) uses a coaxial transmission line (o.d.
0.062"; MgO dielectric material; Inconel 1 sheath) on which a SS-316
cone is silver soldered. The sealing cone can be made of copper
because no electrical insulation from the closure plug is needed. The
fact that we do not have to use an insulated sealing cone (SP-1 in our
original design) improves the reliability of this specific probe design, because depending on operating temperature, the SP-1 cones have
only a limited lifetime of approximately 20 pressure cycles. The high
temperature limit of $200^{\circ}C$ is presently determined by the epoxy resin
used in the feedthrough (see Fig. 5) but changing to ceramic adhesive
would raise the high temperature limit up to $300^{\circ}C$. The current
pressure range is 1 to 5000 bar but an increase of wall thickness of
the titanium alloy high pressure vessel (current o.d. = 5.7 cm; i.d. =
1.4 cm) would allow pressures up to 9000 bar to be reached without
modification of the probe itself. The superconducting magnet used in
these experiments was the wide-bore (12.7 cm) magnet made by Oxford
Instruments, Inc. operating at proton frequency of 180 MHz. The working bore diameter is 11.0 cm when using the room temperature shims.
The high homogeneity of the magnetic field over the sample volume when
using the superconducting and room temperature shims was excellent and
allowed us to obtain resolution 5×10^{-9} for 6 mm samples. In view of
the performance of this probe, we are certain that it will be possible
to increase the sample size without deterioration of the resolution.
The spectrometer system used is analogous to the system which has been
described in detail elsewhere (1).

In summary, we emphasize several features of the probe design
which are important for future applications of this specific NMR
technique: i) wide pressure and temperature range; ii) high resolution; iii) large sample volume; iv) reliable high pressure RF
feedthrough; v) contamination-free sample cell; vi) suitability for
superconducting magnets.

2.2 High Pressure Probe for NMR Studies of Homogeneous Catalysis

A promising application of high pressure NMR spectroscopy, still
little exploited, is in the study of homogeneous catalysts (25-27).
The use of NMR methods of structure determination, and the large body
of spectroscopic data of known organometallic compounds, can be expected to extend considerably the mechanistic understanding of these
systems. Initial experiments in this laboratory (25,26) showed the
need for a probe specifically designed to carry them out. Its important new features are: (i) a built-in stirring mechanism to mix
catalyst solutions with reactive gases, necessary because diffusion
across the gas-solution interface is otherwise extremely slow, and
(ii) much improved sensitivity, achieved by incorporating commercial
coaxial transmission line into the probe, making it possible to study
dilute solutions by ^{13}C NMR.

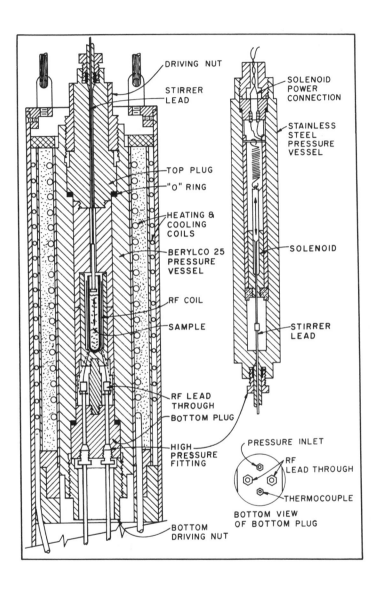

Figure 6. Schematic drawing of the high resolution probe for NMR studies of homogeneous catalysis.

A schematic drawing of the probe is shown in Fig. 6. As in a previous design (22), temperature control (-50 to 150°C) is provided by circulating a liquid through copper tubing coiled on a metal cylinder tightly fitted to the pressure vessel. An outer set of coils carry water to protect the magnet bore from extremes of temperature. The Berylco-25 pressure vessel (o.d. 2.5", i.d. 1.25"), chosen for its resistance to hydrogen embrittlement, is closed at each end with a Berylco plug and driving nut; the pressure seal is made with a Viton or Teflon o-ring. The vessel has a calculated bursting pressure of ca. 5 kbar, and has been pressure tested to 2 kbar.

The NMR insert is mounted on the lower closure plug. It contains an 8 mm sample tube and crossed Heimholtz observe coils. (For clarity, the coils are not shown in Fig. 6.) The coils are connected to leadthroughs made of cryogenic coaxial 50Ω cable with Teflon dielectric and 316 stainless steel sheath (Precision Tube, Co., North Wales, PA). A ground connection is made to a wire soldered directly to the cable sheath. The leadthroughs are sealed at the end with a Vespel SP-1 cap (DuPont, Wilmington, DE) and Stycast resin (Emerson and Cuming, Canton, MA), and sealed in the plug with Swagelok male nuts and ferrules (Crawford, Fitting Co., Solon, OH).

The stirring assembly is contained in a separate pressure vessel located outside of the magnet. It consists of a solenoid (Guardian Electric, Chicago, IL) and a spring which retracts the stirring rod when the solenoid is off. The stirrer is connected to the probe by a 3-foot length of stainless steel pressure tubing; a Nichrome wire inside the tubing drives the stirring rod, which enters the pressure vessel through the top plug. The stirring rod is in two pieces. The larger piece is fitted with a brass disk with a diameter close to that of the sample tube; it remains inside the pressure vessel. The upper piece, attached to the solenoid, is threaded into the lower piece during assembly. A separate power supply energizes the solenoid, moving the stirring rod up and down by one inch, as shown by the double-headed arrows in Fig. 6. An optimum resolution of 1 part in 10^8 was obtained in ^{13}C and ^1H spectra.

As mentioned, a high pressure sapphire tube suitable for homogeneous catalysis studies has recently been described in the literature (21). In comparison with the design presented here, the sapphire insert has the advantages of simplicity and compatibility with conventional, high-resolution probes. The pressure vessel-based design used here has the advantages of a much larger sample volume (which could be further increased), a much better coil filling factor, and an increased safety margin for operation at pressures above those employed in this work or studies using the sapphire tube that have been published to date (27).

3. Experimental High Pressure Laser Scattering Techniques

Since the experimental techniques for infrared and Raman spectroscopy at high pressure are well established and were discussed in detail by Whalley (28) at the last NATO ASI on research at high pressure, the present discussion will be limited to a few comments on pressure

induced polarization scrambling by optical windows. As already mentioned in Sect. 1., this overview is limited to experiments in the
pressure range 1 bar to 10 kbar, and therefore no discussion of the
diamond anvil cell is given. There are many excellent reviews dealing
with the diamond anvil cell and, e.g., a recent monograph by Ferraro
(29) on vibrational spectroscopy at high internal pressures presents
an extensive coverage of both the experimental techniques and the
actual data obtained by the diamond anvil cell.

     To introduce the discussion of the stress induced birefringence,
Fig. 7 gives a schematic drawing of a high volume optical cell which
can be used for laser scattering experiments on fluids at high pressure.

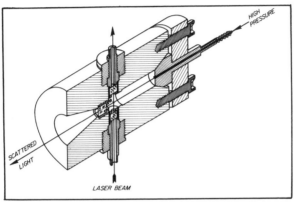

Figure 7.   Schematic drawing of high volume optical cell for light
scattering experiments at high pressure.

     As indicated in the introduction, Raman bandshapes contain
information on the orientational and vibrational dynamics of the
molecular system. The theoretical basis by which this information may
be extracted from the experimental spectra depends on the use of
linearly polarized incident radiation and the ability to measure
accurately the scattered light in two polarization directions. Lasers, of course, are an excellent source of intense, linearly polarized radiation. However, if the optical cell window material is
birefringent, the linear polarized light will become elliptically
polarized, and the results of a bandshape analysis will become questionable. Fortunately, since the most frequently used cell materials
such as fused silica and glass are not birefringent, this is usually
not a problem for measurements at atmospheric pressure. However, the
stress applied to the cell windows in a high pressure experiment can
lead to strain-induced anisotropies which result in a pressure-depend-
ent scrambling of the polarization. Even though the scrambling is
small, it can have a large effect on the I_{VH} band, especially for
cases when the depolarization ratio is small. The effect has been
known for several years and was studied earlier in this laboratory by
Cantor et al. (30).

Ikawa and Whalley (31) have recently analyzed in a great detail the problem of polarization scrambling by glass windows of a Raman cell up to pressures of 18 kbar. Their results showed that the polarization scrambling of the laser beam passing along the cylinder axis of the window can be very small provided one carefully controls the design of the optical cell windows and uses a specific scattering geometry.

Since neglect of the effects of polarization scrambling by optical windows during a high pressure experiment can lead to erroneous results, the following example (32) stresses the importance of a careful analysis of each specific high pressure light scattering experiment where polarization measurements are important. In principle, any compound exhibiting a totally polarized band with the depolarization ratio equal to zero can be used as a test liquid to measure the polarization scrambling of the windows. We found particularly useful spherical molecules of T_d symmetry for which the totally symmetrical modes have no anisotropic spectrum and therefore $\rho = 0$. There are many possible choices for the test liquid, but in view of our experiments (33) we have used tetramethyltin (TMT) as an illustrative example. The spectrum of TMT contains the totally polarized ν_3 (Sn - C symmetrical stretch, A_1) band at 509 cm^{-1} as well as the depolarized ν_{18} (Sn - C asymmetrical stretch, F_2) band at 530cm^{-1} which are both very intense and easy to measure.

Since TMT is a spherical molecule of the T_d point group, symmetry demands that the depolarization ratio equal zero for A_1 vibrations such as ν_3. Therefore any intensity in the ν_3 VH spectrum is due to polarization scrambling. The effects of pressure on the VH spectrum of tetramethyltin is shown in Fig. 8. The ν_3 peak can be seen to grow from a small shoulder on the low frequency side of ν_{18} at 500 bar to a peak nearly as intense as ν_{18} at 4000 bar.

4. Applications of NMR at High Pressure

In the following sections 4.1 to 4.4 the results of several specific NMR experiments at high pressure will illustrate what problems can be studied by this approach.

4.1. Conformational Isomerization of Cyclohexane at High Pressure

Since the high resolution, high pressure NMR techniques show particular promise for studies of chemical and biochemical systems the first example deals with our work (34) on pressure effects on ring inversion of cyclohexane in several solvents. Fig. 9 gives the schematic diagram of conformational isomerization of cyclohexane which has been a subject of a great many NMR studies at ambient conditions. This is not surprising in view of the fact that the problem of cyclohexane inversion represents a seminal problem in conformational analysis.

In addition to the simple goal of investigating the pressure effects on conformational isomerization, the main motivation of our study was to provide the first test of the predictions of the stochastic models (35) for isomerization reactions. In order to make clear the significance of our NMR experimental results, the main

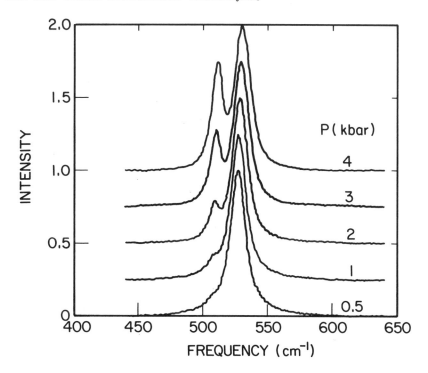

Figure 8.   Depolarized spectra of tetramethyltin at different pressures showing the growth of the ν_3 intensity (509 cm^{-1}) produced by polarization scrambling of the float glass window in the high pressure cell.   The strong band at 530 cm^{-1} is the depolarized ν_{18} stretching mode.   All measurements were carried out at 90°C.

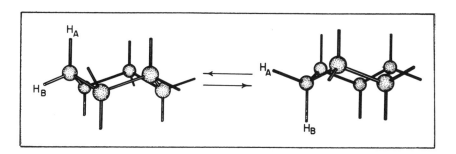

Figure 9.   Schematic diagram for conformational isomerization of cyclohexane.

features of the stochastic models for isomerization reactions have to be briefly introduced.   These models predict the existence of dynamic effects on isomerization reactions because the reaction coordinate is expected to be coupled to the surrounding medium.   In contrast to

classical transition theory (TST) where the transmission coefficient κ
is assumed to be unity and independent of the thermodynamic state,
the stochastic models (35) propose that κ is dependent on the "collision frequency", α, which in the absence of electrostatic effects
reflects the actual coupling of the reaction coordinate to the surrounding medium. The stochastic models introduce κ as

$$k(\Delta t) = \kappa k_{tst} \quad , \tag{5}$$

where $k(\Delta t)$ is the observed isomerization rate and k_{tst} represents the
reaction rate as defined in the classical transition-state theory.
Then it follows that

$$\Delta V_{obs}^{\neq} = \Delta V_{coll}^{\neq} + \Delta V_{tst}^{\neq} \quad , \tag{6}$$

where $\Delta V_{obs}^{\neq}$ is the observed activation volume, $\Delta V_{coll}^{\neq}$ is the collision contribution to the activation volume defined as

$$\Delta V_{coll}^{\neq} = - RT(\partial \ln \kappa / \partial P)_T \quad , \tag{7}$$

and $\Delta V_{tst}^{\neq}$ is the transition-state activation volume. According to
the stochastic model, one would expect a large pressure dependence of
$\Delta V_{coll}^{\neq}$ because κ is a strong function of the collisional frequency α.
The theory predicts that with weak coupling an increase in pressure
will lead to an increase in the isomerization rate, whereas with
strong coupling an increase in pressure will decrease the isomerization rate. There is a nonmonotonic transition between the weak and
strong coupling regimes.

We selected cyclohexane for our study for several reasons. First,
cyclohexane has no dipole moment and the coupling to the surrounding
medium is given by the collision frequency, α, which is approximately
proportional to the shear viscosity, η, of the solvent. Second, the
ring inversion in cyclohexane is a relatively simple process, such
that to a first approximation one can compare the experimental results
with the predictions of the stochastic models on the basis of a one-
dimensional bistable potential.

Because one can relate α to viscosity, η, of the medium by using
simple hydrodynamic arguments, one can select different solvents in
order to cover the various regions of the κ vs. α dependence. Ace-
tone-d_6, carbon disulfide, and methylcyclohexane were selected. Using
standard line shape analysis, we studied the pressure dependence of
ring inversion of cyclohexane at several temperatures in these three
solvents.

In our experiments (34) we found that, as predicted by the stochastic models, the $\Delta V_{obs}^{\neq}$ was strongly pressure and solvent dependent as were the collisional contributions to the activation volume
$\Delta V_{coll}^{\neq}$. Our results are shown in Figure 10 which gives the reduced
transmissions coefficient $\kappa(P)/\kappa(1 \text{ bar})$ as a function of pressure, P,
for the three solvents studied. In acetone-d_6, κ increased by 30% of
its initial value at 1 bar and leveled off at about 2.5 kbar. In
carbon disulfide, κ initially increased but only by about 10% and then

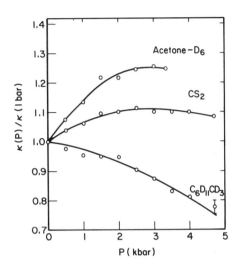

Figure 10.   Reduced transmission coefficient κ as a function of pressure and solvent for conformational isomerization of cyclohexane at 225K.

decreased slightly at the highest pressures. In contrast, κ in cyclohexane-d_{12} showed an 80% decrease from its initial value at 1 bar. From these results we readily see that in the acetone and CS_2 solvents κ is in the weak coupling region (κ proportional to α) whereas in deuterated cyclohexane κ is in the strong coupling region ($\kappa \propto 1/\alpha$). An even better representation of the experimental results is given in Figure 11 which shows the dependence of the reduced transmission coefficient upon the solvent viscosity, η. As predicted by the stochastic models, κ shows a nonmonotonic dependence upon η (approximately proportional to the collision frequency α) and indicates the weak and strong coupling regions. Our results provided the first experimental proof of the theoretical predictions of the stochastic models.

4.2   Coupling Between Rotational and Translational Motions in Supercooled Viscous Liquids

Considerable attention has been devoted to the relationship between the microscopic reorientational correlation time obtained by NMR and the macroscopic viscosity of the medium (36). The adequacy of the hydrodynamic Stokes-Einstein model to describe the microscopic molecular dynamics of liquid systems has been studied extensively. Modifications of this model have been summarized in a recent review article (37).
       A modification based on the concept of anisotropy of intermolecular potentials was introduced by McClung and Kivelson (38), according to which the Debye equation takes the form

$$\tau_\theta = \kappa V_H \frac{\eta}{kT} + \tau_H, \qquad (8)$$

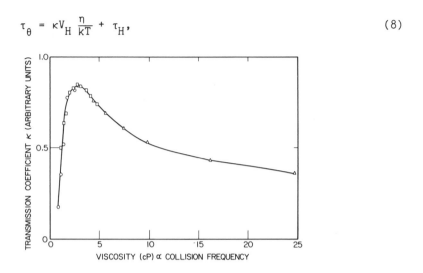

Figure 11. Relation between transmission coefficient κ and viscosity η (viscosity proportional to collision frequency) for conformational isomerization of cyclohexane at 225K. Solvents: (o) acetone-d6; ($\square$) carbon disulfide; (Δ) methylcyclohexane-d14.

where τ_θ is the reorientational correlation time, η is the shear viscosity, V_H is the hydrodynamic volume and τ_H is the zero viscosity correlation time, often associated with the free rotor correlation time. The McClung and Kivelson constant, κ, represents the ratio of the mean square intermolecular torques on the solute molecules to the mean square intermolecular forces on the solvent molecules. The total intermolecular potential energy determines both the torques and the forces, but only the anisotropic part of this potential gives rise to the torques. Therefore, the McClung and Kivelson constant reflects the extent of the coupling between the rotational and the translational motions, and in addition it provides information about the degree of anisotropy which characterizes these motions.

Several studies (37) employing NMR (39,40), ESR (41), Raman (42), and light scattering (43,44) techniques demonstrated that the coupling parameter is independent of temperature and pressure conditions but depends upon the molecular structure of the solute molecule and the nature and size of the solvent molecule. This conclusion was supported by linear τ_θ vs. η/T isotherms, isobars and isochores. The shear viscosity was generally varied by changing the conditions of temperature and pressure or selecting solvents of widely differing viscosities. However, since these studies focused their attention on non-viscous, non-associative liquid systems, the range of viscosities achieved by changing these parameters was fairly limited, often not in excess of one order or magnitude. The narrow range of viscosities was associated with a correspondingly limited variation of the correlation time, τ_θ.

We believed that the linear τ_θ vs. η/T isotherms, isochores and isobars that are thus obtained were in fact only segments of more general nonlinear curves that would be observed if a sufficiently large range of viscosities and correlation times was examined. This belief was also shared by Zager and Freed (45) in their ESR study of the molecular dynamics of the probe molecule PD-Tempone in toluene-d_8 solvent. As pointed out, erroneous conclusions may be drawn if sufficiently large intervals of viscosity and correlation times are not examined. Our study (36) expanded the range of viscosities and correlation times by extending measurements from the liquid into the supercooled state with a proper choice of model compounds known to resist crystallization below the melting point. Because the temperature and density dependence of the coupling parameter was found to be most dramatic at high viscosities, we focused our attention on the viscosity range 1 cP - 1000 cP, which was naturally accomplished by supercooling.

The primary goal of the study was therefore to investigate the nature and the extent of the dependence of the coupling parameter κ upon factors such as temperature, density and molecular structure. Since previous studies demonstrated that molecular symmetry rather than permanent dipole moment plays the decisive role in determining the rotational-translational coupling, the main criteria for selecting the compounds for our study (36) was molecular structure. These compounds were cis-decalin, sec-butylcyclohexane, sec-butylbenzene, n-butylbenzene, isopropylbenzene and toluene. For a more direct comparison, it was desirable to examine the reorientation of the ring segment only, which was achieved by selective deuteration of the ring protons.

The reorientational correlation times were obtained from a direct measurement of the deuterium spin lattice relaxation times,

$$(T_1)^{-1} = \frac{3}{8} (e^2 qQ/\hbar)^2 (1 + \frac{\eta^2}{3}) \tau_\theta \tag{9}$$

where $(e^2 qQ/\hbar)$ is the quadrupole coupling constant and η is the electric field gradient assymmetry parameter. In addition, the shear viscosities and molar volumes of the six model compounds have been measured as a function of both temperature and pressure.

The validity of the modified Debye equation and consequently the density dependence of the rotational-translational coupling parameter is best illustrated by isothermal plots of τ_θ vs. η/T. Since the parameter κ is represented by the slope of such plots, a negative curvature would imply a decoupling of the rotational-translational motions with increased density, whereas the opposite would be true for positive curvatures. Fig. 12 gives the isothermal plots of the reorientational correlation time, τ_θ, as a function of η/T for sec-butyl

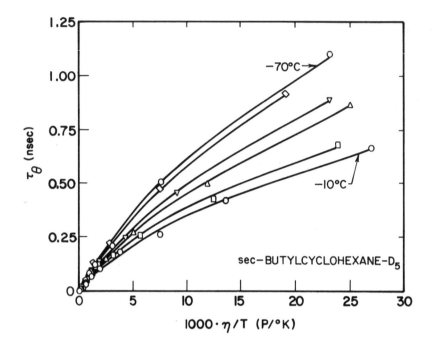

Figure 12. Reorientational correlation time as a function of
viscosity/temperature for sec-butylcyclohexane-d5 at several
isotherms: (o) -10°C; (□) -20°C; (Δ) -40°C; (∇) -50°C; (◇)
-60°C; (o) -70°C.

cyclohexane-d_5. A negative curvature of the τ_θ or η/T plots is clearly observed.

As stated earlier, the coupling parameter gives the ratio of the
torques on the solute molecules to the forces on the solvent molecules. If the molecular symmetry of the solute is such that a relatively free rotation about some axis α is allowed, than by increasing
the pressure (density), the intermolecular torques increase at a
slower rate than the intermolecular forces, resulting in a reduced
parameter κ. Physically this means that a molecule which is able to
rotate relatively unhindered about a given axis will have a negligible
effect on the local structure of the liquid, although its translational ability will be drastically reduced by the increased pressures.
The net result is a decrease of the coupling between the rotational
and translational motions with increased density. After all, viscosity is a measure of linear momentum transfer and is known to have a
strong pressure dependence, whereas the reorientational correlation
time only depends upon angular momentum transfer which may become
nearly pressure independent in highly symmetric molecules.

The behavior of the relatively symmetrical model molecules of
toluene-d_8 and cis-decalin-d_{10} also showed that the τ_θ vs. η/T isotherms displayed non-linear negative curvatures. A very contrasting

behavior was observed for the model molecules which lack an axis of symmetry about which the reorientation is relatively unhindered. The results of our study have shown that for molecules which tumble aniso-tropically and lack an axis of symmetry for relatively unhindered reorientations, the coupling parameter increases with increasing density, in complete contrast to the previous case of symmetric liquid molecules. τ_θ vs. η/T isotherms yielded positive curvatures which is a consequence of the reorientational correlation times exhibiting a stronger pressure dependence than the viscosity. Isopropylbenzene-d_5 and sec-butylbenzene-d_5 fell into this category. Fig. 13 shows the τ_θ vs. η/T plot for isopropylbenzene-d_5 at several isotherms. A physical explanation of this phenomenon is that as density increases, molecular reorientations become increasingly more hindered and anisotropic in nature due to the closer proximity of the neighboring molecules, resulting in enhanced rotational-translational coupling.

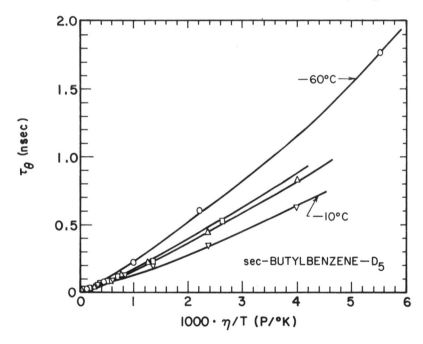

Figure 13. Reorientational correlation time as a function of viscosity/temperature for sec-butylbenzene-d5 at several isotherms: (▽) - 10°C; (△) -20°C; (□) -40°C; (o) -60°C.

The fact that the pressure dependence of the τ_θ is greater than the pressure dependence of viscosity can be understood in physical terms if one considers that the relaxation process is no longer domi-nated by simply collisions as in the case of non-viscous liquid sys-tems, but rather by anisotropic and cooperative reorientations of a more complex nature as well as by possible cluster formations. Such complex reorientational processes are then responsible for the observed behavior of τ_θ vs. η/T plots under these conditions.

Purely kinetic effects are best examined under constant density conditions. In order to investigate the kinetic effects and determine to what extent they affect the coupling parameter, isochoric plots of τ_θ vs. η/T are required. It was possible to obtain such plots for all compounds studied. A representative plot is shown in Fig. 14 for

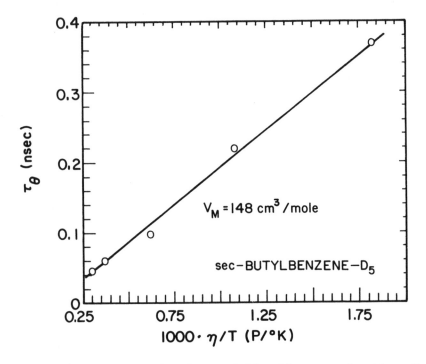

Figure 14. Reorientational correlation time as a function of viscosity/temperature for sec-butylbenzene-d5 under isochoric conditions Vm = 148 cm^3/mole.

isopropylbenzene-d$_5$. The remarkable feature of this plot is the absence of any type of curvature, regardless of the density used. Linear τ_θ vs. η/T isochores were obtained for all densities and all compounds measured. The constant slopes naturally suggest that the rotational-translational coupling parameter is independent of kinetic effects. Given a constant molar volume available for molecular motion, changing the kinetic energy of the molecules may change the frequency of reorientations, but the actual nature of the reorientations which couple to the translational motions remains unaffected. Therefore, volume rather than kinetic energy plays the decisive role in determining the degree of the coupling. This was an important result because one can draw a general conclusion about the major role of volume effects on various dynamic processes in liquids.

4.3 NMR Studies of Disordered Solids at High Pressure

Naturally, high pressure NMR techniques are readily applicable to
investigation of relatively compressible disordered solids. NMR experiments limited only to 5 kbar pressure range can provide unique
information both about dynamic and static structure of disordered
solids. Two experiments, one dealing with the kinetics of solid-solid
phase transformation (46) and the other describing the pressure effects on the quadrupole coupling constant in lithium graphite (47)
show convincingly the high information content of high pressure NMR
experiments on disordered solids.

4.31 Kinetics of Solid-Solid Phase Transformation. Ross and Strange
(48) have introduced the pulse NMR technique to determine the phase
diagrams of plastic crystals. In their study, the abrupt changes in
T_1 or T_2 at the phase transition provided the means of detecting the
precise phase-transition pressure. In our high pressure study we
reported a NMR relaxation study of the kinetics of the plastic phase
transition in adamantane. The high-pressure phase transition of
adamantane has been shown (49) to be structurally analogous to the
low-temperature transition at one bar. The use of pressure as an
experimental variable allowed us to bring rapidly the sample to a
nonequilibrium state to drive the phase transition towards completion.
At the phase-transition pressure, the T_1 of adamantane decreases
abruptly (50) by a factor of about 40 in passing from the plastic (α)
phase to the brittle (β) phase. This large discontinuity allowed us
to monitor the NMR signal during the phase transition and separate the
contributions from each phase as a function of time, since one may
write

$$M_{obs} = x_\alpha M_\alpha + x_\beta M_\beta \; , \tag{10}$$

where M_i are the magnetizations, and x_i are the mole fractions of each
phase contributing to the observed magnetization at any time.
     The linear pressure dependence of T_1 in the pure α and β phases
(50) was measured for each sample and extrapolated into the phase-
transition region as illustrated in Fig. 15 to provide the values of
$T_{1\alpha}$ and $T_{1\beta}$. This measurement also allowed the determination of the
equilibrium transition pressures P_o for the forward ($\alpha \to \beta$) and the
reverse transitions.
     For the kinetic measurements, to ensure that each transition
starts from a pure parent phase, the forward transitions were approached from an initial pressure of 1 bar and the reverse transitions
are approached from 5 kbar. From this initial point, the pressure was
brought stepwise over 30 to 90 min to an equilibrium pressure P_{eq},
which was typically within 50 bar of the equilibrium transition pressure P_o. After spending 15 to 30 min at P_{eq}, the pressure was rapidly
(less than 10 sec) changed to the nonequilibrium, driving pressure, P_o
+ ΔP, at which the kinetic measurement is to be made. This pressure
was carefully maintained within 5 bar over the duration of the measurement.

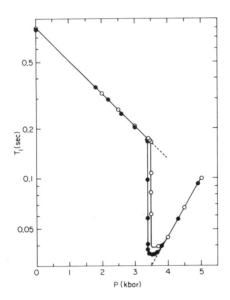

Figure 15. Pressure dependence of the NMR spin lattice relaxation
time, T_1, for adamantane dispersion in D_2O at $0^\circ C$ for compression (o)
and decompression (•) measurements. Dashed lines represent the linear
extrapolation of T_1 into the phase transition region.

The periodic measurement of χ_α began immediately before the
pressure is changed. The time dependence of χ_α at $0^\circ C$ for a number of
driving pressures P_o + ΔP is shown in Fig. 16. Each curve is reproduc-
ible within 5 mole % over the entire time range after the pressure has
stabilized (less than 1 min). In this example, the equilibrium phase
transition occurs at 3475 bar for the forward ($\alpha \rightarrow \beta$) transition and
at 3400 bar for the reverse ($\beta \rightarrow \alpha$) transition. The sample is surrounded within the cell by D_2O as an internal pressure medium to
ensure hydrostatic pressure and to ensure that the phase transition
goes to completion.
It is clear that, as ΔP increases, the transition proceeds at a
much faster rate. In an analogous study, Mnyukh et al. (51) found
that the rate of growth of the daughter phase in other systems increases as $\Delta T = T_o - T$ increases, where T is the ambient temperature
at which a phase change is being observed, and T_o and ΔT are analogous
to P_o and ΔP. Furthermore, the qualitative features of the curves in
Fig. 18 are predicted by the kinetic order-disorder phase-transition
theory of Honig (52). In particular, the transition rate undergoes a
sharp decrease as the phase change approaches completion and, in
addition, the nature of the transformation (growth) curves changes
from exponential to sigmoidal as ΔP decreases for transitions in both
directions.
We demonstrated (46) the applicability of the pulse NMR technique
to the measurement of the kinetics of phase transition in plastic

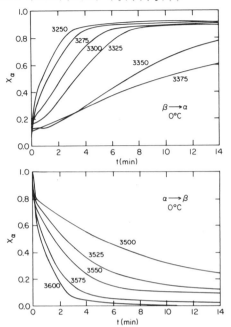

Figure 16. Growth curves for the mole fraction of the plastic phase of adamantane dispersion in D_2O at $0°C$ at various pressures. Pressure is given in bar.

crystals. The proposed method requires that the T_1 exhibits a large discontinuity at the phase transition, and that the rate of phase change be slow compared to the longer T_1 of the two phases involved.

4.32 <u>Pressure Effects on the Quadrupole Coupling Constant of ^7Li in First Stage Lithium Graphite.</u> Graphite intercalation compounds constitute ideal systems for fundamental studies of electronic properties, the dynamics of the intercalated species and the phase transitions of pseudo-two-dimensional systems. The interest in these highly anisotropic compounds arises not only because of their unusual electronic and chemical properties, but also because of their technological importance. Among the donor graphite intercalation compounds the alkali ones are the most well known. Lithium graphite (LiC_6) is structurally the simplest and provides an attractive system to be studied by high pressure NMR techniques. Electronic, optical, thermal and transport properties have been reported in the past (53). In our laboratory (47) we reported the first observation of the pressure effects on the ^7Li NMR lineshape. The main purpose of this experiment was to study the pressure effects on the quadrupole coupling constant (e^2qQ/h), which is closely related to the layer spacings and the symmetry of the Li site as well as to the electronic charge distribution.

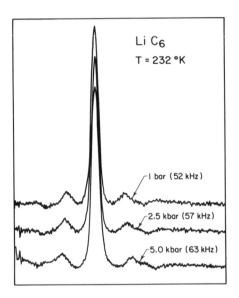

Figure 17. ^{7}Li NMR spectra obtained from stage one lithium graphite
at 1 bar, 2.5 kbar, and 5.0 kbar pressures at T = 232K. Absolute
values of the lithium quadrupole coupling constant e^2qQ/h are given in
parentheses.

The NMR spectra of ^{7}Li obtained from stage one lithium graphite
(LiC_6) at T = 232K and at various pressures are shown in Fig. 17. It
is apparent that the satellite splitting increases with an increasing
pressure. In powders, this frequency difference for spin I = 3/2 is
given by (54)

$$\Delta\nu = \frac{1}{2}\frac{e^2qQ}{h}(1 - \eta) .$$ (11)

Owing to the hexagonal symmetry of Li^+ site in LiC_6, the asymmetry
parameter η of the electric field gradient (EFG) tensor is η = 0 (54)
and the quadrupole coupling constant e^2qQ/h at the site of Li nucleus
can be calculated as

$$e^2qQ/h = (1 + \gamma) . V_{zz} . eQ/h .$$ (12)

Here, Q is the quadrupole moment of the Li nucleus, γ is the
Steinheimer antishielding factor, and V_{zz} is the zz component of the
electric field gradient tensor produced by all other charges except
the Li^+ ion under consideration. We used γ = -0.256 (54).

Conventional NMR experiments do not yield the sign of the quadrupole coupling constant, only e^2qQ/h can be determined. Our experimental results showed that e^2qQ/h for ^7Li in LiC_6 increased linearly with pressure. The rate of increase is 1.5 kHz/kbar, and it was found to be constant over the entire range of temperatures studied (230K ≤ T ≤ 300K). In order to explain this pressure dependence in the stage one lithium graphite, a simple point charge model was proposed. This pressure dependence can be understood qualitatively by realizing that the electric field gradient tensor depends strongly on the interatomic distances. In an anisotropic crystal structure such as LiC_6, one would expect the compressibilities parallel and perpendicular to the layer planes to be very different. The coplanar covalent bonds of carbon atoms are so strong that the in-plane contraction at pressures less than 10 kbar can be neglected. Therefore the dependence of e^2qQ/h on pressure can be explained by considering only contraction of the c-axis. The details of the calculations were reported in our original work (47), and therefore, it is sufficient to emphasize that, by carrying out a high pressure study, it was possible to give a quantitative explanation of the value of the quadrupole coupling constant of ^7Li in LiC_6. In contrast to regular NMR experiments the negative sign of e^2qQ/h could be determined. The average location of the transferred charge was determined to be about 1 A below and above the carbon layers, in an excellent agreement with the theoretical calculation (55). Finally, the model explained well the change of e^2qQ/h with pressure.

4.4 Technological Relevance

Aside from the basic research value of NMR measurements at high pressure, the technological relevance of such work is evident in the application of high-resolution FT NMR at high pressure to the study of homogeneous catalytic processes. Several other aspects of our NMR work bear directly on specific problems in applied fields. For example, a detailed understanding, at the molecular level, of the dynamic processes and intermolecular interactions in supercritical dense fluids will aid in the development of new, highly selective extraction and separation procedures with supercritical fluids as the solvent medium. The work on supercritical compressed steam is relevant to energy-related problems, and the extension of these studies to more complex systems, such as H_2O-CO_2-electrolytes, should be of great interest to geochemists and geophysicists.

In a general sense, a better understanding of the liquid state of matter will have a far-reaching impact on technology. Since most engineering materials are in liquid form at some stage of their fabrication or production, more information about this intermediate state may lead to the development of better engineering materials. Experiments at very high pressures and temperatures should provide valuable new information on the behavior of materials under extreme conditions. The increasing demands of modern technology for materials that perform better under extreme operating conditions are well recognized. One could list a number of additional technological areas which will

benefit from the results obtained by the NMR studies of liquids at
high pressure. For example, new information on highly viscous liquids
at the atomic and molecular level will help workers in the applied
field of lubrication.

Three specific examples of high pressure NMR experiments are
discussed in the following sections 4.41 - 4.43 to illustrate the
direct relevance to several technological areas.

4.41 NMR Measurements of Naphthalene Solubility in Supercritical
Carbon Dioxide. Research on the properties of supercritical fluids
and supercritical fluid mixtures has become very important in recent
years due to the great promise of supercritical fluid extraction
techniques. These techniques and their applications have been reviewed by several authors (56,57). There are many advantages of using
supercritical fluid extraction over conventional extraction techniques. Many low volatility molecular solids show greatly enhanced
solubilities in supercritical dense fluids. Solvent recovery is
easily accomplished by manipulating the density, and therefore the
solvating power, of the supercritical fluid to precipitate the solid.
In addition, although the densities of the supercritical fluids are
comparable to liquid densities, the viscosities are generally an order
of magnitude smaller, and diffusivities an order of magnitude larger
than liquids. A more efficient separation can therefore be achieved.

The NMR method (58) we have recently developed gives a direct, in
situ determination of the solubility and also allows us to obtain
phase data on the system. In our study we have measured the solubilities of solid naphthalene in supercritical carbon dioxide along
three isotherms (50.0, 55.0, and 58.5°C) near the UCEP temperature
over a pressure range of 120-500 bar. We have also determined the
pressure-temperature trace of the S-L-G phase line that terminates
with the UCEP for the binary mixture. Finally, we have performed an
analysis of our data using a quantitative theory of solubility in
supercritical fluids to help establish the location of the UCEP.

The solubilities for naphthalene in supercritical carbon dioxide
(58) were measured at 60 MHz using the NMR spectrometer described
elsewhere (1). The high pressure, high temperature NMR probe and gas
compression system were the same as that used in our other experiments
(1). The solubility sample call was of cylindrical design with 0.250
in. inner diameter and was machined from a high temperature polyimide
plastic (Vespel, DuPont Co.). An excess of solid naphthalene was
loaded into the cell before a solubility determination and the cell
was closed with a close-fitting piston. Pressurized CO_2 entered the
sample region through two small holes (0.016 in.) drilled through the
sample cell walls. To assure that equilibrium solubilities were obtained, enough solid naphthalene was initially placed in the sample
cell so that an excess would be present after dissolution. This made
it necessary to separate the contribution to the NMR signal from the
dissolved naphthalene and the remaining solid. This separation is
easily accomplished due to the radically different spin-spin relaxation rates of dissolved and solid material (T_2, solid $\ll T_2$, dissolved). We used the $90°-\tau-180°$ spin-echo sequence with a pulse

separation of τ = .007 s; this ensured that no contribution to the NMR
echo signal could result from the quickly relaxing protons of solid
naphthalene. In this way we were able to monitor the NMR signal from
the naphthalene dissolved in the supercritical solution exclusively.
This experimental approach for separating the signal from mobile and
immobile nuclei has been used previously in our laboratory (59).

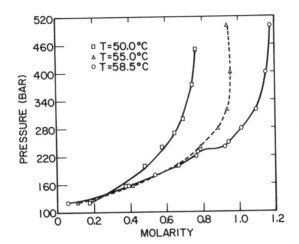

Figure 18. Experimental solubilities for solid naphthalene in
supercritical carbon dioxide determined by the NMR technique.
Concentration expressed as molarity of naphthalene in CO_2.

     The experimental solubility data for solid naphthalene in super-
critical carbon dioxide, given as moles naphthalene dissolved per
liter, are shown in Fig. 18. Qualitatively the three pressure-compo-
sition isotherms show characteristic behavior for a solid-super-
critical fluid system. Each isotherm initially shows a large increase
in solubility with increasing pressure, and then a limiting value is
reached at higher pressures.
     Our NMR technique for the determination of the solid-liquid-gas
phase line that ends at the UCEP again makes use of the fact that the
NMR signal from nuclei with different relaxation rates can be easily
separated. In this case we are distinguishing between the signal from
the naphthalene-rich liquid phase formed when the S-L-G line is
crossed and solid naphthalene. Since the spin-spin relaxation of the
liquid phase naphthalene protons is much longer than that of the solid
naphthalene protons, by once more using the spin-echo sequence, we are
able to monitor the NMR signal from the liquid phase only. Our results showed that our novel NMR method can yield both solubility data
and phase information when studying equilibria in supercritical fluid
mixtures.

4.42. ^{29}Si NMR Study of the Sol-Gel Process for Preparing Glasses.
Sol-gel processes for preparing highly homogeneous oxide glasses at
temperatures significantly lower than the melting temperatures required in the conventional fusing technique have recently stimulated
considerable research activity (60,61). The preparation of oxide
glasses involves first the preparation of a wet monolithic gel by
hydrolysis and polymerization of a silicon-alkoxide solution. It is
followed by heat treatment to remove all residual organics and water
from the porous gel, and finally by the densification of the dry gel
to form a monolithic dense glass. The physical and chemical properties of the resulting oxide glass are highly dependent upon the degree
of homogeneity achieved during the gelling stage. However, the optimum solutions conditions to achieve this state necessitate almost
prohibitively long gelation times, often of the order of days, even
months.

The object of our (62,63) studies was to investigate the role of
pressure on the polymerization kinetics of the $Si(OCH_3)_4$ sol-gel
transition. Using high resolution ^{29}Si NMR spectroscopy the possibility of reducing gelation times by elevated pressures has been explored
at a molecular level. This represented the first study to carry out
^{29}Si NMR measurements under high pressure. The resulting ^{29}Si NMR
spectra were analyzed to monitor quantitatively the time evolution of
all condensed species in solution during the initial stages of the
polymerization reaction.

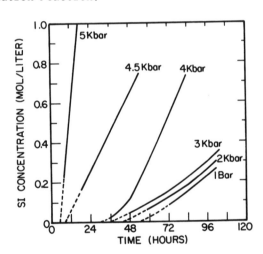

Figure 19. Time evolution of the concentration of the higher
polymeric species as a function of pressure as determined by the ^{29}Si
NMR for the sol-gel process based on tetramethylsilane. For details,
see the text.

An excellent illustration of the dramatic effect of pressure on
the sol-gel process is given in Fig. 19 which shows the time evolution
of the concentration of higher polymer species as a function of pres-

sure. The extent of the condensation rate enhancement was quantitatively evaluated using kinetic principles. Transition state theory
was employed to provide a detailed mechanism based on activated volumes for the pressure-induced acceleration of the gelation process.
In order to ensure that the accelerated kinetics under pressure do not
originate from associated pH changes, but rather from volume contraction effects, an experimental comparison had been sought between
changes produced by pressure by pH separately. The results confirmed
our interpretation.

Our study has shown that high pressures have a dramatic effect on
the polymerization kinetics of sol-gel processes. Although other
parameters produce similar accelerating effects, they generally in-
flict structural modifications in the polymer gel networks. For
instance, the fast polymerization of base catalyzed sols produces
easily sintered gels of high porosity but low purity with respect to
unreacted organic compounds. The slow acid catalyzed process forms
gels of relatively high purity but poor sintering properties. The
results presented here suggest that by using pressure as an experimental variable it might be feasible to accelerate the polymerization
process without altering the structural properties of the resulting
glass-precursor gels.

4.43. <u>Kinetics of Crystallization of Polyethylene</u>. It is well known
that NMR relaxation or linewidth measurements are sensitive to the
degree of crystallinity in solid polymers. Protons in the crystalline
lattice experience strong dipole-dipole interactions which cause fast
spin-spin relaxation and line broadening. Linewidth and relaxation
studies have been used to measure crystallinity in a broad range of
polymers (64).

In our study (59) we extended the NMR technique to follow polymer
crystallization under high hydrostatic pressure. The formation of
extended-chain crystals in polyethylene was observed and a variety of
questions were considered. The dependence of the rate of growth on
pressure and temperature, the value of the Avrami coefficient, the
mechanism of extended-chain crystallization, and the effect of pressure upon the surface energies of the crystal nuclei were examined.
Comparison with previous measurements showed the potential of NMR to
reveal new details about polymer crystallization.

As an example of some of the results of our study (59), Fig. 20
gives the log time dependence of a fraction of the amorphous component
$1 - \chi(\tau)$, where $\chi(\tau)$ is the crystalline fraction at time τ, for polyethylene at 3 kbar. According to the theory discussed in the original
reference (59), the important feature of Fig. 20 is the observation
that the isotherms are superimposable upon shifting the time axis with
the possible exception of the lowest undercooling. This indicates a
consistent mechanism for the formation of extended-chain crystals;
this mechanism must be independent of the density of the amorphous
melt from which the crystals are formed.

In conclusion, we have developed a successful new method for
measuring the crystallization kinetics of polymers under high pressure, sensitive on a molecular scale to the dynamics of chain motion.

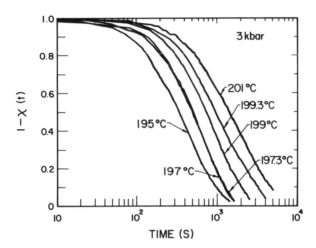

Figure 20. Time evolution of the fraction of the amorphous component in polyethylene at 3 kbar as determined by NMR in the study of kinetics of crystalization of polyethylene under high pressure.

The very rapid formation of extended-chain crystals of polyethylene has been directly observed. A decrease in the surface energy of the crystal nucleus with pressure was indicated. The technique has much promise for future experiments including polymer crystallization, melting, and annealing.

5. Applications of Laser Raman Scattering at High Pressure

In view of our own interest and expertise the following sections 5.1. -5.3. will briefly discuss several laser Raman and Rayleigh scattering experiments on liquids at high pressure. In the Introduction (Sect. 1.1.) we pointed out the unique information laser Raman scattering experiments provide about reorientation and vibrational processes in liquids.
     The broadening of isotropic Raman bandshapes may be influenced by several mechanisms. The two dominant ones involve energy relaxation vis inelastic collision and phase relaxation via quasieleastic collisional processes. Both mechanisms have been investigated (65), and it has been found that phase relaxation (dephasing) in liquids always occurs much faster than energy relaxation. In all the cases discussed in the following section 5.1,the broadening of the isotropic Raman line shapes arises from a dephasing process.
     The section 5.2 will briefly mention the importance of using pressure as an experimental variable in studies of reorientational processes in liquids, and in particular, will focus on the density behavior of reorientational second moments with density. The observed density dependence of the rotational second moments has important theoretical implications and leads naturally to investigation of collision induced phenomena in liquids. Section 5.3 mentions several

experiments dealing with pressure effects on collision induced scattering in fluids.

5.1. Vibrational Dephasing in Liquids

Many different theories dealing with dephasing have recently been developed (for review, see ref. 65), examples of which are the isolated binary collision (IBC) model (66,67), the hydrodynamic model (68), the cell model (69), and the model based on resonant energy transfer (70).

In the measurements of isotropic line shapes of the C-H, C-D, and C-C stretching modes in a variety of simple molecular liquids (3), we found that the IBC model (66,67), which considers only the repulsive part of the intermolecular potential in calculating the dephasing rate, reproduces the general trends of the experimental data observed. It should be pointed out that this model and other models predict that the dephasing rate (line width) will increase with density at constant temperature; therefore, it was interesting to observe a very different behavior of the dephasing rate in liquid isobutylene (71).

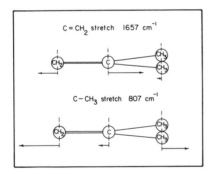

Figure 21.   Schematic representation of the vibrational amplitudes for the C = CH_2 and C - Ch_3 stretching modes of isobutylene.

In order to understand this behavior, one has to realize the different character of these two vibrations of interest. First, the C=CH_2 vibration is strongly infrared active, while the C-CH_3 vibration is inactive. Secondly, the classical vibrational amplitudes show that the repulsive forces of the bath may affect more the C-CH_3 vibration. We calculated these amplitudes in a simplified way by modeling isobutylene as a linear molecule. Using this procedure, we found the amplitudes of the molecular groups proportional to the lengths of the arrows as shown in Fig. 21. The shortest arrow represents 2 X 10^{-3} A. The calculated frequency ratio differs from the experimental one by only 3%, and therefore we estimate that the amplitudes are correct to within 10%.

The Raman line shapes of the ν_4 (A_1) symmetry C=CH_2 stretching mode of 1657 cm^{-1} and the ν_9 (A_1) symmetric C-CH_3 stretching mode of

807 cm^{-1} in isobutylene (71) were measured as a function of pressure
from vapor pressure to 0.8 g cm^{-1}, over the temperature range from -25
to 75°C. Figure 22 shows the density dependence of the experimental
half width $\Gamma = (2\pi c\tau_v)^{-1}$, where τ_v is the dephasing time and Γ is
proportional to the dephasing rate. From Fig. 22 we see that increas-
ing density, at constant temperature, affects the bandwidths of the
two vibrations in a qualitatively different way. The C-CH$_3$ stretching
bandwidth increases with increasing density, a behavior found for many
other modes of liquids. This basic trend can be predicted in terms of
a IBC model based on rapidly varying repulsive interactions. The most

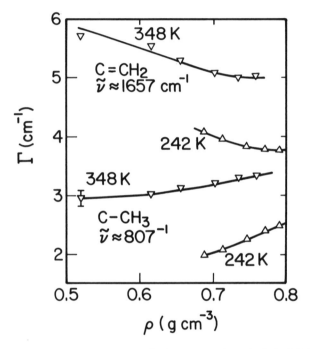

Figure 22. Density dependence of the experimental halfwidth for C =
CH$_2$ and C - CH$_3$ stretching modes in liquid isobutylene for the
temperature range from -25°C to 75°C.

interesting result of this study was the observed decrease of the
bandwidth of the C=CH$_2$ stretching mode (strongly infrared active band)
with increasing density. To our best knowledge, this was the first
experimental observation of a decrease in dephasing rate with increasing density in liquid. It appears that this band is inhomogeneous
broadened as it is affected by environmentally induced frequency
fluctuations. These fluctuations are due to dispersion, induction, and
electrostatic forces that depend on the dipole (0.5 D) and polarizability of the molecule. The decay of the inhomogeneous environment
around a molecule results in motional narrowing. The correlation
function modeling (72), which uses the Kubo stochastic lineshape

theory (73), was in agreement with our experimental data. In further studies we found a similar decrease in bandwidth for the C=O stretching mode in liquid acetone (74). Again, the attractive interactions influence the dephasing process and are responsible for the density behavior of the bandwidth.

The Raman experiments on isobutylene and acetone have been discussed in a theoretical study of dephasing by Schweizer and Chandler (75), who analyzed in detail the relative role of slowly varying attractive interactions and rapidly varying repulsive interactions on the frequency shifts and dephasing in liquids. Their theoretical model correctly predicts the isothermal density dependence of the $C=CH_2$ bandwidth in isobutylene and the C=O bandwidth in acetone.

5.2. Reorientational Motions in Liquids

During the past decade reorientational motions of molecular liquids and gases has been extensively studied by vibrational Raman lineshape analysis (3). High pressure laser Raman scattering experiments provided conclusive evidence for the need of separating of the effects of density and temperature on reorientational processes in liquids. At this point we limit ourselves to showing Fig. 23 which illustrates the pressure effects on the reorientational correlation function of propyne in acetone solution (76,77). Propyne (methyl acetylene) is a symmetric top molecule and the lineshape of the ν_3 C≡C stretching mode at 2142 cm^{-1} was analyzed to obtain the correlation functions given in Fig. 23. Here again the changes with pressure are typical for liquids as with increasing pressure to 4000 bar the rotational process slows by a factor of about 5. Analogous behavior has been observed for other molecular systems.

However, according to our opinion the most interesting observation in this study was the finding that the rotational second moment is strongly pressure dependent. According to classical interpretations, the second moment $M_R(2)$ is related to the rotational kinetic energy of the molecule and should not change with density. Its value is constant at a given temperature, $M_R(2) = 6kT/I_\perp$ ($I_\perp$ is the molecular moment of inertia perpendicular to the symmetry axis). In contrast to this classical prediction we found that $M_R(2)$ is strongly density dependent in liquid chloroform (78) and propyne (76,77). In these cases we found that $M_R(2)$ decreases with increasing density.

For example, for the ν_1 (A_1) C-H stretching mode in liquid chloroform (78) at 303K there is a decrease of $M_R(2)$ from 510 cm^{-2} to 223 cm^{-2} with a density increase from 1.47 to 1.78 g cm^{-3}. The decrease of the second moment with increasing density can be explained in terms of collision induced scattering (CIS) which contributes to the wings of the Raman band. The fact that the CIS affects only the far wings of the band explains why this effect was very often neglected; the correlations functions and the correlations times are affected only slightly. However, second moment analysis, which describes the short time behavior of a correlation function and thus reflects the collision induced high frequency contributions to the spectral line rather strongly, is the most reliable method of investi-

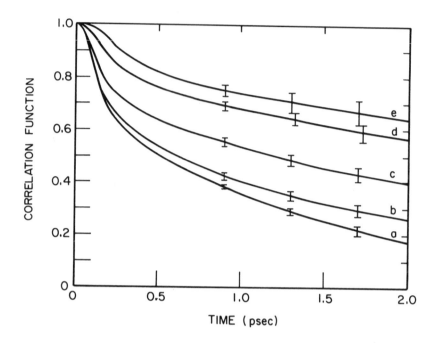

Figure 23. Reorientational correlation function for propyne in
acetone solution at 50°C as a function of pressure: (a)   500; (b)
1000; (c)   2000; (d)   3000; (e)   4000 bar.

gating short time intermolecular interactions. The reason why the
second moment decreases with increasing density is related to the
decrease of collision induced scattering with increasing density.
Clearly, dynamic effects are not as inefficient, over short time intervals, as had been previously assumed. Evidently, three- and four-
particle correlations must be included in any description of the
scattering process. In their molecular dynamics calculations Alder et
al. (79) investigated the effects of density on bandshapes of depolarized light scattering from atomic fluids, taking into account not only
two-particle but also three- and four-particle correlations. At low
densities the pairwise term dominates and increases with density
whereas at higher densities cancellations occur between the pairwise,
triplet and quadruplet terms with the result that the total scattering
intensity decreases. We proposed (3,78) that an analogous process is
responsible for the decrease of CIS with increasing density in our
liquid studies. The observed density dependence of the rotational
second moment makes questionable studies using Raman bandshapes to
calculate torques in liquids.

5.3. Collision Induced Spectra in Fluids

The overwhelming majority of Raman studies of the dynamic processes in liquid dealt with the investigation of the properties of individual molecules. The reorientational and vibrational relaxation reflect only indirectly the influence of intermolecular interactions. In the past decade, the problem of collision induced scattering (CIS) has attracted both theoretical and experimental interest (80,81). It has been observed that collisions in dense liquids or gases produce depolarized Rayleigh spectra in fluids composed of atoms or molecules of spherical symmetry. Collision induced Raman spectra forbidden by selection rules (symmetry) have been investigated in polyatomic molecular liquids as well as collision induced contributions to the allowed Raman bands. The origin of these collision induced spectra lies in the polarizability changes produced by intermolecular interactions. It is clear that studies of CIS can provide direct information about intermolecular interactions. The CIS represents though a very difficult theoretical problem because the scattered intensity depends on the polarizability change in a cluster of interacting molecules, and the time dependence of this change is a function of the intermolecular potential.

In analogy with the Raman studies of vibrational dephasing and reorientational motions in dense liquids, one may expect that studies of CIS at high pressure will contribute in a major way towards a better understanding of these phenomena and will help to establish a sound experimental basis for further theoretical work. Since few studies of CIS in polyatomic molecular liquids have so far appeared, we have started systematic high pressure experiments dealing with CIS in dense liquids (82). Our interest focuses on three main areas: first, for molecules of spherical symmetry, we follow the effects of density and temperature on the depolarized Rayleigh scattering. Second, we investigate the CIS contribution to allowed Raman bands. Third, we study the density and temperature behavior of collision induced forbidden Raman bands.

Depolarized Rayleigh spectra (DRS) in particular have attracted much experimental and theoretical interest. Of course, atomic fluids represent the simplest systems available for the study of DRS because all of the polarizability anisotropy for atomic fluids may be attributed to multi-body interactions (i.e. the entire DRS is collision induced). As in the case of atomic fluids, the DRS for molecules with spherical symmetry are also entirely collision induced. Methane is one of the simplest spherical molecules and the DRS has been studied by a number of investigators.

To illustrate the importance of high pressure experiments in studying the collision induced scattering a selected result taken from our study (83) of DRS of methane is presented. In the high pressure study of CIS in methane we investigated DRS over a wide range of densities and temperatures. The density ranged from 180 to 500 amagat and the temperature range was from -25 to $50^{\circ}C$. Fig. 24 shows the depolarized Rayleigh spectrum of methane and one has to realize that the spectrum is entirely due to collision induced scattering. The

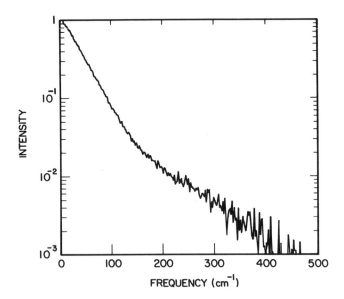

Figure 24.   Stokes depolarized Rayleigh spectrum for methane at -25°C and 454 amagat.

experimental spectra were characterized in terms of the exponential decay constant Δ (cm^{-1}) for the 20 cm^{-1} to 85 cm^{-1} wavenumber region

$$I(w) \propto e^{-\omega/\Delta},\tag{13}$$

where ω is the frequency.

Only one specific result related to the density and temperature dependence of the exponential decay constant Δ for the 25 to 85 cm^{-1} region corresponding to the dipole-induced-dipole mechanism will be mentioned. Figure 25 gives the density dependence of Δ in methane over a range of temperatures and compares the experimental data to the theoretical prediction using the theoretical model developed by Ballucani and Vallauri (84). In comparing the theoretical predictions of Δ using Ballucani and Vallauri's expressions, one can see that the theoretical estimates are 5 - 7 cm^{-1} greater than the experimental Δ's. However, the density dependence of Δ is reproduced quite well.

To conclude the overview of the experiments on collision induced scattering, it is interesting to mention the density effects on the intensity of the DRS for gaseous oxygen (85,86). These quantitative results have recently been obtained in our laboratory.

Analogously as for linear triatomics, the depolarized Rayleigh spectrum for O_2 reflects contributions from rotational motions, collisional effects and cross terms between rotational motions and induced polarizabilities. Unfortunately, for oxygen these spectral contributions occur on the same time scale and therefore cannot be

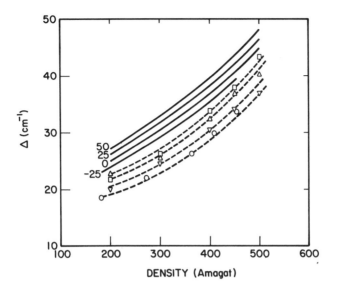

Figure 25. Density dependence of the exponential decay constant d for the depolarized Rayleigh spectrum of methane in the temperature range -25 to 50°C. The full lines denote the theoretical prediction using the theoretical model by Ballucani and Vallauri (84).

separated. Before a detailed comparison with molecular dynamics calculations is carried out the present discussion was necessarily limited only to qualitative aspects of the density dependence of the various spectral moments of fluid O_2.

Figure 26 shows the density dependence of the $M(0)/\rho$ for fluid O_2 at 300K. With increasing density, ρ, the ratio $M(0)/\rho$ first increases up to about 110 amagat, and a substantial decay is observed at higher densities. This is a typical observation for the collision induced spectra, and in the dipole-induced-dipole approximation it can be explained in part as a cancellation effect due to three-body collisions and in part due to the increasing role of the negative cross term, as the density is increased.

From the few results mentioned above and the other studies (3,82,85), one can conclude that studies of collision induced scattering in polyatomic molecular liquids at high pressure represent a very promising field. Again, one can expect that the high pressure experiments will prove to be decisive in unravelling the complex problem of collision induced scattering.

Acknowledgements

This work was partially supported by the National Science Foundation under Grants NSF CHE 85-09870 and NSF DMR 83-16981, the Air Force

Office for Scientific Research under Grant AFOSR 85-0345, and the
Department of Energy under Grant DOE DEFG 22-85-PC80503.

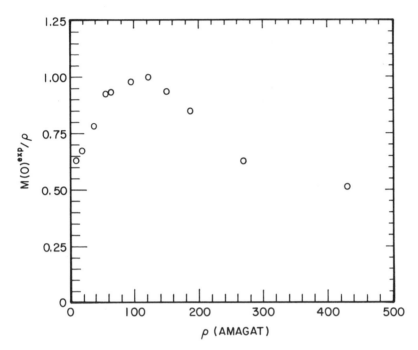

Figure 26.   Density dependence of M(0)/ρ for the depolarized Rayleigh
band for fluid oxygen at 300K.M(0) denotes total intensity of the
Rayleigh spectrum.

References

1. J. Jonas, NATO ASI **41**, 65 (1978).
2. J. Jonas, Science, **216**, 1179 (1982).
3. J. Jonas, Acc. Chem. Res., **17**, 74 (1984).
4. A. E. Merbach, this volume.
5. H. G. Drickamer and C. W. Frank, Electronic Transitions and the
    High Pressure Chemistry and Physics of Solids, (Chapman & Hall,
    London, 1973).
6. C. P. Slichter, Principles of Magnetic Resonance (Springer-
    Verlag, New York, ed. 2, 1978); A. A. Abragam, The Principles of
    Nuclear Magnetism (Clarendon, Oxford, 1961); J. S. Waugh, Ed.,
    Advances in Magnetic Resonance (Academic Press, New York, 1965-
    1976), vols. 1-8.
7. L. A. Nafie and W. L. Peticolas, J. Chem. Phys. **57**, 3145 (1972).
8. S. Bratos, B. Guillot and G. Birnbaum, NATO ASI, Series, B,
    Physics **127**, 363 (1985).
9. P. A. Madden, NATO ASI, Series B, Physics **127**, 399 (1985).

10. W. Schindler, T. W. Zerda and J. Jonas, J. Chem. Phys. 81, 4306 (1984).
11. M. Bradley, T. W. Zerda and J. Jonas, Spectrochim. Acta. 40A, 1117 (1984).
12. W. J. Lamb and J. Jonas, J. Chem. Phys. 74, 913 (1981).
13. D. J. Wilbur and J. Jonas, J. Chem. Phys. 55, 5840 (1971).
14. J. Jonas, Rev. Sci. Instr. 43, 643 (1972).
15. H. Yamada, Chem. Lett. 747 (1972).
16. H. Yamada, Rev. Sci. Instr. 45, 640 (1974).
17. J. G. Oldenziel and N. J. Trappeniers, Physica 82A, 565 (1976).
18. H. Vanni, W. L. Earl and A. E. Merbach, J. Magn. Resonance 29, 11 (1978).
19. G. Volkel, E. Lang and H. D. Ludemann, Ber. Bunsenges. Phys. Chem. 83, 722 (1979).
20. W. L. Earl, H. Vanni and A. E. Merbach, J. Magn. Resonance 30, 571 (1978).
21. D. C. Roe, J. Magn. Resonance 63, 388 (1985).
22. J. Jonas, D. L. Hasha, W. J. Lamb, G. A. Hoffman and T. Eguchi, J. Magn. Resonance 42, 169 (1981).
23. S. Shimokawa and E. Yamada, J. Magn. Resonance 51, 103 (1983).
24. D. G. Vander Velde and J. Jonas, J. Magn. Resonance, 71 (1987), in press.
25. B. T. Heaton, J. Jonas, T. Eguchi and G. A. Hoffman, JCS Chem. Comm. 331 (1981).
26. B. T. Heaton, L. Strona, J. Jonas, T. Eguchi, and G. A. Hoffman, JCS Dalton Trans. 1159 (1982).
27. P. J. Krusic, D. J. Jones and D. C. Roe, Organometallics 5, 456 (1986).
28. E. Whalley, NATO ASI, Series C, 41, 127 (1978).
29. J. R. Ferraro, Vibrational Spectroscopy at High External Pressures; The Diamond Anvil Cell, Academic Press, New York, 1984.
30. D. M. Cantor, J. Schroeder and J. Jonas, Appl. Spectrosc. 29, 393 (1975).
31. S. Ikawa and E. Whalley, Rev. Sci. Instrum. 55, 1273 (1984).
32. S. Perry, P. T. Sharko and J. Jonas, Appl. Spectrosc. 37, 340 (1983).
33. S. Perry, Ph.D. THesis, University of Illinois, Urbana, 1981.
34. D. L. Hasha, T. Eguchi, J. Jonas, J. Chem. Phys. 75, 1571 (1981); J. Am. Chem. Soc. 104, 2290 (1982).
35. J. L. Skinner and P. Wolynes, J. Chem. Phys. 69, 2143 (1978); J. A. Montgomery, Jr., D. Chandler, B. J. Berne, ibid. 70, 405 (1979).
36. I. Artaki and J. Jonas, J. Chem. Phys. 82, 3360 (1985).
37. J. Dote, D. Kivelson and R. Schwartz, J. Phys. Chem. 85, 2169 (1981).
38. R. E. McClung and D. Kivelson, J. Chem. Phys. 49, 3380 (1968).
39. D. L. VanderHart, J. Chem. Phys. 60, 1858 (1974).
40. R. Stark, R. L. Vold and R. R. Vold, Chem. Phys. 20, 337 (1977).
41. J. S. Hwang, J. V. S. Rao, J. H. Freed, J. Phys. Chem. 80, 1490 (1976).

42. K. Tanabe and J. Hiraishi, Mol. Phys. 39, 493 (1980).
43. G. R. Alms, D. R. Bauer, J. I. Brauman and R. Pecora, J. Chem. Phys. 59, 5310 (1973).
44. D. R. Bauer, G. R. Alms, J. I. Brauman and R. Pecora, J. Chem. Phys. 61, 2255 (1974).
45. S. A. Zager and J. H. Freed, J. Chem. Phys. 77, 3360 (1982).
46. M. Fury, S. G. Hwang and J. Jonas, J. Magn. Resonance 33, 211 (1979).
47. C. Marinos, S. Plesko, J. Jonas, J. Conard and D. Guerard, Solid State Commun. 47, 645 (1983).
48. S. M. Ross and J. H. Strange, Mol. Cryst. Liq. Cryst. 36, 321 (1976).
49. T. Ito, Acta Crystallogr. B29, 364 (1973).
50. N. Liu and J. Jonas, Chem. Phys. Lett. 14, 555 (1972).
51. Yu. V. Mnyukh, N. A. Panfilovna, N. N. Petropavlov and N. S. Uchvatova, J. Phys. Chem. Solids 36, 127 (1975).
52. J. M. Honig, in Kinetics of High Temperature Processes (W. D. Kingery, Ed.), Chap. 25, Technology Press/Wiley, New York, 1959.
53. P. Lauginie, M. Letellier, H. Estrade, J. Conard and D. Guerard, Proc. Vth London Carbon and Graphite Conf., London (1978).
54. M. H. Cohen and F. Reif, Solid State Physics, 5, Eds. F. Seitz and D. Turnbull (Academic Press, New York, 1957).
55. L. A. Girifalco and N. A. W. Holzwarth, Mater. Sci. Eng. 31, 201 (1977).
56. M. E. Paulaitis, V. J. Krukonis, R. T. Kurnik and R. C. Reid, Rev. Chem. Eng. 1, 179 (1983).
57. G. M. Schneider, Ber. Bunsenges. Phys. Chem. 88, 841 (1984).
58. D. M. Lamb, T. M. Barbara and J. Jonas, J. Phys. Chem. 90, 4210 (1986).
59. D. R. Brown and J. Jonas, J. Polym. Sci., Polym. Phys. Ed. 22, 655 (1984).
60. S. P. Mukherjee, J. Non-Crystalline Solids 42, 477 (1980).
61. B. J. J. Zelinskii and D. R. Uhlmann, J. Phys. Chem. Solids 45, 1069 (1984).
62. I. Artaki, S. Sinha and J. Jonas, Mat. Lett. 2, 448 (1984).
63. I. Artaki, S. Sinha, A. D. Irwin and J. Jonas, J. Non-Crystalline Solids 72, 391 (1985).
64. B. Wunderlich, Macromolecular Physics, Vol. 1, Crystal Structure, Morphology, Defects (Academic, New York, 1973).
65. D. W. Oxtoby, Ann. Rev. Phys. Chem. 32, 77 (1981).
66. S. F. Fischer, A. Lauberea, Chem. Phys. Lett. 35, 6 (1975).
67. D. W. Oxtoby, S. A. Rice, Chem. Phys. Lett. 42, 1 (1976).
68. D. W. Oxboty, J. Chem. Phys. 70, 2605 (1979).
69. D. J. Diestler, J. Manz, J. Mol. Phys. 33, 227 (1977).
70. D. Doge, Z. Naturforsch., A 28, 919 (1973).
71. W. Schindler, J. Jonas, J. Chem. Phys. 72, 5019 (1980) and W. Schindler, J. Jonas, ibid. 73, 3547 (1980).
72. W. G. Rothschild, J. Chem. Phys. 65, 455 (1976).
73. R. Kubo, in Fluctuations, Relaxation, and Resonance in Magnetic Systems," Ter Haar, D. Ed.; Plenum Press, New York, 1962.
74. W. Schindler, J. Jonas, Chem. Phys. Lett. 67, 428 (1979).

75. K. S. Schweizer, D. J. Chandler, J. Chem. Phys. **76**, 1128 (1982).
76. S. Perry, T. W. Zerda and J. Jonas, J. Chem. Phys. **75**, 4214 (1981).
77. T. W. Zerda, S. Perry and J. Jonas, J. Chem. Phys. Lett. **83**, 600 (1981).
78. J. Schroeder and J. Jonas, J. Chem. Phys. **34**, 11 (1978).
79. B. J. Alder, J. J. Weiss and M. L. Strauss, Phys. Rev. **47**, 281 (1973); B. J. Alder, H. L. Strauss and J. J. Weiss, J. Chem. Phys. **59**, 1002 (1973).
80. W. M. Gelbart, Adv. Chem. Phys. **26**, 1 (1974).
81. G. Birnbaum, Collision-Induced Vibrational Spectroscopy in Liquids in Vibrational Spectroscopy of Molecular Liquids, 1980.
82. J. Jonas, NATO ASI, Series B, Physics **127**, 525 (1985).
83. K. H. Baker, Ph.D. Thesis, University of Illinois, Urbana, Illinois, U.S.A., 1984.
84. V. Ballucan, and R. Vallauri, Mol. Phys. **38**, 1099, 1115, (1979).
85. J. Jonas, Disc. Faraday Soc., in press (1986).
86. T. W. Zerda, X. Song, B. M. Ladanyi and J. Jonas, Manuscript in preparation.

SOLID STATE HIGH PRESSURE RESEARCH : X-RAY AND NEUTRON SCATTERING

J. Voiron[1] and C. Vettier[2]
[1]Laboratoire Louis Néel, C.N.R.S., 166X, 38042 Grenoble cedex
France
[2]Institut Laue-langevin, 156X, 38042 Grenoble cedex, France

ABSTRACT. A brief account of solid state high pressure physics is presented. Special emphasis is given to investigations by X-ray and neutron scattering.

The lectures are primarily devoted to a very basic description of high pressure studies in solid state physics. Indeed, pressure is a powerful tool for research in condensed matter physics : it is used as an external parameter to induce small but well-defined changes in physical properties, which can give a better understanding of the zero pressure phase. It can also create entirely new phases. These investigations have stimulated many theoretical studies, because application of pressure provides critical test of models. Moreover, so many improvements have been done in the high pressure techniques for the last twenty years that the field of high pressure research in solid state physics is wide open. The aim of this paper is to discuss some experimental results to illustrate the use of pressure in solid state physics.
      In the first part, we give examples of various effects of pressure, which include compressibility studies, structural phase transitions induced by pressure and all electronic changes such as metal-insulator transitions, superconductivity, valence changes or magnetic order. Secondly after a brief comparison of neutron and X-ray scattering, we will present some high pressure cells designed for these scattering techniques. Finally, selected experimental investigations will be presented. They deal with the effects of pressure on magnetic properties of materials and, as a consequence, they have been performed using neutron scattering techniques.

R. van Eldik and J. Jonas (eds.), High Pressure Chemistry and Biochemistry, 237–262.
© *1987 by D. Reidel Publishing Company.*

I. INTRODUCTION TO SOLID STATE PHYSICS AT HIGH PRESSURE

The essential effect of pressure is to reduce interatomic distances,
which leads to modifications in lattice constants of crystalline
materials as well as to changes in atomic positions within crystallographic cells. In general, reductions of lattice spacings and force
constants between atoms induce modifications in the band structures of
solids and thus in electronic properties.

I.1. Compression of solids

The primary effect of applied pressure is to reduce the lattice
volume. The softness of the lattice response may be expressed in terms
of linear compressibility $K = -(1/V)(dV/dP)$ (or bulk modulus, $B = 1/K$),
which depends on temperature and pressure. Pressures, in the range of
interest, are expressed in kbar or gigapascal (1 GPa = 10 kbar).
Variations of volume versus pressure can be presently determined by
X-rays techniques at pressures higher than 1 Mbar (100 GPa), which
indicates the order of magnitude of experimentally available static
pressures. Typical values of initial compressibility (K.at P = 0 kbar)
are given in table I for some solids.

TABLE I : Compressibility of selected solids.

Elements	He	Cs	K	Na	Li	U	Cu
K_o(Mbar^{-1})	1168	32	23	13	6.3	1.0	0.73

Compressibility depends on pressure and this variation can be described
by an equation of state as, for example, the first-order equation of
state of Murnaghan :

$$P = \frac{B_o}{B'_o} \left((\frac{V_o}{V})^{B'_o} - 1 \right)$$

based on a linear variation of the bulk modulus with pressure
$B(P) = B_o + B'_o P$.
Compressibility depends mainly on the outer electron configuration and
on chemical bonding. Many studies have been carried out to evaluate
the cohesive energy of noble gases, ionic crystals, covalent or
metallic crystals and their associated compressibility and equilibrium
lattice constants. For example Yamashita et al (1) have calculated
theoretically the P-V relations of a number of alkali halides, oxides
or sulfides within a self-consistent KKR band calculation and compare
them to the experimental compression curves of Drickamer up to 300
kbar (2). The theory is able to reproduce fairly well the observed
curves. Figure 1 shows the pressure-volume curve for MgO, as
determined experimentally

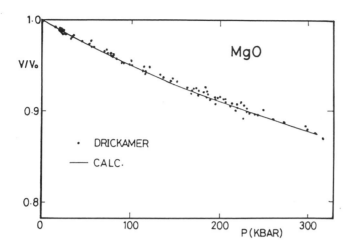

Figure 1 : Pressure-volume relation of MgO (from ref. 1).

by X-ray diffraction in comparison with the theoretical curve. This
model leads to a lattice constant "a" for the P = 0 kbar rocksalt
structure of 4.217 A, in very good agreement with the experimental
value 4.21 A. However, such calculations cannot predict whether the
P = 0 kbar ground state is of the NaCl- or CsCl-type, as we will see
in the case of KCl.

I.2. Structural phase transitions induced by pressure

KCl exhibits a first-order transition from NaCl-type phase at low
pressure to CsCl-type structure at high pressure (3) (figure 2). Band
calculations, as for MgO, are able to reproduce the observed P-V
relations in each phase and thus predict the volume change at the
phase transition but they do not predict the structure of the
P = 0 kbar phase and therefore cannot provide information on the
transition pressure. In fact the cohesive energies of the various
possible pressure induced crystallographic structures are usually very
close to each other, which makes it difficult to predict induced
crystallographic phase transitions. Of course the density of the high
pressure phase is larger than the density of the low pressure phase.
So when increasing the pressure, a sequence of structural phase
transitions can be expected corresponding to an evolution towards
higher density.
        A good illustration of this type of work is given by
theoretical and experimental studies of phase transitions of germanium
induced by pressures up to 1.25 Mbar (4). Ge is found to have the
following structural transition sequence : cubic diamond to the
tetragonal β-Sn structure at 10.8 GPa, β-Sn to a simple-hexagonal (sh)
phase at 77.7 GPa and finally the sh phase to a double-hexagonal close-
packed structure at 102 GPa. The Ge samples were studied in a diamond
anvil cell by energy-dispersive X-ray diffraction techniques using a
high-energy synchrotron source. In figure 3, the pressure variation of

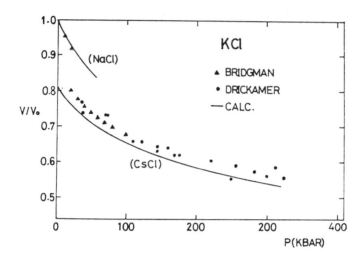

Figure 2 : Pressure-volume relation of KCl. The upper curve is that for the NaCl structure and the lower curve is that for the CsCl structure (from ref. 1)

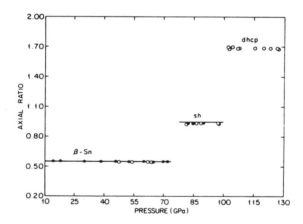

Figure 3 : Axial ratio for β-Sn (c/a), sh (c/a) and dhcp (c/2a) phases of Ge. The solid circles represent the experiment to 90 GPa. The open circles represent the experiment to 125 GPa. The solid lines represent the theoretical calculations (from ref. 4).

the axial ratio (c/a) is given for various phases of Ge. Total-energy pseudopotential calculations predict the various transitions respectively at 9.5, 84 and 105 GPa (4) which is in remarkable agreement with experimental data.

Crystallographic phase transitions can occur discontinuously or continuously. Continuous crystallographic phase transitions have been seen in CsI (5) from a cubic phase at atmospheric pressure to a tetragonal phase and to an orthorhombic one at high pressure (figure 4) but the compression curve of CsI (figure 5) up to 100 GPa does not show any anomaly within the experimental resolution at phase transitions observed at 38 and 56 GPa.

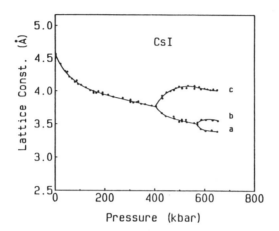

Figure 4 : Pressure induced structural transition in CsI (from ref. 5) from cubic to tetragonal and from tetragonal to orthorhombic.

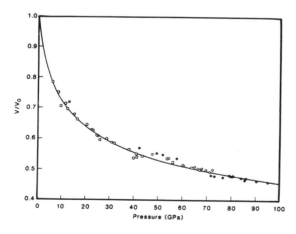

Figure 5 : The measured equation of state for CsI to 95 GPa. Various symbols represent different high-pressure experiments on the same sample (from ref. 6).

I.3. Pressure-induced electronic changes

By reducing interatomic spacings, applied pressure produces changes in
band structure, which can drive drastic changes in the electronic
properties of solids and induce electronic transitions (e.g.
metal-insulator transitions, superconductivity, magnetic order, valence
changes.

Metal-insulator transitions
Numerous theoretical and experimental works have been devoted to
metallization induced by pressure in insulators. The most famous
example is the prediction of a metal insulator phase transition in
solid hydrogen under a few Mbar. The most recent calculations (11)
predict this transition at 1.7 ± 0.2 Mbar and also, at a much higher
pressure (4 ± 1 Mbar), a structure phase transition to an atomic
metallic hcp phase with a high superconducting transition temperature.
Such pressures of 1.7 Mbar are now accessible with the diamond anvil
cell technique that we will present in the second part. Effectively,
infrared reflectivity measurements have been carried out on the alkali
halide CsI, up to 1.7 Mbar (12), which leads to an estimated
transitions pressure of 1.1 Mbar in agreement with theoretical
estimates.
        The case of solid hydrogen can be compared to the situation
of iodine which forms a molecular crystal. Iodine was known to undergo
a continuous pressure-induced transition to a metallic state, the
energy gap disappearing around 180 kbar. The interesting question was
to know whether this transformation to the metallic state occurs in
the molecular crystal due to a continuous conduction-valence band
overlap, or whether iodine dissociates to become a monoatomic metal,
the metallic character being the consequence of the unfilled 5p band.
High pressure X-ray measurements (13,14) have provided answers to
these questions. According to these results, the space group is the
same at 206 kbar as that at atmospheric pressure and iodine still
retains its molecular character. However, iodine transforms to a
monoatomic lattice around 210 kbar (14) with about 4 % change in
volume. Thus the metallization in iodine occurs by band overlap prior
to the dissociation. Figure 6 shows the projections of the structure
of iodine at three different pressures.

Superconductivity
It has been seen in the last section that pressure induces
crystallographic transitions in germanium. Actuallly, resistivity
measurements under pressure (7) have also shown that the diamond to
β-Sn crystallographic transition is also associated with a semiconductor to metal transition. Another similar and interesting case is
silicium which exhibits several transitions under pressure. Si was
found to have successive structural phase transitions from cubic
diamond to tetragonal β-Sn, then to sh to an intermediate structure
and finally to a hcp structure above 40 GPa (8,9). But in agreement
with theoretical prediction, superconductivity was evidenced in both
phases β-Sn and simple hexagonal (10). In the β-Sn phase of silicium,

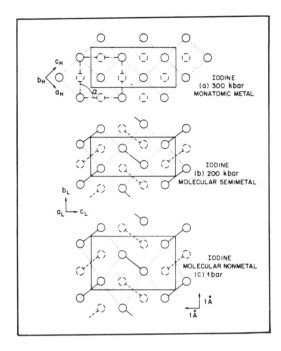

Figure 6 : Projections of the crystal structure of iodine at three different pressures. The atoms represented by solid and dashed lines lie, respectively, at the basal plane and halfway above in a perpendicular plane (from ref. 14)

the superconducting transition temperature T_c = 6.3 K is not sensitive to pressure. For the sh phase, T_c goes through a maximum of 8.2 K at 15.2 GPa (figure 7) : this critical temperature is among the highest found for simple elements.

Valence changes

Pressure induces also electronic transitions in lanthanides where magnetic 4f electrons are involved. The most typical example is the γ-α transition of cerium at 7.6 kbar : the transition is isostructural (fcc) with a volume collapse of about 16 %. At higher pressures two phase transitions take place at 50 kbar and 120 kbar towards a body-centered monoclinic (α"-Ce) and a tetragonal phase. Figure 8 shows the volume of cerium as a function of pressure up to 460 kbar as determined by the most recent studies (15) using a diamond anvil cell and synchrotron radiation. The γ-α transition is associated with the delocalization of the single 4f electron of cerium and consequently it results in a partial valence change, from nearly trivalent state towards a tetravalent state, with a decrease of the susceptibility at the transition. Several compounds of cerium have a susceptibility which decreases with increasing pressure as a consequence of an unstable 4f shell.

Samarium sulfide exhibits a similar isostructural transition at 6.5 kbar from a semi-conductor state to a metallic state with a ~ 10 % volume decrease. Valence change in SmS has been evidenced by X-rays absorption technique (16). L_{III} edge absorption spectra, shown in figure 9, exhibit two mean peaks characteristic of the Sm^{2+} ($4f^6$) and Sm^{3+} ($4f^5$) configurations. The Sm^{3+}/Sm^{2+} intensity ratio is found to increase with pressure. Analysis of the intensities gives a mean valence of Sm which varies from 2.12 at ambient pressure to a value of 2.42 at 7.5 kbar.

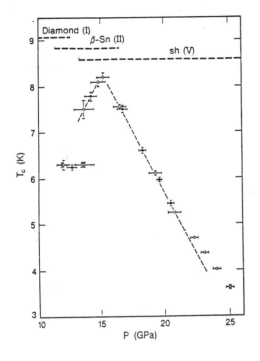

Figure 7 : The superconducting transition temperature T_c as a function of pressure in Si. The points represent the measured values with experimental error bars ; the two types of points denote two samples. The dashed line is used as a guide (from ref. 10).

Valence transitions can also occur continuously as it has been observed in SmSe, SmTe, SmB_6 or YbO. The pressure-volume data up to 35 GPa for YbO as measured using a diamond anvil cell and the energy-dispersion X-ray technique (17) are presented in figure 10. The pressure-volume relationship clearly indicates abnormal compressibility above 8 GPa.

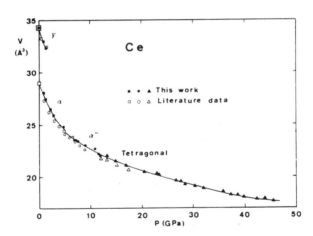

Figure 8 : Atomic volume of cerium metal as a function of pressure. The curve through the point for α-Ce, α"-Ce and tetragonal cerium has been calculated from Murnaghan's equation of state using B_0 = 20 GPa and B_0' = 5.5 and the extrapolated value V_0 = 29.0 $Å^3$ (from ref. 15).

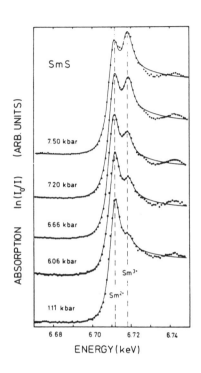

Figure 9 : Sm L_{III}-edge
absorption as a function of
photon energy for SmS at
various hydrostatic pressures
(from ref. 16).

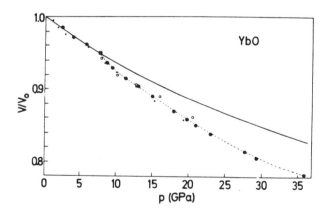

Figure 10 : Pressure-volume relationship for YbO. The solid line was
calculated using the Murnaghan equation of state with B_0 = 130 GPa and
B_0' = 4. The experimental points deviate after 8GPa due to valence
change. Various symbols represent different high pressure media (from
ref. 7).

This anomaly is attributed to a gradual and isostructural valence change of ytterbium from a divalent state ($4f^{14}\ 5d^0\ 6s^2$) towards a trivalent state ($4f^{14-x}\ 5d^x\ 6s^2$). A change of color from black at low pressure to yellow at high pressure, associated with a semiconductor to metallic transition, has been observed and gives a further evidence for a valence change in YbO. This example shows that valence change can be spread over a large pressure range.

Onset of magnetic order

Structural and electronic transitions are correlated and all energies must be taken into account to determine the nature of magnetic ground state. As an illustration, figure 11 presents the pressure temperature phase diagram of CeZn determined by resistivity and magnetization measurements under pressure (18). The low pressure phase is cubic CsCl and exhibits a magnetic transition from a paramagnetic state at high temperature to an antiferromagnetic state at low temperature. Pressure induces a structural transition above 8 kbar. The high pressure phase is tetragonal and the magnetic order at low temperature is ferromagnetic. Lattice instability, crystalline electric field, Kondo interactions and exchange interactions should be considered to explain the values of the magnetic moment of cerium and the variations of the magnetic ordering temperature with pressure.

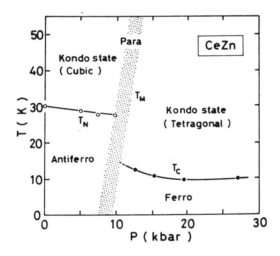

Figure 11 : Pressure-temperature (p-T) magnetic and structural phase diagram in CeZn. The shaded part denotes the width of the thermal hysteresis of the structural transition (T_M) (from ref. 18).

II. TECHNOLOGY OF HIGH PRESSURE NEUTRONS AND X-RAY CELLS

II.1 X-rays versus neutrons

In the past two decades major advances have been made in the design of high pressure equipment for neutron and X-ray scattering. These techniques contribute to the large development of high pressure studies, since they give essential informations on crystallographic and magnetic structures of solids in the high pressure state. Several reviews describe the technologies and the physics of neutrons (19-21) and X-ray (22,23) scattering under pressure and in this paper we will give a rapid survey of these techniques in order to emphasize their characteristics and limitations.

   The advantage of neutron scattering is the ability to investigate magnetic properties and to perform inelastic scattering to probe elementary excitations. In addition, neutrons have low absorption cross-sections and constant atomic scattering factors. However, due to low fluxes of neutron sources in comparison with X-ray sources, larger sample volumes are necessary (~ 0.1 cm^3) which limits the highest available pressure at about 50 kbar. X-ray scattering under pressure has focused renewed interest with the steady development of diamond anvil cell, which can be used in the megabar region. Such experiments are even more powerful now with the widespreading use of synchrotron radiation sources. The synchrotron radiation has enormous advantages : extreme spectral intensity, highly collimated beam and well-defined polarization. Figure 12 illustrates the very high resolution obtained with synchrotron beam in comparison with conventional sources such as X-rays produced by a rotating anode or neutrons (21). The high brightness of synchrotron radiation allows also new types of measurements such as absorption measurements and kinetics studies of time-dependent phenomena (34).

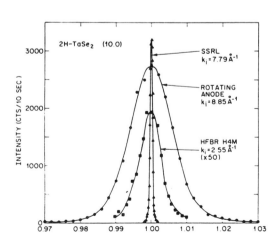

Figure 12 : comparison of (10.0) reflection of 2H - $TaSe_2$ in high pressure cells using a synchrotron source (SSRL), a rotating anode or a neutron source (HFBR) (from ref. 21).

II.2. Neutron scattering

The design of a pressure cell for neutron diffraction is a balance
between absorption of neutrons against the strength of the walls. The
choice of materials is confined to aluminium alloys, titanium-zirconium
alloys, copper alloys, maraging steels or sintered alumina. At
pressure below 6 kbar, the unsupported aluminium cell (24) given
figure 13 has been intensively used at I.L.L. (Grenoble) from 1.5 to
300 K, for 2-axes elastic scattering, 3-axes inelastic scattering as
well as 4-circles elastic scattering experiments. The sample has a
volume as large as 5 cm^3. The pressure medium is helium gas which
offers good hydrostaticity even at low temperature. A beryllium copper
clamp device can also be used more conveniently for experiments under
magnetic field (25) up to 10 kbar.

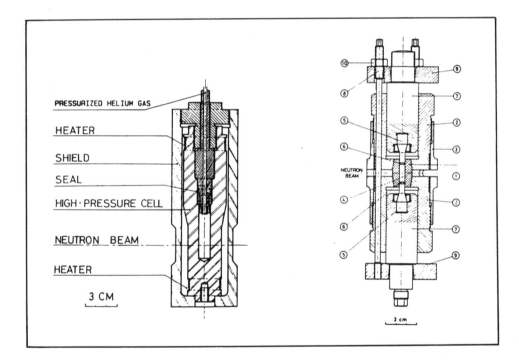

Figure 13 : Aluminium cell   Figure 14 : Alumina cell
for neutron diffraction.   1   alumina part,
                                               5   tungsten carbide
                                                   piston.

At higher pressures, the cells need to be externally supported. Sintered alumina (Al_2O_3) is chosen as a material for the pressure cylinder because of its relative transparency to neutrons and its very high compressive strength. In that case, the restricted height of the window due to supporting parts limits the volume of the sample accessible to neutrons (0.1 - 0.2 cm^3). This type of cell shown in figure 14 produces pressures of 40 kbar (26-28).

Some attempts are done actually to extend the available pressure range by using large-diamond anvil cell : the volume of the sample is only $\simeq$ 0.3 mm^3 and measurements of neutron diffraction under pressure of 210 kbar have been reported (29).

Many experimental devices have also been developed to perform neutron scattering experiments under uniaxial stress (30). In that case, the maximum stresses which can be achieved depend mainly on the strength of the material being studied.

II.3. X-ray scattering

After the era of classical Bridgman-anvil high pressure devices with two sintered tungsten carbide anvils, the improvements in high pressure cells for X-ray scattering have been obtained by using diamond anvil cell (DAC). The reader is referred to the very complete review on the diamond anvil cell by A. Jayaraman (23). Figure 15 shows the basic principle of the DAC. The pressure is generated in a hole drilled in a metal gasket which is tightened between two opposed diamond anvils. Pressure increases inside the sample chamber when the gasket is compressed as shown in the upper part of the figure. Many DAC (31) have been designed with different ways to apply the load and to ensure alignment of the diamonds, which is of prime importance to generate very high pressures without breaks of the diamonds. The DAC designed by Basset et al (31b) and shown in figure 16 is one of the most simple device and has been extensively used for X-ray powder diffraction studies. A pressure of about 400 kbar can be obtained with this cell. The Mao-Bell (31c) presented in figure 17 is capable to generate pressures higher than 1 Mbar for X-ray studies. The highest pressure attainable depends, of course, on the area of the diamond flat which varies from 0.3 to 0.7 mm, and on the optimization of the size and shape of diamond anvils. The volume of the pressure chamber is fixed by the diameter of the hole, which is typically 200 μm and by the thickness of the metallic gasket which is around 0.25 mm before compression. Thus the total amount of sample is very small. Many pressure media have been experimented to obtain the best hydrostatic environment for the sample : A 4:1 methanol:ethanol mixture (32) remains hydrostatic up to 10^4 kbar at room temperature. The pressure is determined from the lattice parameter of internal markers (NaCl, CsCl, Ag...) mixed with the sample or more and more from the shift of the red ruby fluorescence line (33), which is now the most convenient method. Indeed, a tiny chip of ruby 5-10 μm in dimension is sufficient to do measurements of fluorescence excited by a laser.

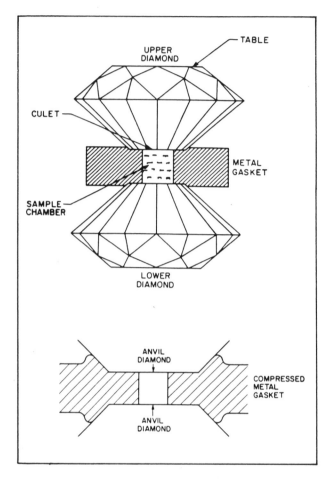

Figure 15: Above: opposed diamond anvil configuration, with a metal gasket for sample confinement in a pressure mechanism. Below: the shape of a compressed gasket supporting the diamond edges (from A. Jayaraman, ref. 23)

For powdered samples, both angle- and energy-dispersive diffraction techniques are used with high pressure cells. As an example, figure 18 shows the energy-dispersive powder diffraction spectrum of cerium at 42 GPa obtained with a synchrotron source (15).
For single crystal studies, many high pressure cells have been adapted to be mounted on a goniometer for four circle diffractometry.

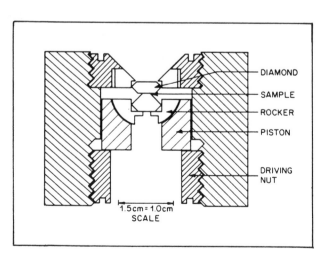

Figure 16: Basset-type cell with half-cylindrical rockers for diamond alignment for high-pressure X-ray scattering (from Basset et al, ref. 31b).

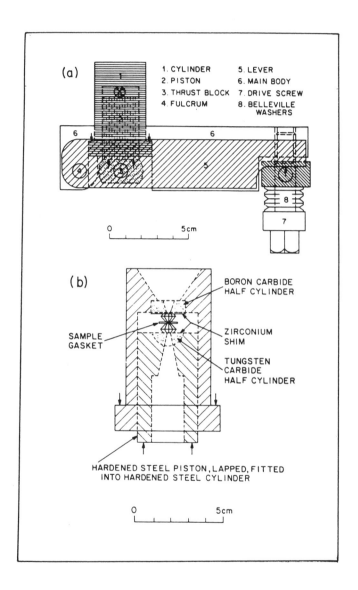

Figure 17: Mao–Bell cell for high pressure X-ray scattering (from ref. 23).

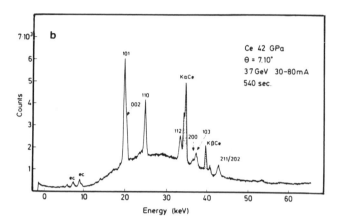

Figure 18 : X-ray synchrotron source energy-dispersive powder
spectrum of cerium at the pressure 42 GPa in the body-centred tetragonal
phase. Escape peaks are denoted by ec (from ref. 15).

III. SOME EXAMPLES OF HIGH PRESSURE STUDIES IN SOLID STATE PHYSICS

III.1. Magnetic phases of CeSb

The case of CeSb is a typical example which shows the irreplaceable
role of neutron diffraction techniques to investigate magnetic
properties. CeSb has the very simple NaCl crystallographic structure
and orders at zero magnetic field with a first-order transition at
T_N = 16.2 K. Magnetization measurements on a single crystal (35)
indicate a very large magnetic anisotropy and reveal the existence of
many magnetic phases versus magnetic field and temperature. However,
these experiments cannot provide the exact nature of the magnetic
ordering. The magnetic phase diagram has been determined by neutron
diffraction experiments (36,37) performed as a function of magnetic
field and temperature. This H-T magnetic phase diagram at zero pressure,
the most complex never reported, exhibits fifteen phases as shown on
figure 19. These different phases are characterized by a propagation
vector $\vec{k}$ = (0, 0, k) always parallel to a fourfold axis and by a strong
magnetic anisotropy which confines the moment direction along the
propagation vector. They consist of a periodic packing of paramagnetic
planes and ferromagnetic planes with magnetization parallel (up) or
antiparallel (down) to the magnetic field. Thus three types of
magnetic structures can be distinguished in the H-T phase diagram : 1)
the antiferro-ferromagnetic arrangement (AFF) with only up and down
planes, 2) the antiferro-paramagnetic arrangement (AFP) with the
coexistence of paramagnetic up and down planes, and finally 3) the
ferro-paramagnetic arrangement (FP) with only up and paramagnetic
planes. The existence of such paramagnetic planes is quite unusual and
the existence of such arrangements indicates a very weak coupling
between planes to allow the coexistence of ferromagnetic and
paramagnetic planes.

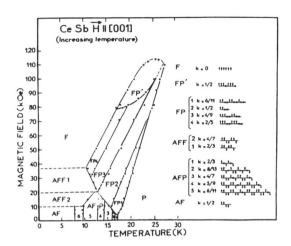

Figure 19 : Magnetic phase diagram of CeSb as a function of
temperature and magnetic field (from ref. 37).

        The unusual properties of CeSb are well explained by the p-f
mixing model, which is based on hybridization between p and f bands
and takes into account the symmetry of the 4f states due to crystalline
electric field (CEF). In this model properties should be sensitive to
pressure via the p-hole concentration.
        Magnetization measurements under pressure have first shown
the sensitivity of the phase diagram of CeSb to applied pressures
(38). However, only high pressure neutron diffraction experiments
(37,39) were able to determine the P-T phase diagram up to 21 kbar in
zero field. Figure 20 shows typical thermal variation of the intensity
of superlattice used to determine the magnetic phase diagram (figure
21). In the low pressure range (P < 9 kbar), the ordering temperature
T_N increases linearly with pressure (at the rate dT_N/dP = +0.65 K/kbar).
The phase transition at T_N remains first-order as at ambient pressure.
All the AFP phases remain but their stability ranges become narrower
with increasing pressure. However, the stability range of the low
temperature phase (AFIA) becomes wider as the pressure increases.
        In the intermediate pressure range (9 kbar < P < 13 kbar) T_N
increases strongly with pressure and a new ordered phase appears below
the transition temperature T_N, which becomes second-order. This new
phase at T_N corresponds to a type I antiferromagnetic ordering.
        In the high pressure range (P > 13 kbar), two completely new
phases are observed between the antiferromagnetic phase (AFI) at high
temperature and the (AFIA) always stable at low temperature. The
characteristics of the behaviour at high pressures is the disappearance
of the magnetic structures containing paramagnetic planes.

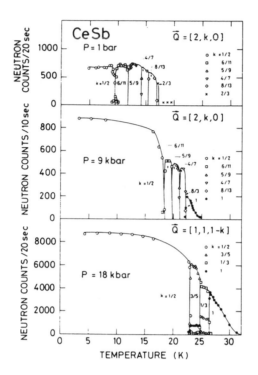

Figure 20 : Thermal variation of the intensity of the superlattice
peak [2, k, 0] at P = 1 bar and P = 9 kbar and [1, 1, 1-k] at
P = 18 kbar in CeSb (from ref. 39).

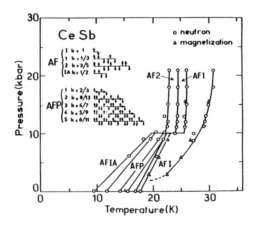

Figure 21 : Magnetic phase diagram of CeSb as a function of temperature
and pressure (from ref. 39).

The main pressure effect being to modify the p-hole concentration, it was concluded that the existence of the non-magnetic phases results from a delicate balance between the CEF-splitting and the p-f mixing term via the p-hole concentration. The large pressure dependence of magnetic properties can be understood in terms of non-linear effects in the p-f mixing.

III.2. Magnetic ordering induced by uniaxial stress in praseodymium

Ions submitted to electric fields present in solids may have properties largely different from that of the free ion. Therefore these properties may be severely modified through the variation of hydrostatic pressure or uniaxial stress. We have chosen as an illustration the Pr^{3+} ion which possesses a moment $g_J J = 4/5 \times 4 \ \mu_B$. When embedded in Pr or PrSb the $2J+1 = 9$ fold degeneracy of the free ion is lifted which gives rise to a singlet ground state. The crystal field levels can be determined using neutron spectroscopy.

Many investigations have been devoted to those compounds where interatomic exchange J and crystal field splitting Δ are comparable in magnitude, and specially to those with a non-magnetic single crystal field ground state. For a two singlets levels with energy separation Δ :

$$\chi = \chi_c/(1 - J\chi_c)$$

where χ_c, the Van Vleck susceptibility, is inversely proportional to Δ. A spontaneous polarization can occur when the product $J\chi_c$ passes through 1. High pressure can be used to tune the value of $J\chi_c$ through 1, and therefore to study the T = 0 magnetic-non-magnetic transition in homogeneous materials. This kind of phase transition also occurs in a large variety of physical problems when a spontaneous polarization results from two opposing mechanisms. Such T = 0 non-magnetic to magnetic phase transitions have been observed via the effect of hydrostatic pressure in PrSb (40) and Pr_3Tl (41) and through the action of stress in Pr (42) where an uniaxial stress of 800 bar creates a modulated antiferromagnetic structure with $T_N = 7.5$ K. In this last case hydrostatic pressure does not lead to a formation of a magnetic moment.

Neutron scattering and magnetization studies have established that the ions on the hexagonal sites in dhcp praseodymium metal ($4f^2$, J = 4) have singlet ground states and that the exchange/crystal field splitting ratio is just below the threshold value for magnetic ordering. The wavefunctions of the ground state singlet and the first excited doublet are ($J_z = 0$) and $| \pm 1\rangle$, respectively. Transitions between these states are readily observable by neutron scattering : at T = 5 K, and ambient pressure conditions, the dispersion of the lowest energy magnetic excitations exhibits a broad minimum around a wave vector of 0.10 to 0.15 τ_{100} demonstrating that the exchange coupling J(q) is at a maximum in this region.

The magnetic behaviour of single crystal Pr under uniaxial stress was examined using neutron scattering techniques. Neutron elastic-scattering studies of Pr have shown that intense magnetic satellites are observed around ($\pm Q$, 0, $\mp Q$, 3) with $Q = 0.128$ under the application of an 800 bar-stress in the a direction. This induced magnetic ordering corresponds to a sinusoidally modulated structure with amplitude of magnetic moments m $\simeq$ .88 μ_B at 970 bar and T = 1.3 K. At this pressure the corresponding temperature ordering is T_N = 9.7 K. This situation is well understood theoretically : the effect of a stress applied in the a-direction is to remove the degeneracy of the $|\pm 1\rangle$ doublet and also to couple $|\pm 2\rangle$ components of higher energy crystal field states to the $|0\rangle$ ground state, resulting in an induced magnetic ordering.

The most striking changes in the dispersion relation of magnetic excitations as a result of the applied stress have been observed by neutron inelastic scattering measurements (42, 43). Figure 22 shows typical inelastic scans at (0.127, 0, 1) near the satellite wave vector at different stresses. From such inelastic spectra and from the temperature dependence of these excitations below and above T_N, the magnetic relation dispersions are determined.

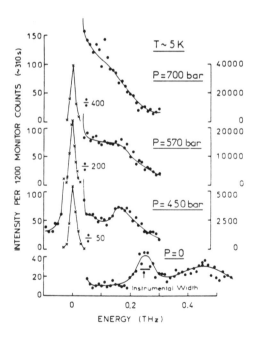

Figure 22 : Neutron inelastic scattering spectra showing the softening of the LO mode as a function of uniaxial stress in Pr. The intense elastic scattering arises from the magnetic satellite. (from ref. 43).

The dispersion along [1, 0, 0] (ΓM) of the lowest two modes
of the magnetic excitations propagating on the hexagonal sites is
shown in figure 23 for an applied stress of 970 bar and compared with
equivalent measurements at zero stress. At P = 0, the lower and upper
modes have longitudinal and transverse optic characters respectively,
providing direct evidence of anisotropic exchange. The splitting
between these two modes increases substantially as stress is applied.
Most dramatically, the broad minimum in the lower mode softens
completely and the excitations appear to emanate with a linear
dispersion from the satellite reflection at a wave vector of about
$Q = 0.13 \tau_{100}$.

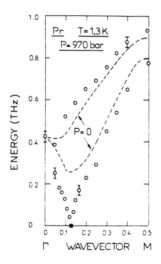

Figure 23 : Dispersion along ΓM of the LO and TO magnetic exciton
modes in Pr at 1.3 K for an applied stress of 970 bar. For comparison,
zero stress results are shown as dashed lines. The closed square
denotes the magnetic satellite position (from ref. 43).

III.3 Pressure effects in concentrated Kondo lattice of $CeNi_xPt_{1-x}$

One of the most fascinating subject of these last few years in
magnetism concerns Kondo effect. Kondo interaction arises from a
negative exchange interaction J between a localized magnetic moment
and the spin of the conduction electron, which tends to give a non-
magnetic ground state. The same negative interaction leads to indirect
exchange interactions, between localized magnetic moments, via the
R.K.K.Y. (Ruderman-Kittel-Kasuya-Yosida) spin density oscillations,
which lead to long range magnetic ordering. It results a competition
between these R.K.K.Y. exchange interactions and Kondo effect,

depending on the strength of J. The Kondo temperature T_K depends
exponentially on J, whereas the magnetic ordering temperature
increases as $T_0 \simeq J^2$. So for small values of J, R.K.K.Y. interactions
dominate and magnetic ordering is possible. For large values of J,
Kondo fluctuations, which are the most important, give reduction of
magnetic moments and may suppress long range magnetic ordering.

Kondo effect can be observed in intermetallic compounds of
"anomalous" rare earths, like cerium, where the 4f level is not too far
below the Fermi level, since Kondo coupling constant is given by
$J \propto |V|^2/(E_{4f} - E_f)$, where E_{4f} and E_F are respectively 4f and Fermi
energy levels and V the hybridization parameter between 4f and
conduction (s-d) band states.

The pseudo-binaries CrB-type orthorhombic $CeNi_xPt_{1-x}$ compounds
(44) give a good example to illustrate the competition between R.K.K.Y.
and Kondo interactions, and application of pressure give informations
about the applicability of the Kondo model.

CePt and compounds with $x \leq 0.9$ order ferromagnetically with
their Curie temperature T_C going through a maximum versus nickel
content. Spontaneous magnetization M_S decreases continuously against
increasing Ni content as shown figure 24. These results are consistent
with the Kondo lattice model (45) which predicts such variations of
T_C, T_K and M_S versus $|Jn(E_F)|$, where $n(E_F)$ is the density of states
at the Fermi level (figure 25).

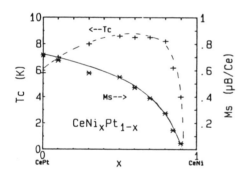

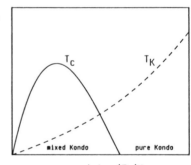

Figure 24 : Curie temperature
and spontaneous magnetization
versus Ni content x in the
$CeNi_xPt_{1-x}$ compounds
(after ref. 44).

Figure 25 : Schematic representation
of the ordering temperature and of
the Kondo temperature as a function
of the product $|Jn(E_F)|$ after Lavagna
et al (from ref. 45).

The variations of the magnetization M_S at 4.2 K and the Curie
temperature T_C as a function of pressure have been determined for some
ferromagnetic alloys (46) and relative variations of T_C and M_S are
reported figure 26. They exhibit a large variety of behaviours :
increase of T_C for CePt, no pressure effect on T_C for x = 0.5,
decrease of T_C for x = 0.8 and x = 0.85 and decrease of M_S when
pressure is increased with larger rates for larger Ni concentrations.
In fact, these various behaviours are completely consistent with the
Kondo lattice model also evidenced at room pressure.

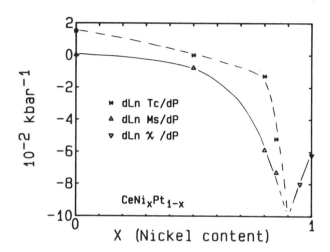

Figure 26 : relative variations with pressure of the Curie temperature (T_c), of the spontaneous magnetization (M_s) for ferromagnetic alloys and of the susceptibility (χ) for non magnetically ordered alloys versus Ni content.

The main effect of pressure is to increase $|Jn(E_F)|$, and the increase of $|Jn(E_F)|$ (or pressure) brings out an increase of the Kondo temperature and a variation of T_c passing through a maximum. Effectively in CePt where the 4f level is well below the Fermi level, the J parameter is small and Kondo interactions are weak. Consequently M_s is not sensitive to pressure and T_c increases with pressure. Increasing the nickel content shifts down the Fermi level and leads to a larger J parameter. For x = 0.5, which corresponds to the maximum of T_c versus $|Jn(E_F)|$ expected in the model, larger pressure effects are observed in M_s and T_c is pressure independent. At last for larger concentrations in nickel, larger decreases of both T_c and M_s are expected. Effectively, the decrease of M_s with pressure becomes huge ($\approx$ 45 % in 6 kbar for x = 0.85) whereas T_c is found to decrease faster and faster when Ni concentration increases above x = 0.5. Thus pressure effects in this system are a good confirmation of the Kondo lattice model.

CeNi and $CeNi_{0.95}Pt_{0.05}$ do not order magnetically and the thermal variation of their susceptibility presenting broad maxima at 140 and 60 K, respectively, are typical of intermediate valence states or non-magnetic Kondo systems with high Kondo temperatures. Figure 27 shows the thermal variation of the susceptibility of CeNi and $CeNi_{0.95}Pt_{0.05}$ at P = 0 and P = 4 kbar (47). The maximum susceptibility temperatures increase with pressure whereas the susceptibility itself decreases. The rates of variations are very large and indicative of strong 4f instability. These results are also in agreement with an increase of the Kondo temperature with pressure. Moreover, in CeNi, a pressure of 2 kbar is enough to induce a first-order transition with a large hysteresis between two magnetic states (48) (figure 27). This transition could be explained within the Kondo collapse model proposed by many authors (45,49). Thus first-order transition induced by pressure is presently studied by neutron scattering.

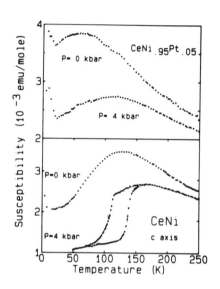

Figure 27 : Thermal variation of the susceptibility of a polycrystalline sample of $CeNi_{0.95}Pt_{0.05}$ at room pressure and at 4 kbar and thermal variation of the susceptibility of CeNi measured along the c-axis at room pressure and at 4 kbar (from ref. 47).

REFERENCES

1 - J. Yamashita and S. Asano, J. Phys. Soc. Japan, **52**, 3506 (1983).

2 - H.G. Drickamer, R.W. Lynch, R.L. Clendenen and E.A. Perez-Albuerne, Sol. Stat. Physics, **19**, 135 (1966).

3 - S.N. Vaidya and G.C. Kennedy, J. Phys. Chem. Solids, **32**, 951 (1971).

4 - Y.K. Vohra, K.F. Brister, S. Desgreniers and A.L. Ruoff, Phys. Rev. Letters, **56**, 1944 (1986).

5 - K. Asaumi, Phys. Rev. B, **29**, 1118 (1984).

6 - Y.K. Vohra, K.E. Brister, S.T. Weir, S.J. Duclos and A.L. Ruoff, Science, **231**, 1136 (1986).

7 - S. Minomura and H.G. Drickamer, J. Phys. Chem. Solids, **23**, 451 (1961).

8 - H. Olijnyk, S.K. Sikka and W.B. Holzapfel, Phys. Letters, **103A**, 137 (1984).

9 - Z. Hu and I.L. Spain, Sol. Stat. Commun., **51**, 263 (1984).

10 - K.J. Chang, M.M. Dacorogna, M.L. Cohen, J.M. Mignot, G. Chouteau and G. Martinez, Phys. Rev. Letters, **54**, 2375 (1985).

11 - B.I. Min, H.J.K. Jansen and A.J. Freeman, Phys. Rev. B, **33**, 6383 (1986).

12 - R. Reichlin, M. Ross, S. Martin and K.A. Goettel, Phys. Rev. Letters, **56**, 2858 (1986).
13 - O. Shimomura, K. Takemura, Y. Fujii, S. Minomura, M. Mori, Y. Noda and Y. Yamada, Phys. Rev. B, **18**, 715 (1978).
14 - K. Takemura, S. Minomura, O. Shimomura and Y. Fujii, Phys. Rev. Letters, **45**, 1881 (1980).
15 - J. Staun Olsen, L. Gerward, U. Benedict and J.P. Itié, Physica, **133B**, 129 (1985).
16 - K.H. Frank, G. Kaindl, J. Feldhaus, G. Wortmann, W. Krone, G. Materlik and H. Bach, in **Valence instabilities**, edited by P. Wachter and H. Boppart (North-Holland, Amsterdam, 1982) p. 189.
17 - A. Werner, H.D. Hochheimer, A. Jayaraman and J.M. Léger, Sol. Stat. Commun., **38**, 325 (1981).
18 - H. Kadomatzu, H. Tanaka, M. Kurisu and H. Fujiwara, Phys. Rev. B, **33**, 4799 (1986).
19 - C.J. Carlile, D.C. Salter, High Temperatures, High Pressures, **10**, 1 (1978).
20 - C. Vettier, in Neutron Scattering, A.I.P. Conf. proc., **89**, 121 (1981).
21 - D.B. McWhan, Revue Phys. Appl., **19**, 715 (1984) and references therein.
22 - W.B. Holzapfel, Revue Phys. Appl., **19**, 705 (1984) and references therein.
23 - A. Jayaraman, Rev. of Modern Physics, **55**, 65 (1983).
24 - J. Paureau and C. Vettier, High Temperatures, High Pressures, **7**, 529 (1975).
25 - D.B. McWhan in **High Pressure Science and Technology,** K.D. Timmerhaus and M.S. Barber Editors (Plenum, N.Y., 1979) p. 292.
26 - D.B. McWhan, D. Bloch and G. Parisot, Rev. Sci. Inst., **45**, 643 (1974).
27 - D. Bloch, J. Paureau, J. Voiron and G. Parisot, Rev. Sci. Inst., **47**, 296 (1976).
28 - U. Walter, Revue Phys. Appl., **19**, 833 (1984).
29 - S.Sh. Shil'shtein, V.P. Glazkov, I.N. Makarenko, V.A. Somenkov and S.M. Stishov, Sov. Phys. Solid Stat., **25**, 1907 (1983).
30 - a - A. Draperi, D. Herrmann-Ronzaud and J. Paureau, J. Phys. E, **9**, 174 (1976).
     b - J. Pretchel, E. Luscher and J. Kalus, J. Phys. E, **10**, 432 (1977).
     c - A. Draperi and C. Vettier, Revue Phys. Appl., **19**, 823 (1984).
31 - a - G. Piermarini and S. Block, Rev. Sci. Inst., **46**, 975 (1975).
     b - W.A. Bassett, T. Takahashi and P.W. Stook, Rev. Sci. Inst., **38**, 37 (1967).
     c - H.K. Mao and P.M. Bell, in Carnegie Institution of Washington Year Book, **77**, 904 (1978) and **78**, 659 (1979).
     d - G. Huber, K. Syassen and W.B. Holzapfel, Phys. Rev. B, **15**, 5123 (1977).
     e - L. Merrill and W.A. Bassett, Rev. Sci. Inst., **45**, 290 (1974).
32 - G.J. Piermarini, S. Block and J.S. Barnett, J. Appl. Phys., **44**, 5377 (1973).

33 - J.D. Barnett, S. Block and G.J. Piermarini, Rev. Sci. Inst., **44**, 1 (1973).

34 - See for examples : **Solid State Physics under Pressure : Recent Advance with Anvil Devices,** Edited by S. Minomura, Terra Scientific Publishing Company (1985).

35 - H. Bartholin, D. Florence, Wang Tcheng-Si and O. Vogt, Phys. Stat. Sol. (a), **29,** 275 (1975).

36 - J. Rossat-Mignod, P. Burlet, J. Villain, H. Bartholin, Wang Tcheng-Si, D. Florence and O. Vogt, Phys. Rev. B, **16,** 440 (1977).

37 - J. Rossat-Mignod, J.M. Effantin, P. Burlet, T. Chattopadhyay, L.P. Regnault, H. Bartholin, C. Vettier, O. Vogt, D. Ravot and J.C. Achart, J. Magn. Magn. Mat., **52,** 111 (1985).

38 - H. Bartholin, D. Florence and O. Vogt, J. Phys. Chem. Solids, **39,** 89 (1977).

39 - T. Chattopadhyay, P. Burlet, J. Rossat-Mignod, H. Bartholin, C. Vettier and O. Vogt, J. Magn. Magn. Mat., **54-57,** 503 (1986).

40 - C. Vettier, D.B. McWhan and F.J. Blount, Phys. Rev. letters, **39,** 1028 (1977).
     D.B. McWhan, C. Vettier, R. Youngblood and G. Shirane, Phys. Rev. B, **20,** 4612 (1979).

41 - R.P. Guertin, J.E. Crow, F.P. Missel and S. Foner, Phys. Rev. B, **17,** 2183 (1978).

42 - K.A. McEwen, W.G. Stirling and C. Vettier, Phys. Rev. Letters, **41,** 343 (1978).

43 - K.A. McEwen, W.G. Stirling and C. Vettier, Physica, **120B,** 152 (1983) ; J. Magn. Magn. Mat., **31-34,** 599 (1983).

44 - D. Gignoux and J.C. Gomez-Sal, Phys. Rev. B, **30,** 3967 (1984).

45 - M. Lavagna, C. Lacroix and M. Cyrot, Phys. Letters, **90A,** 210 (1982).

46 - D. Gignoux and J. Voiron, Phys. Letters, **108A,** 473 (1985).

47 - D. Gignoux and J. Voiron, J. Magn. Magn. Mat., **54-57,** 363 (1986).

48 - D. Gignoux and J. Voiron, Phys. Rev. B, **32,** 4822 (1985).

49 - J.W. Allen and R.M. Martin, Phys. Rev. Letters, **49,** 1106 (1982).

PRESSURE TUNING SPECTROSCOPY

H. C. Drickamer
School of Chemical Sciences, Department of Physics
and Materials Research Laboratory
University of Illinois
Urbana, IL 61801 USA

ABSTRACT. The effect of pressure on electronic excitations is discussed, with emphasis on the balance between <u>intra</u> and <u>intramolecular</u> interactions.

The optical, electrical magnetic and chemical properties (collectively the electronic properties) of condensed phases depend on the characteristics of the energy levels available for the outer electrons on the atoms, ions or molecules which make up the phase; most strongly on the normally occupied levels (the ground state), however, they also depend on the characteristics and energy spacing of states accessible to thermal or electromagnetic excitation. The effect of compression is to increase the overlap of the outer electronic orbitals. Since different types of orbitals differ in their spacial characteristics – in their radial extent, in their angular momentum (shape) and in their diffuseness (compressibility) they are perturbed in different degrees by compression. The study of the change in spacing of these energy levels with pressure is well characterized as "pressure tuning spectroscopy" (PTS). A number of reviews of related work exist.[1-5]
It should be pointed out that there are other very important forms of pressure tuning spectroscopy which involve the perturbations of various kinds of vibrational levels of molecules or of crystal lattices which yield significant information about molecular structure and inter and intramolecular interactions. However, in this presentation we shall limit our discussion to the study of electronic phenomena.
PTS is very broad in its implications and applications and has been used for a variety of problems of interest to physicists and chemists. The physics community has been largely concerned with insulator-metal transitions, the band structure and impurity levels of semiconductors. The electronic structure of the alkali, alkaline earth and rare earth metals, transitions involving a change in the magnetic or superconducting states of matter and behavior of some inorganic phosphors.
In this paper we shall concentrate on optical absorption and luminescence studies of molecules or molecular ions in crystals, polymeric media, or liquids. These problems are of general interest for the chemistry and chemical physics community. In addition they form a background for developments of technological importance.

R. van Eldik and J. Jonas (eds.), High Pressure Chemistry and Biochemistry, 263–276.

A significant aspect of a number of these studies is the balance between intramolecular and intermolecular interactions for particular excitations.

When a molecule is formed from two atoms or molecular fragments one obtains a boundary orbital which is stabilized relative to the fragments and an antibonding orbital which is destabilized. From the strictly intramolecular view point the effect of compression is to destabilize the antibonding orbital further with respect to the bonding orbital. This would result in a shift to higher energy (blue shift) of an excitation from the bonding to the antibonding orbital.

For materials such as we are discussing here, the major intermolecular interaction is of the van der Waals type. This is an attractive interaction of the form:

$$E \sim - \frac{\alpha_1 \alpha_2}{R^6}$$

where α_1 and α_2 are the polarizabilities of the molecule and of the surrounding medium and R is the effective distance between interacting electron clouds. In general (but not always) the polarizability of a molecule is larger when an electron is in an excited (antibonding) state than when it is in the ground state because the electron cloud is more deformable. In this case the volume of the system will decrease when an electron is excited so that pressure will tend to stabilize the excited state vis a vis the ground state and cause a shift to lower energy (red shift) of the excitation. The larger the polarizability of the medium (α_2) the larger this effect will be. In addition, different degrees of mixing of various atomic orbitals in the bonding and antibonding states, different degrees of ionicity etc. can affect the importance of the van der Waals interaction. Finally one must consider the effective value of R. A large molecule can shield the orbitals being studied especially if they are quite localized. A tightly bound solvation shell can also increase the effective R. A number of these factors will come into play in the examples discussed below.

Azulene derivatives

The emission spectra of certain azulene derivatives, e.g. dicarboethoxychloroazulene, provide a convenient illustration of polarizability effects.[6] The normal ground state (S_0) is of π symmetry. There is an excited state (S_2) at $\nu_0 = 24810$ cm^{-1} which is the usual π^* antibonding orbital. As can be seen in Figure 1, the emission $S_2 \rightarrow S_0$ for the molecule dissolved in polymethylmethacrylate (PMMA) shifts to lower energy with increasing pressure as one would expect from our discussion above. There is a second excited state (S_1) at $\nu_0 = 15230$ cm^{-1} which involves a molecular distortion that considerably localizes the π electrons so that the molecule is less polarizable when the electron is in S_1 than in S_0 or S_2. Therefore, as illustrated in Figure 1, the shift of the $S_1 \rightarrow S_0$ emission is to higher energy with pressure.

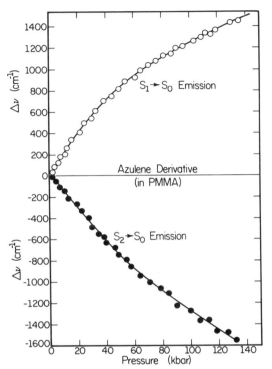

Figure 1. Shift of the $S_1 \rightarrow S_0$ and $S_2 \rightarrow S_0$ emission peaks with pressure for an azulene derivative in PMMA (polymethylmethacrylate).

In Figure 2 we show the normal $S_2 \rightarrow S_0$ emission in PMMA and polystyrene (PS). The polarizability of PS is greater than PMMA and therefore we obtain a larger red shift. This study illustrates clearly the effect of the polarizability of various electronic states as well as that of the medium on electronic excitations.

Figure 2. Shift of $S_2 \rightarrow S_0$ emission peak with pressure for an azulene derivative in PMMA and PS (polystyrene).

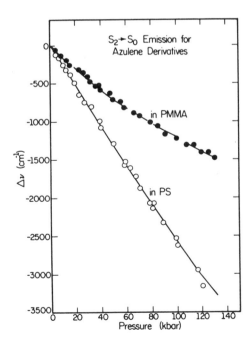

Binuclear metal cluster compounds

Metal cluster compounds provide an illustration of several of the
factors discussed in the introduction. These compounds may involve
anywhere from two to twenty metal atoms connected by metal—metal bonds
stabilized by appropriate ligands. We limit ourselves here to binuclear
clusters where, in first order, the bonding between metal atoms can be
classified as σ, π, or δ according to their angular momentum around the
bond axis. In general, the $\delta - \delta^*$ excitation lies lowest in energy (in
the red part of the spectrum), the $\pi - \pi^*$ excitation is in the blue
region and the $\sigma - \sigma^*$ in the ultraviolet. There is evidence that the σ
orbitals directly associated with metal—metal bond are most localized
while the δ orbitals are most delocalized so one would expect the van
der Waals forces to be least important for the former and most important
for the latter. Indeed $\sigma - \sigma^*$ excitations in metal carbonyls shift blue
(to higher energy) with pressure (7) in the solid and in solution as one
would expect if <u>intra</u> molecular interactions predominated.

We discuss here examples of $\delta - \delta^*$ and $\pi - \pi^*$ excitations.(8)
Figure 3 shows the shift of the $\delta \to \delta^*$
excitation in the crystalline compounds
$(n-Bu_4N)_2Re_2Cl_8$, $(n-Bu_4N)_2Re_2Br_8$ and
$(n-Bu_4N)_2Re_2I_8$. This was the only peak
within range of our equipment for the
solid chloride. In the solid, the
chloride and the bromide $\delta - \delta^*$
excitations both shift red (to lower
energy) at about -10 cm^{-1}/kbar. The
$\delta - \delta^*$ excitation in the solid iodide
is not present above 70 kbar because of
an isomerization effect discussed
elsewhere.(9) The small size of the
$\delta \to \delta^*$ peak near this pressure made it
difficult to fit, hence the increased
scatter in the data. The smaller shift
may indicate shielding (increased
effective R) of the intermolecular van
der Waals interaction by the large
iodine ligands. It is clear from the
red shift of the $\delta \to \delta^*$ excitation that
the van der Waals forces between the δ^*
orbital and the surroundings dominate
over the stabilization of the ground
state relative to the excited state due
to compression. Either the δ^*
orbitals are more polarizable than the
δ orbitals or they are more exposed to
the attractive interaction with the
surrounding molecules. These van der Waals forces may be enhanced by
the greater ionic nature of the first excited state compared to the
ground state which Hay(10) finds in his generalized valence bond
calculation on $(n-Bu_4N)_2Re_2Cl_8$.

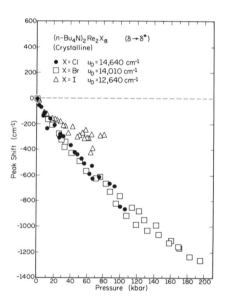

Figure 3. Pressure shifts of
the $\delta \to \delta^*$ excitations in
the solid crystalline
compounds $(N-Bu_4N)_2$ Re_2X_8 (X
= Cl, Br, I).

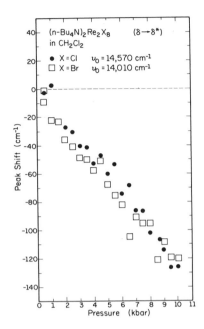

Figure 4. Pressure shifts of the $\delta \to \delta^*$ excitations in the compounds $(N-Bu_4N)_2Re_2X_8$ (X=Cl, Br) dissolved in dichloromethane.

Figure 5. Pressure shifts of the $\delta \to \delta^*$ excitations in the compounds $(n-Bu_4N)_2Re_2X_8$ (X=Cl, Br, I) dissolved in acetonitrile.

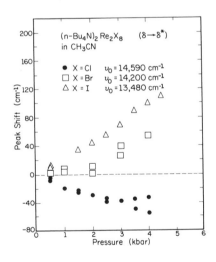

Figures 4 and 5 contrast the shifts of the $\delta - \delta^*$ excitation in the solvents dichloromethane and acetonitrile. In dichloromethane, the $\delta \to \delta^*$ peaks for $(n-Bu_4N)_2Re_2Cl_8$ and $(n-Bu_4N)_2Re_2Br_8$ shift red at -12 cm^{-1}/kbar, approximately the same value as in the solid state. The $\delta - \delta^*$ peak of the iodide could not be adequately resolved in dichloromethane. Acetonitrile freezes at 4.5 kbar at 25 C so we present

data only to 4 kbar. In this solvent we observe very different behavior
for the $(Re_2X_8)^{2-}$(X=Cl, Br, I) ions. The $\delta \to \delta^*$ excitation for the
chloride shifts slightly red for the first 2-3 kbar and then levels.
For the bromide the $\delta - \delta^*$ excitation shows very little shift below 2
kbar, and then shifts increasingly blue. The $\delta - \delta^*$ excitation for the
iodide shifts blue over the entire range of pressure. These ions in a
high dielectric medium should be strongly solvated. This solvation
shell increases the effective R in the van der Waals interaction and
decreases the magnitude of the attractive interaction. The shift data
indicated that this effect is largest for the iodide (the largest halide
used here) and smallest for the chloride (the smallest halide used
here). It should be noted (see Figures 3 and 5) that the chloride and
bromide exhibit only small changes in initial peak location in the
various media, while the iodide blue shifts by ~ 900 cm^{-1} (13,480 cm^{-1}
vs. (12,640 cm^{-1}) in the acetonitrile compared to the solid. This may
indicate some specific axial solvation of the iodide.

To test our hypothesis about the effects of solvation we studied
$Re_2(piv)_2Cl_4$ and $Re_2(piv)_2Br_4$ in acetonitrile. These are neutral
molecules and will not be strongly solvated. As can be seen from Figure
6 these molecules exhibit a red shift of ~ -16 cm^{-1}/kbar, a behavior very
similar to that of the (unsolvated) ions in the crystal or in CH_2Cl_2.

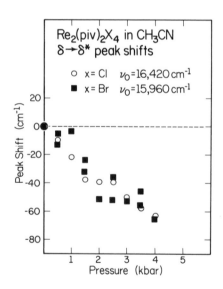

Figure 6. Pressure shifts
of the $\delta \to \delta^*$
excitations in the
compounds $Re2(piv)2X_4$
(X=Cl, Br) dissolved in
acetonitrile.

The $\pi \to \pi^*$ transitions for two different sets of ligands are
compared in Figure 7. The $\pi \to \pi^*$ excitation in crystalline
$(n-Bu_4N)_2Re_2Br_8$ shifts little in the first 30 kbar and then shifts blue
at about 5 cm^{-1}/kbar above this pressure. The $\pi \to \pi^*$ excitation in the
iodide shifts blue at about 10 cm^{-1}/kbar. The dotted line shows the
shift of the $\delta - \delta^*$ excitation in the bromide for comparison.

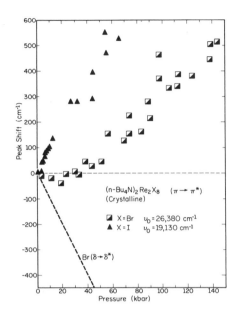

Figure 7. Pressure shifts of the $\pi \to \pi^*$ excitations
in the solid crystalline compounds $(n-Bu_4N)_2Re_2X_8$
(X=Br, I). The dashed line gives the shift of the
$\delta \to \delta^*$ excitation in $(N-Bu_4N)_2Re_2Br_8$ for
comparison.

A calculation by Mortola et. al.[11] on the $\pi - \pi^*$ excitation in
$(Re_2Cl_4)^{2-}$ indicates that there is a larger admixture of polarizable
ligand in the ground state than in the excited state. This tends to
increase the polarizability of the ground state relative to the first
excited state and to reduce the effect of the van der Waals forces.
Since iodine is more polarizable than bromine, the effect on the
polarizability should be greater for the iodide than the bromide, hence
the larger blue shift in the iodide. Because of the much greater
compressibility of intermolecular distances relative to intramolecular
distances, any change in ligand mixing under pressure should be a second
order effect.

In addition, as mentioned above, the π orbitals associated with the
metal-metal bond apparently have a smaller radius than the corresponding
δ orbitals. This would increase the effective van der Waals R and
decrease the relative importance of the van der Waals interaction. As
with the $\delta \to \delta^*$ excitations in acetonitrile one might expect a larger
effective R for the iodide than the bromide. This could contribute to
the difference in the observed shifts.

In this study we see how various properties of the molecular
orbitals and surrounding medium can affect the relative importance of
intra and intermolecular interactions.

Metal chain crystals - intensity borrowing

A subject of considerable interest in spectroscopy is intensity
borrowing between electronic or vibrational states. A special case is
Fermi resonance which involves intensity transfer between coupled
vibrational levels. This has recently been studied as a function of
pressure and the effect of pressure tuning of vibrational levels has
been demonstrated quantitatively in terms of agreement between theory
and experiment.[12-14]   Intensity borrowing between two different
electronic excitations is more complex and less susceptible to simple
quantitative analysis. Nevertheless, it can be well illustrated from
high pressure studies.

A variety of glyoxime ligands form complexes with d^8 metals. Many
of these crystallize in layers with the complexes stacked with alternate
molecules rotated 90° and, for symmetrical ligands, the metals form a
chain.[15,16] In general, the distance between layered molecules is the
same for the Ni, Pd, and Pt metal complexes so that it is established by
the organic ligand. A rather intense absorption peak occurs in the
visible region of the spectrum which is not present for the molecules in
solution. It has been assigned to an interband nd to (n+1)p excitation
so that the excited states of predominately (n+1)p character are

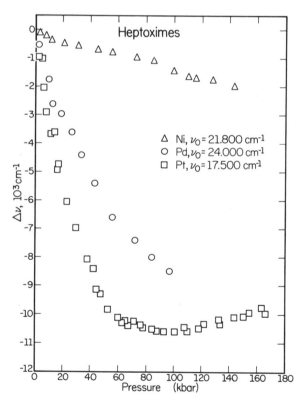

Figure 8. Shift
of metal d → p
excitation
peak with
pressure in
Ni, Pd, and Pt
heptoxime.

stabilized in the crystal with respect to the occupied levels. Figure 8
shows the change in peak location with pressure for Ni, Pd and Pt
heptoxime.[17] The initial effect of the compression is to increase
this this stabilization. This effect intensifies in the order Pt > Pd >
Ni because of the greater spacial extent of the orbitals with larger n
value. At sufficiently high pressure the overlap repulsion of the
platinum p orbitals with the molecules above and below causes a reversal
of this process and a resultant blue shift. Similar shift behavior is
observed for a variety of other metal glyoxime systems.[17]

The intensity of the d → p excitation has been the subject of some
consideration. Ohashi et al (18) ascribe a significant part of the
intensity to intensity borrowing from the intra-atomic metal to ligand
charge transfer peak which lies at higher energies. In Fig. 9 we show
the change in integrated intensity (area under the peak) with pressure
for Pt(HEPT)$_2$. There is a large initial drop in intensity, followed by
a rise from ~ 60 - 90 kilobars and a subsequent decrease. We also
measured the shift of the tail (extinction coeff. = 1) of the charge
transfer peak as a function of pressure. In Figure 10 we plot the
difference in energy between the tail and the excitation on the metal.
As this difference increases the intensity should drop, and vice versa.
As can be seen from Figures 9 and 10 the intensity changes and
differences in energy mirror each other extremely closely. This is,
then a rather elegant demonstration of intensity borrowing effects.
Other glyoxime systems exhibit similar behavior although in a less
dramatic fashion.

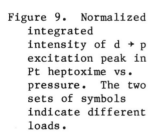

Figure 9. Normalized
   integrated
   intensity of d → p
   excitation peak in
   Pt heptoxime vs.
   pressure. The two
   sets of symbols
   indicate different
   loads.

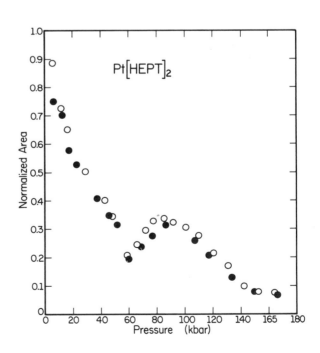

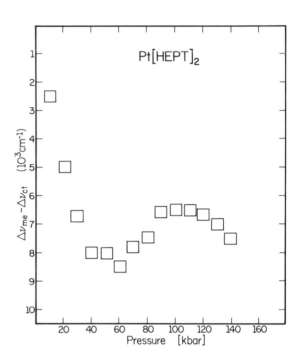

Figure 10. Difference in
shift of the metal d
to p excitation peak
and shift of tail of
charge transfer peak
vs pressure for Pt
heptoxime. Note
agreement with Figure
9.

Energy transfer in rare earth phosphors

Rare earth phosphors are of interest both theoretically and
technologically. They give sharp emission lines and can be induced to
perform as lasers. The problem is to get energy into the lowest excited
state of the rare earth. The direct absorption is inefficient because
it is forbidden on both spin and orbital bases and the peak is very
narrow. One usually surrounds the rare earth ion with ligands which
have a strong absorption and then makes use of energy transfer from the
excited state of the ligand to the rare earth. We present here a study
of this process for europium ion complexed to three dibenzomethide (DBM)
ions so that it is in a cage of six oxygens. This is one of a number of
similar systems studied in our laboratory.[19]

The absorption is to the S_1 state of the ligand from which the
energy must transfer to the lowest excited state of the rare earth
($^5D_0$). The $S_1 \to T_1$ transfer in the ligand is very rapid so that the
transfer to $^5D_0$ is from the triplet state T_1.

There are three possibilities for the location of T_1 with respect
to the rare earth excited states. It may lie above $^5D_1$, the second
excited state of Eu^{+3}. In this case most of the transfer is to $^5D_1$ and
since the coupling between $^5D_1$ and $^5D_0$ is weak, the transfer is poor and
most of the excitation is dissipated as heat. At the other extreme it

may lie just above or just below $^5D_0$. The transfer to $^5D_0$ is rapid, but back transfer to T_1 is very active also. Since T_1 has easy non-radiative paths to S_0, this also leads to weak emission from $^5D_0$. The ideal location for T_1 is between $^5D_1$ and $^5D_0$ – near enough $^5D_0$ for good transfer, but enough above to minimize back transfer.

In this work the DBM complex was studied as a crystalline solid and dissolved in polymethlymethacrylate (PMMA), which is significantly less polarizable than the crystal. In Figure 11 we show on the right the energy levels of Eu^{+3} and on the left those of DBM in the two media. The shaded area represent the pressure tuning in 40 and 80 kilobars. In the polymer the energy levels lie higher initially and shift less because of the lower polarizability of the medium Triplet states shift less than singlets because they are less polarizable.

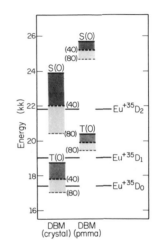

Figure 11. Energy levels of Eu^{+3} and of the DBM ligand. The shaded areas represent the pressure tuning of the ligand $S_0 \rightarrow S_1$ and $T_1 \rightarrow S_0$ excitations in 40 and 80 kilobars in the crystalline compound and in PMMA.

From Figure 11 we would estimate that as pressure increases there should be a marked increase in phosphor efficiency until back donation becomes dominant. Then there should be a drop in intensity. For the polymer all transfer is via $^5D_1$ so that there should be a modest increase and then a drop at high pressure but no dramatic effects. These predictions are confirmed in Figure 12. One must keep in mind that, superimposed on the above

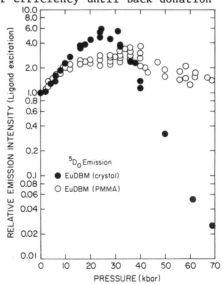

Figure 12. Change in intensity of the Eu^{+3} emission vs pressure in the crystal and in PMMA.

considerations there is a continuous increase in efficiency with pressure
(by a factor of ~ 2 in 80 kilobars) due to an increase in the radiative
rate k_r caused by increasing d-f orbital mixing.

Figure 13 shows another aspect of the energy transfer. One can
excite $^5D_0$ by direct absorption. This is an inefficient process, but
one which is relatively independent of pressure. Using these results as

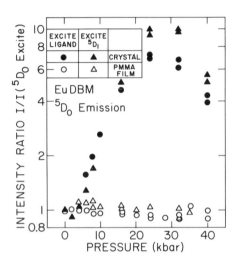

Figure 13. Effect of exciting the Eu^{+3} emission in
the ligand and in the $^5D_1$ level normalized to the
effect of direct absorption in $^5D_0$ vs pressure in
the crystal and in PMMA.

a fiducial marker we can measure the relative efficiency of exciting in
ligand S_1 or by absorption directly in the second Eu^{+3} excited state
$^5D_1$. As we expect, when exciting in S_1, for the crystal there is a
relative increase in efficiency as pressure increases, followed by a
decrease at higher pressure. Exciting in $^5D_1$ exactly mirrors the
relative changes induced by exciting in ligand S_1. This demonstrates
that the most efficient path from $^5D_1$ to $^5D_0$ is via the ligand T_1 state.
When one does the same experiment in the polymer there is no effect of
type of excitation since in any case the process is limited by the
direct $^5D_1 \rightarrow {}^5D_0$ transfer.

Finally we should consider the back transfer from $^5D_0$ to ligand T_1.
Along some configuration coordinate there is an energy barrier between
the potential wells representing the two states. We measure this barrier
as a function of pressure from the temperature coefficient of the
intensity or of the lifetime (both give the same results). In a simple
picture the barrier height should decrease by the amount the T_1 emission
shifts to lower energy with pressure. We measure this from the shift of
the phosphorescence from T_1. The results appear in Figure 14. The
dotted line represents a one to one relationship which holds very well.

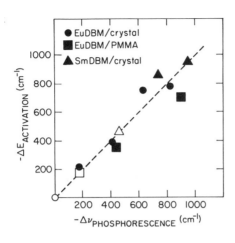

Figure 14. Shift of T_1 phosphorescence vs change in activation energy with pressure for back donation $-Eu^{+3}$ $^5D_0$ to ligand T_1.

These results indicate how pressure can be used to establish parameters for phosphor design.

Summary

The above examples represent only a small fraction of the areas of modern science where pressure tuning spectroscopy is being applied. Other topics of particular interest to chemists include thermochromic-photochromic transitions, rates of radiative and non-radiative decay of phosphors, neutral to ionic transitions in charge transfer complexes, solvent-molecule interactions studied by means of vibrational (infrared and Raman) spectroscopy, vibrational dephasing problems, and rates and mechanism of electron transfer processes in mixed valence compounds.

Acknowledgement

Support from the Materials Science Division of the Department of Energy under contract DE-AC02-76ER01198 is gratefully acknowledged.

References

1. H. G. Drickamer, Ann. Rev. Phys. Chem. **33**, 25 (1982).

2. H. G. Drickamer, Inter. Rev. Phys. Chem. **2**, 171 (1982).

3. H. G. Drickamer and G. Weber, Quart. Rev. of Biophysics **16**, 89 (1982).

4. H. G. Drickamer, G. B. Schuster and D. J. Mitchell in Radiationless Transitions S. H. Lin, ed. 289 (1980).

5. H. G. Drickamer, in Solid State Physics **17**, pg. 1, F. Seitz and D. Turnbull ed., Academic Press, N.Y. (1965).

6. D. J. Mitchell, H. G. Drickamer and G. B. Schuster, J. Am. Chem. Soc. **99**, 7489 (1977).

7. T. L. Carroll, J. R. Shapley and H. G. Drickamer, Chem. Phys. Letters **119**, 340 (1985).

8. T. L. Carroll, Ph.D. Thesis (University of Ilinois) 1986.

9. T. L. Carroll, J. R. Shapley and H. G. Drickamer, J. Am. Chem. Soc. **107**, 5802 (1985).

10. P. J. Hay, J. Am. Chem. Soc. **104**, 7007 (1982).

11. A. P. Mortola, J. W. Moscowitz, N. Rosch, C. D. Cowman and H. B. Gray, Chem. Phys. Lett. **32**, 283 (1975).

12. W. Schindler, T. W. Zerda and J. Jonas, J. Chem. Phys. **81**, 4306 (1984).

13. T. W. Zerda, M. Bradley and J. Jonas, J. Chem. Phys. Lett. **117**, 566 (1985).

14. M. Bradley, T. W. Zerda and J. Jonas, J. Chem. Phys. **82**, 4007 (1985).

15. L. E. Godycki and R. E. Rundle, Acta Chryst. **6**, 487 (1953).

16. G. V. Ganks and D. Barnum, J. Amer. Chem. Soc. **80**, 3579 (1958).

17. M. Tkacz and H. G. Drickamer, J. Chem. Phys. **85**, 1184 (1986).

18. Y. Ohashi, I. Hanazaki and S. Nagakura, Inorg. Chem. **9**, 2551 (1970).

19. A. V. Hayes and H. G. Drickamer, J. Chem. Phys. **76**, 114 (1982).

SECTION II:

CHEMICAL ASPECTS

KINETICS OF ORGANIC REACTIONS AT HIGH PRESSURE

W. J. le Noble
Department of Chemistry
State University of New York
Stony Brook, NY   11794   USA

INTRODUCTION. With the privilege of participating as a repeater in this second Advanced Study Institute on High Pressure comes the duty not only to state one's objectives, but also to outline the differences with the Chapters written for the earlier occasion. Both of those Chapters dealt with kinetics at high pressure, the first with "model reactions"[1] and the other with "problem reactions".[2] The model reactions included, especially, protic acid base equilibria. The reaction volumes of large numbers of such equilibria are known, and these provide a wealth of information on which kinetic studies can be securely founded. However, the models also included a number of data on rate constants under pressure. The reactions discussed comprised those in which the mechanism is exceedingly simple and dominated by a single feature, and in which alternative studies had left essentially no doubt about the pathway. The second chapter described the application of knowledge thus gained, to reactions the mechanisms of which were still unsettled, or even controversial.

The present Institute gives me the opportunity to discuss how well the high pressure kineticist in Organic Chemistry has fared with his views and conclusions. In the second session, I shall deal with applications in organic synthesis. In this unusual sequence, I follow historic precedent: mechanistic investigations came first in the high pressure field, and synthetic studies later.

Logic and continuity dictate that while the fundamental considerations must be stated, lengthy iterations of earlier writings need not. Students intending to make future plans in high pressure chemistry should consult those earlier writings, as well the several books and reviews with which this field has been blessed.[3] As in 1978, it appears that the purposes of this Institute are best met by letting these monographs serve as the basic references, and by avoiding the long lists of quotations usually seen at the end of formal and and exhaustive reviews. Exceptions are made for the rather recent advances.

R. van Eldik and J. Jonas (eds.), High Pressure Chemistry and Biochemistry, 279–293.

1. FUNDAMENTALS

1.1 Purpose

What the kineticist studying a given chemical reaction hopes to learn is the structure profile of the process. The structure profile is a detailed description of the states that the initial molecule(s) must traverse to arrive at the product molecules. The features of special interest include the energy, the bond lengths and angles, the distances between non-bonded atoms, and the charge distributions not only of the species appearing in the stoichiometry, but of the nearby solvent molecules as well. Some of the features of the profile are far outside the scope of this discussion; they include the structure of the initial and final states, the overall thermodynamics, and the determination of the rate law. This latter result is always based on an examination of the effect of the concentrations of all species conceivably involved in the reaction (except the solvent). One principal piece of information that seems hidden to kineticists concerns the concertedness of the reaction of interest. If a reaction is concerted in the lowest energy mode, the atoms move without interruption to their final positions, and the energy profile exhibits but a single maximum; if the reaction is stepwise, there are at least two maxima, and the minima between them represent semi stable states that have relatively long life times.

Inextricably interwoven with the determination of the rate law is that of the rate constant (or constants) appearing in it. After chosing a set of standard conditions for his rate constants, the kineticist systematically alters one after the other of all of the variables in his bag of tricks; these include the substituent effect, solvent effect, temperature effect, isotope effect and many others. By means of the changes in rate produced thereby, one can determine the energy and entropy differences between transition and initial states, and inform oneself about the transition state location, i.e., whether the activated complex resembles the initial state or the final state.

Our present concern is with one of these variables: the external pressure. The effects of pressure on the rate and equilibrium constants tell us about the difference in volume between the transition and initial states, and about the change in volume overall, respectively. In turn, it seems reasonable to anticipate that volume changes and differences will convey information about the evolution of structure along the reaction coordinate.

In the sections to follow, we shall briefly discuss certain experimental aspects of high pressure kinetics, the formalisms used, the caliberation of kinetic effects by means of equilibrium effects, and the principal structural features that can be discussed on the basis of the volume. We shall then be in a position to evaluate the status of high pressure chemistry, the opinions its practitioners have based on it, and the reservations that should be kept in mind.

1.2. Apparatus

The generation, maintenance, control and measurement of pressure in
the range of 0–20 kbar are all well known, and laboratory equipment
for these activities can be obtained commercially. The price tag
is not only a function of the pressure to be reached, but also and
even especially of the volume to be compressed; the needs in both of
these quantities should be carefully assessed since an exaggerated
desire for flexibility may lead to exorbitant costs. The basic
reason for this is that the strength of vessels is not governed by
the wall thickness but rather by the ratio of external and internal
diameters. Those whose interest is primarily in kinetics can almost
always do their work with small space requirements and in modest
pressure ranges (1–3 kbar); chemists hoping to make synthetic
applications usually opt for vessels of higher capacity, in both
volume and pressure (to 20 kbar). On the other hand, kineticists
will more often need to consider the need for continuous analysis of
their samples while they are compressed. For most of the common
techniques (conductometry, dilatometry, UV–visible and IR spectroscopy, NMR, ESR, etc.) ways and means have been invented, and reported
in the literature.

   Safety does not place constraints upon this type of work that
are not applicable to any kind of chemical laboratory, provided that
the samples compressed are liquids and not gases. Reactions that
produce gases should be avoided unless the concentration can be
held to or below solubility thresholds at ambient pressure. One should
also be aware that viscosities and melting points rise with pressure,
often rather steeply, and that the solubilities of solids as a rule
decrease.

1.3. Formalisms

The reaction volume ΔV can in many instances be obtained from measurements of the equilibrium constant as a function of pressure; then:

$$\Delta V = - RT \ \frac{\partial \ln K}{\partial P}$$

The dependence of ln K on p is always found to be a non-linear one: ΔV
is a function of the pressure. The curvature is such that the absolute
value of ΔV tends to asymptotically approach zero, which means that the
more dense side of the reaction equation is also the less compressible.
To compare and discuss reaction volumes, we therefore have to choose a
standard pressure; the custom is that the value at zero pressure (not
significantly different from that at ambient pressure) serve as that
standard. Accordingly, we need to know the slope at one extreme and of
the series of data.

   To find this slope, the best procedure is to fit the data to an
empirical expression, and calculate the slope from the derivative at
P = 0. Several such expressions have been proposed. The most common
has been ln K = a + $bP + cP^2$, so that $\Delta V_o = - bRT$. It is not a very

satisfying solution to the problem, however. Thus, this relation has
an extremum at P = - b/2c. One can locate this maximum generally at
pressures a little above the highest pressure used, and the consequence of trying to force the data into this parabolic straight-
jacket is a flattening of the curve at zero pressure. Thus, the
absolute value of ΔV is almost always underestimated, and the problem
is worst in those cases in which they most extensive pressure range
is employed. One should not have to pay a price in terms of accuracy
for an extensive data collection! Fortunately, a new expression is
now available in the form of the Asano equation[4]:

$$\ln k_p/k_1 = aP + b \ln (1 + cP)$$

It is found to provide much superior fits, and is strongly recommended.
In most high pressure literature, the original raw data were recorded,
permitting recalculation when desired; but it may be added that the
error introduced by other evaluations is in general not large.

A rather more embarrassing difficulty is that many authors have
mistreated their data in another way: via the misuse of units. It
goes without saying that to calculate ΔV from data on K versus P, the
equilibrium constant must be expressed in terms of pressure independent units such as mole fractions, molality units, or molarity units
at a standard pressure (again: zero or ambient is preferred). In
many reports, one finds assertions that the data were corrected for
compressibility effects because molarity units were used, but in
cases where such corrections were uncalled for and, indeed, wrong.
This situation has been discussed repeatedly and will not be reiterated
here; suffice it to say that this literature should be consulted by all
newcomers to whom the usual needlessness of the compressibility
correction is not clear.[5]

The expression for the activation volume is

$$\Delta V^\star = - RT \frac{\partial \ln k}{\partial P}$$

The evaluation of ΔV* is entirely the same as that of Δ V, and subject
to the same caution and reservations.

One additional problem may be mentioned here because the question
is sometimes raised: is it possible that the ln K (k) vs. P plots hide
a narrow region at and near zero pressure where the curvature is much
greater than the data reveal? This would mean of course that the $\Delta V_o^\star$
(or ΔV_o) term is grossly underestimated. A similar problem often
plagues chemists studying conductance extrapolated to high dilution.
Fortunately, this question can be answered with a resounding No. The
reaction volume can be measured in two alternative ways. One of these
depends on the measurement of the partial volumes of all of the species
participating in the reaction and combining them in the way demanded
by the stoichiometry: $\Delta V = \Sigma V_{product} - \Sigma V_{reactant}$. The other
method is the very direct experimental approach known as dilatometry.
Neither of these alternatives depends on pressure at all. For many

instances, one or the other -and in some instances, both- have been
applied, and in all instances, the results agree well with those obtained
from high pressure data. There is consequently no demonstrated justification for this concern.

1.4 Role of Equilibria: Caliberation.

The fact that these three methods of measuring reaction volumes are
available means that we can readily obtain the overall volume changes
in any type of equilibrium, in the widest chemical sense of the word.
Included are not only equilibria as students usually understand the
phrase, with reactants and products coexisting and detectible, but
also "complete" reactions in which the reactant totally vanishes in
favor of product, and even hypothetical reactions that cannot be
induced to take place. Even a brief contemplation of these three
methods will quickly convince anyone of the correctness of this
statement.

It is a time-honored practice in chemistry to use equilibria
as measuring sticks with which to judge kinetic variables. Acid base
equilibria in particular can be quoted in support of every effect
that has ever been invented to explain kinetic variations. This is
as true for pressure effects and for the derived volume parameter as
it is for any of the other, perhaps more familiar effects and parameters.

1.5 Molecular Variations, and their Effects on the Volume

1.5.1. Effect of Bonding

It is axiomatic that bound atoms are closer together than non-bonded
ones, and unless one or more of the other features conspire against
it, the proposition can be offered that covalent bond formation
reduces the volume of the system, and hence, it is promoted by
pressure. The difference in volume between covalently bound and non-
bonded first row atoms may on the basis of known bond length and
van der Waals radii be estimated to be in the range of A^3/molecule,
or cm^3/mole. Values up to 10 or 20 cm^3/mole have been encountered.
The hope is, of course, that activation volumes will now tell us some-
thing about the length of breaking bonds in the transition state.
Eyring, on the basis of such the relatively few data then available
estimated that the bond extension of breaking bonds was about 10%, and
this became a rule of thumb.[6] Later Hamann, using much more extensive information but of the same sort, expressed the belief that such
extensions often amounted to about 50%.[7] Quotations of this new
value often brought objections from chemists who knew the old rule
but were not aware of the nature of its origin!

It is important to realize that no such rule can be expected to
have general validity. Transition states may be "early" or "late",
i.e., resemble substrates or products, and accordingly, extensions
of breaking bonds may be modest or substantial. A further complication is, as Newman has often emphasized,[8] is the fact that bond

cleavage cannot be detected if return occurs; in such cases, the
volume change reflects a totally broken bond plus whatever volume
requirement there may be for the fragments to diffuse apart. Thus,
while it is certainly heartening to see that kinetic pressures effects
lead to volume changes that appear to be in the right ballpark by
structural standards, care must be exercised if one wishes to base
geometric appraisals on them.

The volume changes are smaller if weaker bonds are considered.
Hydrogen bond formation, for example, leads to a volume decrease of
4 cm^3/mole, on the average, and charge transfer complexation is
similar in that respect. Thus, while these changes are smaller, they
are much larger than one might expect on the basis of their energies.

Changes in the nature of the bonds of course are also common,
particularly interchanges between single and multiple bonds. While
the π components of multiple bonds bring the nuclei closer together,
the electrons occupy a region of space outside of the internuclear
line, and these two features taken together lead to a net increase in
volume. Thus, to mention one example, the net substitution of double
bonds by single bonds in polymerization leads to a substantial
reduction in volume. A similar argument can be made for single bonds
that have a significant π component, i.e., bent bonds, as in
cyclopropanes.

The simple packing of non-bonded molecules can have dramatic
volume consequences. It is known, for example, that when aqueous
β-cyclodestrin is allowed to interact with t-butylbenzene, the
complexation leads to a net expansion of about 15 cm^3/mole. This
increase is due to the transfer of 6 or 7 water molecules out of the
cyclodextrin cavity to the bulk solvent, and that of the ill-fitting
t-butyl group into this crevice. By contrast, when the same experiment is done with the tight-fitting cylindrical ferrocene, a contraction of about 15 cm^3/mole occurs.[9]

Another related feature is that of crowding. The enforced close
proximity of two functional groups have the effect of reducing the
overall volume, usually by several cm^3/mole. One observes this in
the densities of Ø-disubstituted benzenes vs. their m- and p- isomers,
those of cis-olefins vs. their trans-isomers, trans- to gauche-
shifts in conformational equilibria, and so on.

1.5.2. Effect of Ionic Charge.

One often encounters statements to the effect ionization reactions are
exceptional in the sense that they are dissociations, A - B → A$^+$ + B$^-$
characterized by volume reductions. These reductions are described as
electrostriction, the contraction due to the solvation of charged
particles.

In fact, this phenomenon is not an exception at all. To view it
as such is to be misled by the symbolism. The equation should really
be written as (for example): (a + b) H$_2$O + A-B → A (H$_2$O)$_a^-$ + B$^+$(OH$_2$)$_b$.
While we have indeed lost one covalent bond, we have been compensated
by (a + b) electrostatic bonds. The volume reductions ascribed to

covalent bonds in the previous section in fact characterize all attractive interactions, whether covalent, dipole - dipole (e.g., H-"bonds"),
or ion - dipole. The latter are quite substantial, especially because
a and b may be large. Even as rewritten, the symbol is inadequate in
that it neglects the strengthening of hydrogen bonds between first and
second waters, and beyond. As a result, ionogenic processes are always
characterized by shrinkage, even if concomitant covalent bond cleavage
occurs.

    The Drude-Nernst equation describes this shrinkage, occurring upon
the immersion of a spherical charge q (radius r) out of the vacuum and
into a dielectric:

$$\Delta V = \frac{-q^2}{2rD} \frac{\partial \ln D}{\partial P}$$

The effects of radius, charge and solvent properties are all nicely
anticipated on this basis. Thus, ionization volumes of acids and
bases are reduced when ions of larger radius form or when oxyanions
with delocalized charge are produced, and strongly increased when poly-
valent ions are at stake.

2. MECHANISTIC APPLICATION

2.1. Known Applications

Several types of applications were known at the time of the first Corfu
Conference; new examples in these areas are presented in the next four
sections. New types of applications are mentioned in the sections to
follow.

2.1.1. Concertedness in Multiple Bond Reorganizations

The question of concertedness in reactions in which several bonding
changes occur continues to excite chemists, and at least one of the
reactions to which high pressure kineticists have applied their tool
remains shrouded in controversy.[10] The basis of the application is
of course simply the thought that if the transition state foreshadows
the reaction volume when only a single structural change occurs, it
should mirror all of them when several features operate. Thus, the
near identity of the large and negative activation and reaction volumes
in Diels-Alder cycloadditions should be taken as an indication that
both of the new bonds are well on the way toward completion when the
transition state is reached.

    While the evidence in the case of cycloadditions is little short
of overwhelming, much less is known in the case of other pericyclic
reactions, sigmatropic shifts among them, and hence one of the new
applications mentioned here is in that category. It concerns the
1,4-shift of the pyridine oxides shown:

This reaction involves an array of six π electrons, and hence may occur in concerted, suprafacial fashion. It apparently does so when R = H; at least, the resolved deuterio analog reacts with clean retention of configuration.[11] The alternative radical pair mechanism takes over when R is phenyl; in this case, the NMR spectrum of the product exhibits CIDNP during the reaction, and when R is p-chlorophenyl, total racemization characterizes the process. The change in mechanism is probably related to the greater stability of the benzhydryl radical as compared to benzyl; but whatever the reason be, the activation volumes do not fail to indicate it:[12]

$$R = H \qquad \Delta V^\star = -30 \text{ cm}^3/\text{mole}$$
$$R = \emptyset \qquad \Delta V^\star = +10 \text{ cm}^3/\text{mole}$$

Another controversy of this type has cropped up in the more traditional type of $S_N 2$ displacement reactions. Following the introduction of the curved arrow symbolism by Ingold, most organic chemists simply assumed that the implied concert of the reaction was real, but in 1965 Sneen[13] startled them by suggesting that prior ionization to a pair that is captured by nucleophile later was a possibility not ruled out by kinetics:

In subsequent papers, Sneen expressed the belief that many and perhaps all $S_N 2$ reactions follow this pathway;[14] but others were critical of this point of view and it was not widely accepted.[15] High pressure kinetics cannot undisputably settle this question, because both $S_N 2$ and solvolytic $S_N 1$ reactions are characterized by contraction - in the latter case, due to the solvation of the incipient charges. However, this picture changes if a neutral leaving group is used; heterolysis should then <u>not</u> be characterized by electrostriction, but primarily by the bond <u>lengthening</u>. Indeed, in the reaction:

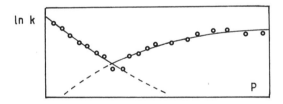

ΔV^{*} is + 19 cm^3/mole: clearly, the process is dominated by bond clevage and pair formation.[16] An unanticipated bonus came our way when we tried to induce the classical displacement mechanism by removing the charge stabilizing <u>p</u>-methoxy group: both mechanisms compete then, giving rise to the minimum shown in the Figure.

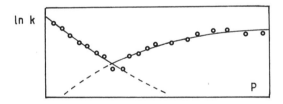

Figure. Minimum in a plot of ln k <u>vs</u>. P; see text

2.1.2. Freeness of Intermediates

When reactive intermediates are definitely implicated, it is often possible to gauge whether they are solvent bound or not from the pressure dependence on the rate constant. Thus, we were able to settle the long uncertain question concerning the nature of the intermediate in the base promoted solvolysis of haloforms: the fact that pressure retards the formation of the intermediate in chloroform hydrolysis can only be squared with the formation of free CCl_2, and not with the water-bound O-ylid $H_2O \overset{+}{\underset{}{-}} CCl_2$. Many applications of this sort have been made, with examples as diverse as carbenes, nitrenes, carbocations and benzyne. A recent addition to this list has been metaphosphate ion, PO_3^-, a species that has stoutly resisted proof ever since

its first hypothesis twenty years ago.[17] Severe doubt about its
reality developed in recent years when Knowles[18] used doubly oxygen
labelled precursors to prove that its products had the inverted configuration. The fact that the activation volume is negative is now
virtual proof that, at least in water, metaphosphate does not exist
as a free species.[19]

An interesting extension has now also come to light in the area
of extended unsaturated carbenes. The pressure insensitivity in the
reaction shown below had led us to suppose that a polarized carbene-
anion pair was the first intermediate past the proton abstraction
stage and support for this was found in terms of the well-known
criteria for internal return: lack of exchange of the leaving group,
and retained stereochemistry.

$$\underset{Me}{\overset{Me_{\text{'''}}}{\searrow}}\overset{\oplus}{C}-C\equiv C^{\ominus} \qquad \underset{Cl^{\ominus}}{} \qquad \longleftrightarrow \qquad \underset{Me}{\overset{Me_{\text{'''}}}{\searrow}}C=C=C \qquad \underset{Cl^{\ominus}}{}$$

An even more extreme case concerns the ethynylog shown next:

$$Me_2CCl-(C\equiv C)_2H \xrightarrow{\ B\ } Me_2\overset{\oplus}{C}-C\equiv C-C\equiv C^{\ominus}$$

For this instance, ΔV^* is large and <u>negative</u>, implying that the
bond cleavage is essentially an ionization unassisted by the ace-
tylenic unshared pair, with the formation of a zwitterion rather than a
polarized carbene.[20]

2.1.3. Sigma-assisted heterolysis and non-classical ions

The twenty five year old controversy over the nature of the norbornyl
ion has disappeared from the organic chemistry journals, with both
sides claiming victory. It may be recalled that the question – which
has broad ramifications in spite of its apparently narrow scope –
concerned the rate of formation and physical properties of the
2-norbornyl ion:

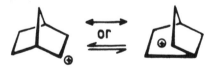

The root of the problem was the initially unappreciated but now
virtually proven fact that cation stabilizing substituents have at
most a modest effect on both the kinetics and thermodynamics of this
and related species.[21] We had at one time supported the reality

of the non–classical ion by showing that the concept correctly
predicted a smaller activation volume for the reaction producing
the delocalized charge. In subsequent experiments, we have shored
up this conclusion by determining the pressure dependence of the NMR
coalescence temperature of a known classical example:

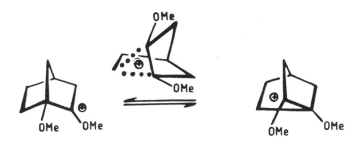

Each time that the classical forms equilibrate via the non–classical
form –in this case, as the transition state– the volume expands by
8 cm^3/mole.[22]

2.1.4. Transition state variation in geometry

The steady approach of $\Delta V^*/\Delta V$ to unity in a family of reactions as
more and more endothermic members are considered, was discussed
previously in terms of the Hammond postulate:

R	$\Delta V^*/\Delta V$
H	0.50
—	—
t–Bu	0.90

This interpreation has now been supported further by measurements
of the chlorine isotope effect, in the reaction of two of these
pyridines with methyl chloride:

R	k_{35}/k_{37}
H	1.00355
Me	1.00384

The isotope effect clearly also supports a later transition state
for the lutidine. The isotope and pressure effects allow comparable
estimates of the transition state shifts as a result of ortho
substitution.[23]

2.2. New Applications

2.2.1. Contraction during Neutralization?

The interpretation of high pressure kinetic results has been so
successful that it has focussed special attention on those cases in
which apparent discrepancies occur. One of these has been the
fading of triphenylmethane dys in aqueous alkaline medium, according
to:

$$CAr_3^+ + OH^- \longrightarrow CAr_3OH$$

The reaction is accelerated by pressure; yet, all of our experience
tells us that ΔV^* should be positive in such reactions as charge
neutralization leads to solvent relaxation and expansion.   Subsequently, we have found that the reaction volume is indeed positive;
the reason for the negative activation volume turned out to be the
fact that a significant fraction of the reaction involves not
hydroxide ion in the rate controlling step, but water![24]   In other
words, the reaction takes place in two steps, and the relaxation does
not materialize until the fast, second step:

$$CAr_3^+ + H_2O \xrightarrow{\text{slow}} CAr_3\,OH_2^+ \xrightarrow[OH^-]{\text{fast}} H_2O + CAr_3OH$$

At the time of the first Corfu Conference, we had –without the
benefit of any experiments– suggested that the slow step might have
involved but not bound hydroxide ion, in the formation of a
tight ion pair. In such a pair, the positive charge might be expected
to be more localized, and hence, to produce more intense solvation.
Since then, Isaacs[25] has found negative activation volumes also in
some solvents that cannot participate as neutral nucleophiles and
where the neutralizing anion must be involved as such; hence our
earlier ad hoc explanation may therefore still be correct, in those
cases.

2.2.2. Contraction during Bond Cleavage

This peculiar phenomenon had been observed by the late Plieninger[26] in
the aromatization of hexamethyl Dewar benzene.

We were able to confirm the direction of this effect, and offer an
explanation in terms of the increase in crowding that must occur as
the six methyl groups are forced into a coplanar array. In support
of this interpretation, we studied the pressure effect also in the
parent compound, and in that case indeed found the volume of activation to be positive.[27]

A few other cases of C–C bond cleavage with contraction are also
known and understood. Retro Diels–Alder reactions are accelerated
when secondary orbital interactions are involved: two non-bonding
centers then engage as part of the activation process.[28] In certain
[2+2] cycloreversions, the bond cleavage is essentially ionic, so
that zwitterionic intermediates cause the contraction.[29]

2.2.3. Ion Transport Through Membranes: With or Without Water?

This question is related to the controversy between the so-called
Chock and Eigen mechanisms of complexation of alkali metal cations
by crown ethers; the views differ on the point whether or not
desolvation of the ion is part of the activation process. We
have found one quite spectacular instance in which dehydration
controls the work term. This is found in the migration of Gd(III)
through vesicle walls of egg albumin. This migration, which can be
observed by the effect the ion has on the esr line width of TEMPONE
(a stable free radical), does not occur unless an ionophore such as
lacalocid is present. The activation volume equals about + 150 cm^3/mole,
which just about equals the known hydration volume in magnitude.[30] A
similar agreement between the two terms was found with Co(II).

3. RESERVATIONS

The instructor must guard against the uncritical presentation of
the virtues of his subject, lest he lead his audience astray. There
are limitations! While a very detailed discussion of the scope of
high pressure kinetics is not appropriate here, some of the limits
should be mentioned.

One of these is certainly the transition state theory itself. The theory's longevity is testimony to its immensely successfuly correlation of huge numbers of data in kinetics with those of thermodynamics. Critics often point to the illogical assumption of equilibrium between the initial and hypothetical transition states, and indeed, one should be on guard in considering reactions in which such a equilibrium seems a particularly far-fetched assumption. It goes without saying that extremely fast processes are especially prone to this weakness. Diffusion controlled reactions and the photophysical steps in photochemical reactions are good examples of this. Thus, the combination of long chain free radicals which is an important termination process in polymerization is known to be strongly pressure retarded, in spite of the fact bond formation should reduce the volume. This step is known to be diffusion controlled, and one could attempt to explain the phenomenon in terms of the need for holes in the liquid lattice into which diffusing particles may move; however, it is clear that the beauty of the simple correlation between bonding and pressure effects is then compromised. Troe[31] has had much success in correlating fast kinetic processes by means of Kramers theory, and it is hard to say where the limits lie; suffice it perhaps to conclude that one must be on guard in the border area.

Similarly, one cannot blindly extend the high pressure empiricisms into the areas of very large molecules, or large organized assemblies of molecules. If a molecule occupying a volume of one hundred cm^3/mole is found to expand by ten, one is justified in concluding that some bond must be breaking, but the same volume increase in a molecule of ten thousand dalton cannot be analyzed with the same confidence. How can one deal with the fact that ΔV for the association of E. coli ribosomal subunits is very large and positive ($\Delta V = + 240$ cm^3/mole)?[32] This is obviously not to say that such information is not useful, or should not be gathered; but one should be aware that the simple conclusions that can be drawn with small molecules and slow processes must not be thoughtlessly carried over into new chemistry in which these ground rules do not apply.

Footnotes and References

1. W. J. le Noble, Proc. NATO/ASI, "High Pressure Chemistry", H. Kelm, Ed., D. Reidel, 325 (1978).
2. W. J. le Noble, ibid., p. 345.
3. For a listing, see W. J. le Noble and H. Kelm, Angew. Chem., Int. Ed. Engl., 19, 841 (1980).
4. T. Asano and T. Okada, J. Phys. Chem., 88, 238 (1984).
5. S. D. Hamann and W. J. le Noble, J. Chem. Educ., 61, 658 (1984).
6. A. E. Stearn and H. Eyring, Chem. Rev., 29, 509 (1941).
7. S. D. Hamann, "High Pressure Physics and Chemistry", Vol. II, R. S. Bradley, Ed., Academic Press, 1963; p. 163.
8. R. C. Neuman, Accounts Chem. Res., 5, 381 (1972).
9. W. J. le Noble, S. Srivastava, R. Breslow, and G. Trainor, J. Am. Chem. Soc., 105, 2745 (1983).

10. M. J. S. Dewar and A. B. Pierini, J. Am. Chem. Soc., 106, 203 (1984). These authors, while attacking the concept of concertedness, considered the activation volumes among "the most convincing evidence" for it, together with Seltzer's isotope effects which they went on to reinterpret.
11. U. Schöllkopf and I. Hoppe, Justus Liebigs Ann. Chem., 765, 153 (1971).
12. W. J. le Noble and M. R. Daka, J. Am. Chem. Soc., 100, 5961 (1978).
13. H. Weiner and R. A. Sneen, J. Am. Chem. Soc., 87, 287 (1965).
14. R. A. Sneen, Accounts Chem. Res., 6, 46 (1973).
15. D. J. McLennan, Accounts Chem. Res., 9, 281 (1976).
16. A. R. Katritzky, K. Sakizadeh, B. Gabrielsen, and W. J. le Noble, J. Am. Chem. Soc., 106, 1879 (1984).
17. W. W. Butcher and F. H. Westheimer, J. Am. Chem. Soc., 77, 2420 (1955).
18. S. L. Buchwald, J. M. Friedman and J. R. Knowles, J. Am. Chem. Soc., 106, 9411 (1984).
19. F. Ramirez, J. Marecek, J. Minore, S. Srivastava, and W. J. le Noble, J. Am. Chem. Soc., 108, 348 (1986).
20. S. Basak, S. Srivastava, and W. J. le Noble, unpublished observation.
21. C. K. Cheung, L. T. Tseng, M.-H. Lin, S. Srivastava, and W. J. le Noble, J. Am. Chem. Soc., 108, 1598 (1986).
22. W. J. le Noble, S. Bitterman, P. Staub, F. K. Meyer, and A. E. Merbach, J. Org. Chem., 44, 3263 (1979).
23. W. J. le Noble and A. R. Miller, J. Org. Chem., 44, 889 (1979).
24. W. J. le Noble, E. Gebicka, and S. Srivastava, J. Am. Chem. Soc., 104, 3154, 6167 (1982).
25. O. H. Abed and N. S. Isaacs, J. Chem. Soc., Perkin Trans. II, 839 (1983). Unfortunately, this valuable paper is marred by a serious misquotation, according to which water could not have offered serious competition to hydroxide ion "in the 0.2 M sodium hydroxide ....used". Our experiments used 0.0004 M base, as did the original experiments by Laidler and Chen; under those conditions, the water contribution is indeed dominant.
26. R. Mündnich and H. Plieninger, Tetrahedron, 32, 2335 (1976).
27. W. J. le Noble, K. R. Brower, C. Brower, and S. Chang, J. Am. Chem. Soc., 104, 3150 (1982).
28. G. Jenner, H. Papadopoulos and J. Rimmelin, J. Org. Chem., 48, 748 (1983); A. V. George and N. S. Isaacs, J. Chem. Soc., Perkin Trans. II, 1845 (1985).
29. W. J. le Noble and R. Mukhtar, J. Am. Chem. Soc., 97, 5938 (1975).
30. J. A. Balschi, V. P. Cirillo, W. J. le Noble, M. M. Pike, E. C. Schreiber, S. R. Simon, and C. S. Springer, Rare Earths in Mod. Sci. Techn., 3, 15 (1982).
31. J. Schroeder and J. Troe, Chem. Phys. Letters, 116, 453 (1985).
32. C. Pande and A. Wishnia, J. Biol. Chem., 261, 6272 (1986).

ORGANIC SYNTHESIS AT HIGH PRESSURE

W. J. le Noble
Chemistry Department
State University of New York
Stony Brook, NY 11794 USA

ABSTRACT. The recommendation is developed that all synthetic chemists
should always consider pressure as one of the routine variables
available to them, along with temperature, solvent, reaction time,
catalysts, and so forth. Ideally, equipment should be available both
in the 1-5 and 10-20 kbar range, the former for the habitual testing
of all reaction steps under increased pressure, and the latter for the
more unusual instances in which gigapascal pressures appear to force
otherwise reluctant reactions. The hope is that synthetic chemists
will learn to optimize yields and conversions with respect to pressure
as well as the other variables, and to report the pressure used just
as routinely as they report reaction times and temperatures now.

INTRODUCTION. The goals of the mechanistic and synthetic chemist are
different in the sense that the former seeks to discover basic features
about chemical transformations and the latter hopes to exploit them.
In the study of pressure effects and how these might be related to the
pathway, it is obviously very necessary that the effects observed be
attributable to the pressure alone; in contrast, when one has successfully prepared a difficult substance, it matters much less to precisely
what factor the success should be attributed. It is simply human, to
wish to spend one's time on taking the next step, rather than an
analyzing the previous one. Still, we must do it, if we are to profit
from discoveries, and to learn from mistakes.
  This paper follows the one on kinetics,[1] and a good understanding
of that chapter would be helpful in following this one. However, not
everyone has the time or interest to read that much preparatory
material, hence the present chapter begins with some general qualitative remarks about kinetics that are hoped to serve as an adequate
background.

R. van Eldik and J. Jonas (eds.), High Pressure Chemistry and Biochemistry, 295–310.
© 1987 by D. Reidel Publishing Company.

1. ON RATES AND RATE CONSTANTS

1.1 The Rate Equation

The tremendous progress is synthetic organic chemistry that occurred
in the past two decades or so can be traced in part to a change in
attitude by its practitioners toward physical organic chemistry.
Disdain gave way to appreciation, and it is fair to say that the most
successful synthetic chemists nowadays are completely at home with
kinetics and spectroscopy, areas traditionally considered irrelevant
to them. In this spirit, let us approach the question of possible
applications of high pressure in synthesis from the viewpoint of
Physical Organic Chemistry.

It is an article of faith that any chemical process can be written
in terms of one or more elementary steps, and that the progress of
each step with time can be described by differential $d[C]/dt$ that
is proportional to all of the concentrations of the components on the
left side of the arrow. Manipulation of these differentials may lead
to a form that can be integrated, and as a result we obtain the experimentally testable rate law, which tells us (a) how the reaction
progress with time will vary with the concentrations and (b) the
magnitude of the proportionality or rate constants.

Establishing the rate law may require quite a bit of effort,
and the synthetic chemist does in practice not have the time or need
for it. Fortunately, the background mechanistic information now
covers so many reactions and is so widely disseminated that one can
at least make reasonable guesses in most cases. It is also very
generally known that the rate constant can be represented by

$$k = \frac{RT}{Nh} e^{-\Delta G^*/RT}$$

which can be rewritten as

$$\ln k = \ln \frac{RT}{Nh} - \frac{\Delta E^*}{RT} + \frac{\Delta S^*}{R} - \frac{P\Delta V^*}{RT}$$
$$\quad\quad\quad (A) \quad\quad (B) \quad\quad (C) \quad\quad (D)$$

The magnitudes of the terms (A) - (D) vary greatly. Term (A) equals
$\ln 2 \times 10^{10} T$; it allows essentially no manipulation. Thus, one should
have to double the absolute temperature to double the rate effect due
to term (A): this is obviously not worth while. The temperature effect
that one so often uses in the laboratory is in the (B) term dependence,
which is more conveniently written as

$$\ln \frac{k_2}{k_1} = \frac{\Delta E^*}{R} \left(\frac{T_2 - T_1}{T_1 T_2} \right)$$

The activation energy often lies between 20 and 40 kcal/mole, and the most readily available temperature range is from 300-500°K; some simple arithmetic will serve to verify the well-known rule of thumb that the rate will increase by a factor of 2-3 for every 10° rise in temperature. The (C) term often controls the chemistry since the entropy of activation may have either sign: values between +30 and -30 gibbs are common. Thus, the entropy term contributes a factor between 10^{-6} and 10^{6} to k. In ionic reactions, we may sometimes hope that a solvent change will lead to a less unfavorable entropy term, but temperature variation will not help: to a first approximation, ΔS^* is independent of temperature, and hence we cannot do very much about it.

Finally, the work term (D) is negligible compared to the others. The product of 1 cm^3 and 1 atm equals about 0.024 cal. Even when the volume of activation is in the normal range of +30 and -30 cm^3/mole, we need to use pressures in the range of kbars to arrive at kcal units. Of course, this is what high pressure in synthesis is all about, but before we review what is known and attempt to forecast what is still to come, some cautionary notes are in order. It is possible, namely, that a number of successes already claimed in the high pressure and synthetic literature are simply illusory.

1.2   The Need to Trace Success

A typical situation one encounters in the synthetic lab is as follows. The reaction:

$$A + B \longrightarrow C$$

in THF as solvent is attempted without success. The chemist raises the temperature to the boiling point of the solvent (80°C), and his patience to an overnight run; if the next day, analysis shows no trace of C, a common dilemma begins. Should we use a higher boiling solvent? This ploy is seldom used; it would sacrifice the convenience of isolation by stripping off the solvent via flash evaporation. Furthermore, one of the reaction components may well be volatile enough to undo the possible advantage of higher temperature. An alternative would be to use a sealed tube, and this is indeed done now and then. However, this also has its disadvantages: sealed glass tubes may rupture, metal vessels may produce unintended chemistry, and stirring of heterogeneous mixtures may be difficult. In addition, the solution may well reach the critical region, with the result that the expected rate increase is reduced, or even cancelled altogether.[2]

In recent years, many synthetic chemists have become aware of the possibility of promoting reactions by means of high pressure. This need not be a hit-or-miss proposition; as will be noted further below, literature listings of pressure effects on reaction rates and equilibria are now so extensive that the reaction of interest or a close analog can very probably be found in them. The data, furthermore, usually cover a pressure range such that reasonable projections

can be made of the accelerations to be expected. However, in all this,
it is assumed that the problem is with the rate constant, and not
with the concentrations.

What often happens is that the chemist, after having taken the
trouble to buy or borrow high pressure equipment realizes that he can
also avail himself of higher temperatures as well, and it may be
thus how he achieves success. The apparatus then serves principally
to hold the components together in the hot liquid medium, and much
more marginally or not at all to increase the rate constant by virtue
of a substantial negative work term. One could then clearly have done
just as well with one of the small, modest-pressure reactors now
commonly available. They can be obtained in a wide range of sizes
and can be used at a wide range of temperatures; they may be glass-
lined and equipped with devices for stirring or shaking, gas introduction and removal, and devices for measuring pressure and temperature.
Anyone who thinks in terms of 20 kbar solutions to synthetic problems
should begin by trying these low pressure reactors which are obviously
far cheaper and easier to acquire and use. This advice is extended
for various synthetic difficulties other than slow rates as well; these
include reactions that take place but produce unreasonable amounts of
unwanted side products, or that take place but to a small extent
because of unfavorable thermodynamics.

1.3 Criteria for Pressure Effects and for Apparatus

As noted, activation volumes for innumerable reactions are known and
listed in reviews. Suppose that the reluctant reaction discussed in
the preceding section or a close analog is known to have an activation
volume of -20 cm^3/mole; that is, the transition state is 20 cm^3/mole
smaller than the initial state. The chemist will then benefit from
doing the reaction at high pressure. From the expression:

$$\Delta V^* = -RT\ \partial \ln k/\partial P$$

one can quickly derive the rule of thumb that a volume change of 20
cm^3/mole in the transition state will halve or double the rate
constant per kbar depending on the sign, and hence the first hope
raised is for a rate increase of $2^{10} \approx 1000$ at 10 kbar, or even 10^6
at 20 kbar. However, it should at once be realized that "the"
activation volume is quite sensitive to factors such as temperature,
solvent and especially: pressure! What one finds reported is ΔV_o^*,
which is the activation volume at zero pressure. The absolute
magnitude of ΔV^* is always reduced at higher pressures. This levelling
of ln k vs. P curves is often already quite noticeable in the neighborhood of 1 kbar; it is generally quite pronounced at 5 kbar, where ΔV^*
may be half of ΔV_o^* or less, and at 10 or 20 kbar is but a small
fraction of ΔV_o^* - perhaps one mL or two. This of course drastically
reduces the first euphoric estimate, but there are two factors in
consolation. The first is that more accurate forecasts can usually
be made by inspecting the original data (which are not in the reviews)

and appraising them for this curvature; see Fig. The second factor
is that the effect is most
pronounced in the low pressure
region. If pressure does not
provide significant benefit
from, say, 1-5 kbar, it is
pointless to go beyond; the
advantage is by far the most
pronounced <u>at</u> <u>low</u> <u>pressure</u>.
This point is emphasized
because essentially all of the
recent contributions from high
pressure synthetic laboratories
mention only the results
obtained at the limits of their
apparatus, often 15 or 20 kbar.
In some instances, ambient
pressure experiments are not
even mentioned. For all the
foregoing reasons, it must be
suspected that virtually
equally good results might in
some cases have been achieved
at the much more easily and
economically reached pressure
of 5 kbar, or even at a few
hundred atmosphere!

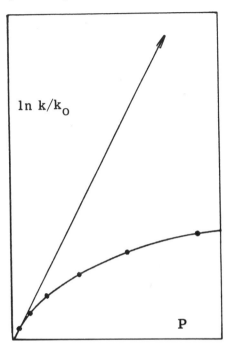

Fig. Fancied <u>vs</u>. actual effects of
pressure on the rate.

When may volume reduc-
tions in the transition state
be expected? Once again,
reference is made to the
reviews,[3] but a brief summary
now follows. Bond formation almost always leads to volume reduction,
typically of 10-20 cm^3/mole. This includes even interactions as
weak as the hydrogen bond (about 5 cm^3/mole). Bond cleavage has
the opposite effect; displacements with no net change in the number
of bonds have transition states of modestly reduced volume (typically
5-10 cm^3/mole). Crowding reduces the volume required: thus,
<u>o</u>-disubstituted benzenes occupy 3-5 cm^3/mole less space than the
<u>m</u>- or <u>p</u>- isomers. Among bond types, π bonds occupy 10-20 cm^3/mole of
space, and small rings (especially cyclopropyl) occupy some extra
space because of the bent bonds. The largest effects are often seen
in charge solvation: ionization of neutral molecules always leads to
drastic decreases in volume, even though a bond is broken. Alterna-
tively, of course, it would perhaps be more accurate to write that
electrostatic interactions are characterized by reduced volumes just
as the covalent ones are. However one prefers to think of it, this
effect usually dominates. Thus, even in ionogenic Grob fragmentations
in which 2 bonds break, concomitant solvation causes a net reduction
of 10 cm^3/mole or more:

Of course, the reverse is true during charge neutralization. It should also be pointed out that these effects are especially large in relatively non-polar solvents, where the range of coulombic forces is larger, and compressibilities greater. Of course, ionic processes cannot be conducted at will in non-polar media; and even when ions are involved, they are virtually always paired with the counter ion.

A large question mark hangs over all of these factors and estimates: the "timing" of the transition state. Only if the transition state is late do these factors apply to maximum effect; if the transition state is early, there will be little change in volume and pressure effects will be minimal. Unfortunately, this information is often not available, and in fact, when it is, the reaction's volume profile is often the basis for it.

2. SYNTHETIC APPLICATIONS

2.1 Pressure Induced Reactions

One of the most dramatic instances of a reaction occurring solely because of high pressure was reported by Okamoto,[4] who found that the Menschutkin reaction shown below occurred in high yield in a few hours at 5 kbar, while no such product was found at ambient pressure (R = t-Bu):

A careful reinvestigation in our labs showed[5] that the product was not the N-methylpyridium salt but the protonated analog; nevertheless, we also learned that the reaction was indeed one with an extremely steep pressure dependence such that it was not observable at pressures below 2 kbar. Several circumstances conspire to lead to the unusually large ΔV^*, estimated to be -50 cm^3/mole; chief among these is a late transition state, with a high degree of bond and charge formation, in a low polarity solvent (acetone).

A major example in this section must be the area of cycloadditions in general, and that of Diels-Alder reactions in particular. The pressure effects in [2+4] cycloadditions has a very long history indeed, the first example being reported in the thirties.[6] Although controversy[7] swirled about the mechanistic significance of the high pressure sensitivity of this reaction, the work of Eckert,[8] Stewart[9] and Jenner[10] has firmly established that the reasons for the large, negative activation volumes is the concurrent formation of two new sigma bonds from the π electrons. It is surprising,[11] indeed, almost disturbing that the volume of the transition state should so nearly equal that of the final state, since the conventional wisdom is that the usually highly exothermic Diels-Alder must surely have an early transition state. In any event, this sensitivity to pressure is seen in all types of cycloadditions, such as the [4+6] type,[12] the [2+3] dipolar type,[13] and the [2+2] zwitterionic type.[14] In the latter case, the acceleration is still large though only one bond is initially formed, but since charges also develop, solvation helps reduce the volume.[15] But the most important area remains the Diels-Alder reaction. Its usefulness in synthesis in general gave high pressure chemistry a dramatic lift, courtesy of Dauben, who applied it in the preparation of a number of well-known natural products.[16]

Some of these applications rested on the large rate increases achievable, and others on improved thermodynamics. Thus, the loss of aromaticity in the cycloaddition of furans to olefins is such that the reaction goes in the opposite reaction at ambient pressure; but by applying high pressure, and then cooling before pressure release, unstable Diels-Alder products can be isolated.[17] Still another angle is the possible improved competition with side reactions. Thus, the cycloaddition of a second molecule of cyclopentadiene to the initial methylquinone adduct is not possible because at the temperature apparently required, this adduct aromatizes:

But pressure accelerates the Diels-Alder reaction much more than the
rearrangement, and we found that the former becomes the principal
reaction at 7 kbar.[18]

Bonding, charge formation and crowding are all features that are
encountered in high pressure applications again and again. Another
good example is that of the oximation of di-t-butylketone, a
reaction that is not observable at ambient pressure presumably because
of steric hindrance. It readily takes place at 9 kbar, giving an
excellent yield.[19] It is one of many examples of such reactions

$$\text{(CH}_3)_3\text{C}_2\text{C=O} + \text{H}_2\text{NOH} \longrightarrow (\text{CH}_3)_3\text{C}_2\text{C=NOH} + \text{H}_2\text{O}$$

with a highly contracted transition state; the C····N bond development
and concurrent charge formation in a crowded molecule all contribute to
this. Further instances include the Robinson annelation reaction:[20]

When the terminal methylene group is highly substituted, the reaction
fails but it can still be carried out then under high pressure conditions. Almost exactly the same remarks apply to certain Michael
additions, that do not occur under conditions of a weak nucleophile
such as nitromethane and the crowded unsaturated ketones encountered
in the steroid field:[21]

Also in the category is the synthetically very useful Mannich reaction, normally limited to formaldehyde in unhindered settings. Under high pressure conditions (10 kbar), methylene chloride can be used[22] together with highly substituted substrates.

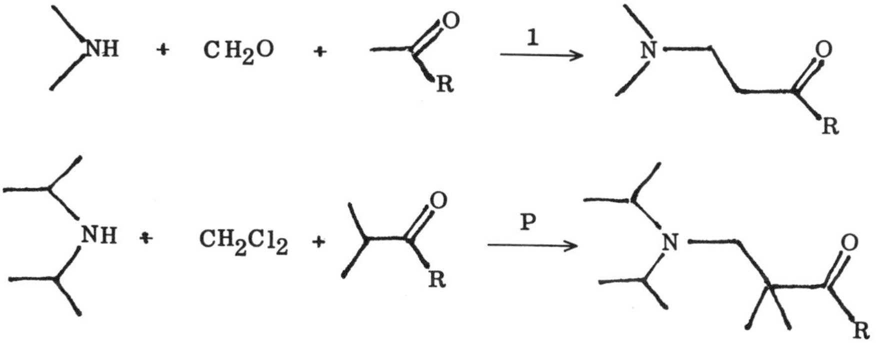

Somewhat different chemistry, but the same principles can be seen at work in hydroboration.[23] Thus:

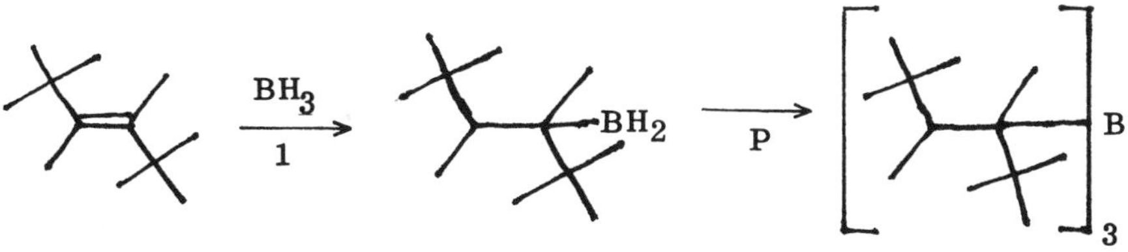

2.2 Pressure Promoted Reactions

The pressure effects are obviously not always so drastic that high yields and conversions are found in reactions unobservable at ambient pressure. What is much more common is a modest change which may or may not be useful.

Thus, the very first reaction we carried out under high pressure at Stony Brook[24] was the allylation of phenoxide ion in aqueous solutions:

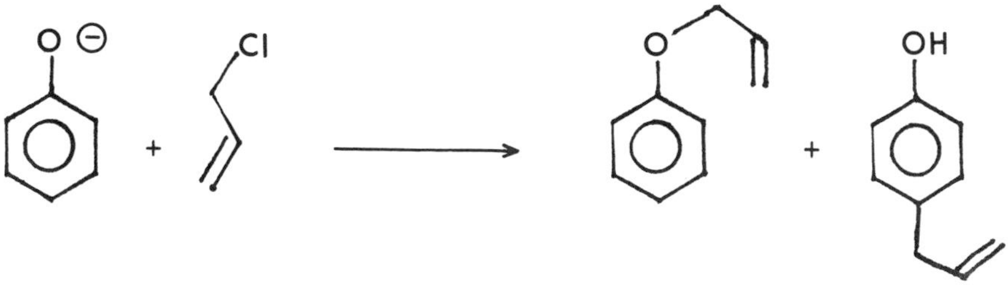

Under ordinary conditions, the ether product is about two thirds of the
product mixture. At 7 kbar, it represents only about one third. If
C-C bond formation had been the goal (in Organic Chemistry, that is
usually the case!), a significant improvement would have been observed.
The reason for the change is that the charge distribution in this
ambident ion heavily favors the more electronegative oxygen atom,
and hence that is where the solvation is concentrated. To allow
displacement, the solvating molecules must vacate their position:
an expensive action in terms of volume.

The important point here is that a small but useful and perhaps not
easily foreseeable effect turned up in a type of reaction that is
routinely used in synthesis. Once again, the potential user will
seriously sell himself short if he begins by assuming that pressure
effects are limited to the spectacular all-or-nothing cases published
already. Numerous further examples could be mentioned in which
competing reactions are shifted in one direction or another by
pressure, sometimes for reasons not easily identified, and yet, only
the surface has been scratched. A few further examples may suffice
to whet the appetite.

Several organic chemists have applied high pressure in the hope
of favorable changes in stereoselection. Thus, Plieninger[25] has
reported that the oxirane shown (An is p-anisyl) isomerizes in
chiral solvents such as (-)-menthol via the isolable achiral alcohol
to a ketone that shows no enantiomeric excess at 1 atm, but asymmetric
induction does occur to the extent of 15% at 10 kbar.

Similarly, pressure has been found to improve the ability of a chiral
center already present in a diene to induce configurational preferences
at the newly forming saturated sites in Diels-Alder reactions:

While rationalizations for such observations are sometimes possible, they have not as yet proved to be generally useful. In other words, success in these experiments is certainly not predictable; but this should not inhibit anyone from trying.

One should be aware, in such explorations, that side reactions are not innocent bystanders: they may introduce new problems even as one solves an old one. Thus, Eckert[27] found that high pressure was helpful in bringing about the conversion of the tosylate shown into the corresponding amine; however, pressure also promoted H-D exchange and racemization.

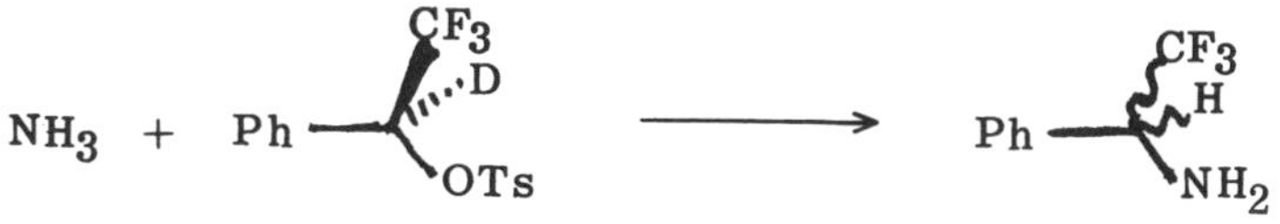

$$NH_3 \quad + \quad Ph \underset{\text{OTs}}{\overset{\text{CF}_3}{\underset{\text{...D}}{\bigg|}}} \quad \longrightarrow \quad Ph \underset{\text{NH}_2}{\overset{\text{CF}_3}{\underset{\text{H}}{\bigg|}}}$$

Still another instance in which selectivities play an important role is in the copolymerization of mixtures of olefins. Jenner[28] has reported that in the copolymerization of maleic anhydride with olefins, pressure tends to increase the incorporation of olefin especially when this component is crowded with substituents.

Yamada et al. have found pressure to be helpful in coupling reactions involved in peptide synthesis.[29] In the reaction shown, the yield increased from 9 to 70% by the application of 10 kbar:

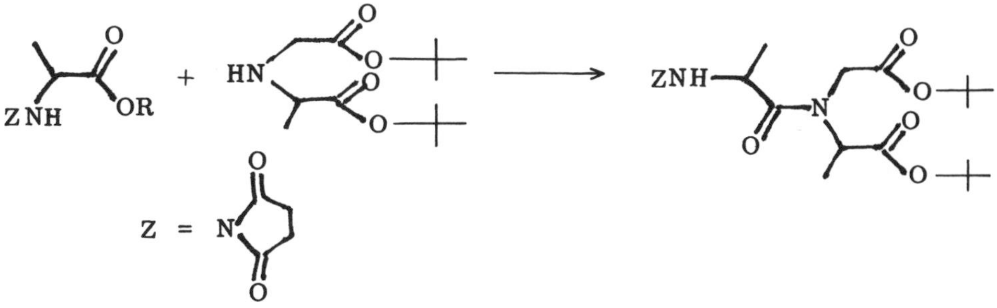

Jurczak[30] used pressure to promote the synthesis of the three-dimensional cryptand shown from a pseudo-planar crown ether.

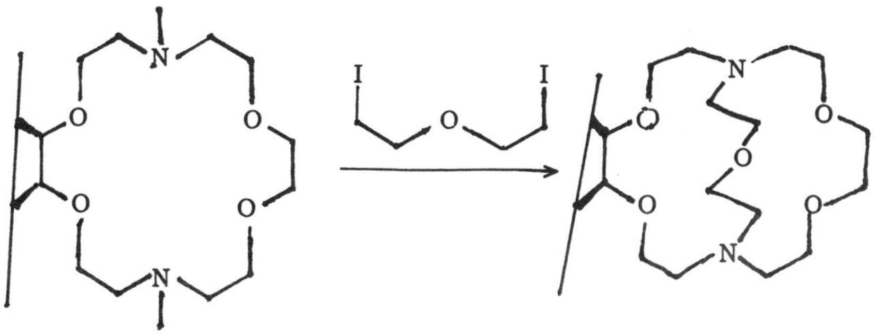

Dauben[31] found that the introduction of OH- protecting silyl groups
into sensitive alcohols can often nicely be achieved under pressure.
In the example shown, the yield improved from 5 to 97%.

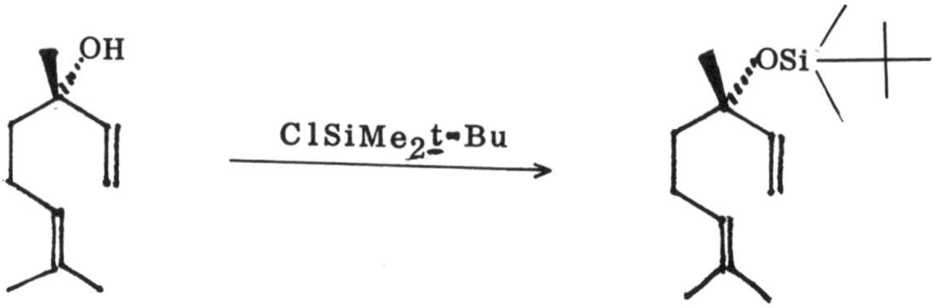

Okamoto has reported[32] that he could raise the yield of the fulvalene
shown below from less than 10 to over 90% by applying pressure.

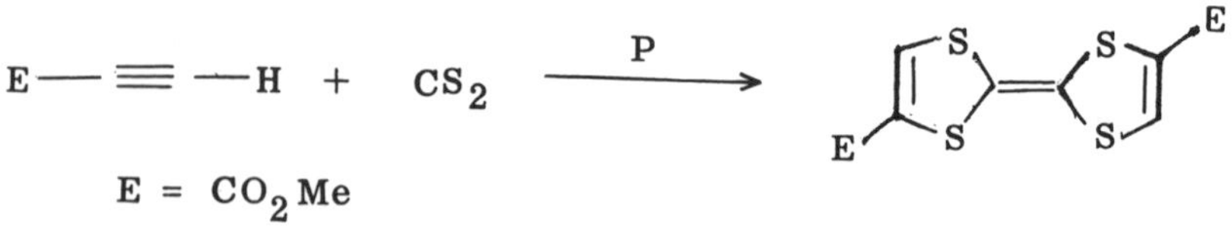

$$E = CO_2Me$$

2.3 Pressure Inhibited Reactions

The assumption abroad appears to be that pressure increases reaction
rates. However, it must be clear by now that retardation can occur, if
expansion characterizes the transition state. This usually happens
if bond cleavage occurs without compensation by solvation. This
phenomenon is also capable of exploitation, although there are as
yet virtually no examples in the literature at all. The use could
be an expended thermal stability range of the molecule in question.
The potential is nicely demonstrated by the following results.
     Triquinacene has long been held out as a stepping stone to
dodecahedrane:

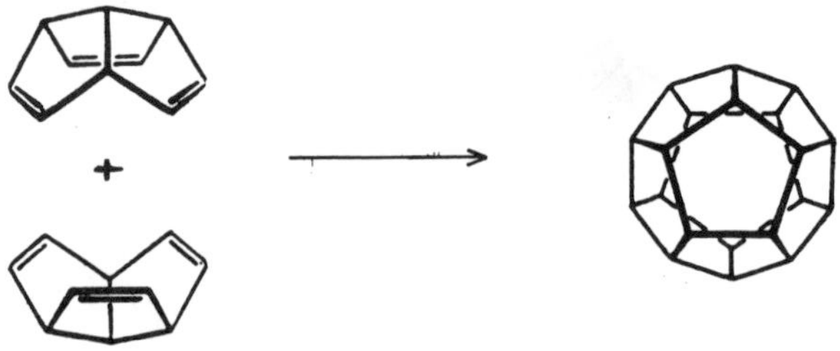

The reaction does not occur; the various synthetic schemes
devised to force it include the clever one by Doering[33] et al.

to install removable dimethylene sulfur bridges to hold the units permanently face to face.

Heating this substance to 150° brought a distressing picture into view: elemental sulfur began to appear. The large increase in volume that must certainly characterize this reaction led to the supposition that pressure should inhibit it. Eventually it turned out that at 50 kbar, we could heat the sample to over 500°C overnight, without significant decomposition![34] Although the hoped-for closure did not occur even at that temperature, the experience dramatically revealed the potential of this approach to extend a thermal stability range. With so many types of highly labile molecules accessible for study now, this potential is ripe for exploitation. One thinks of cyclophanes, dibenzenes, aromatic valence isomers, crowded olefins and innumerable other classes of strained compounds as candidates for high pressure stabilization.

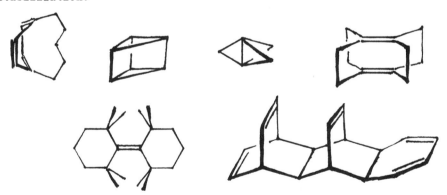

3. CONCLUDING REMARKS

While a number of spectacular cases are known in which pressures of 15-20 kbar or more permit reactions to occur that are unobservable at ambient pressure, and while more of such examples will surely be forthcoming, these are not the principal awards that await the synthetic chemist willing to experiment with high pressure apparatus. To begin with, one should make sure that the problem is not one of

holding the components together in the liquid at the desired temperature: in that case even the simplest two-or-three-hundred-atmosphere
pressure vessels may solve it. But even when this is not the problem,
before one thinks of GPa (10 kbar) range apparatus, one should realize
that equipment to achieve pressures in that range is not inexpensive,
and it is not trivial to operate or maintain - especially if it is not
routinely used. It will for the vast majority of synthetic chemists
prove much more worthwhile - at least, in the beginning - to obtain
equipment in the 1-5 kbar range, which is inexpensive, and simple
to operate and maintain, and to routinely study all of their reactions
at some pressure in addition to atmospheric. That will help identify
those cases in which larger scale, higher pressure experiments might
be worthwhile. It should be noted in this connection that the
steepest accelerations will always occur at the lower pressures.
Although there is of course no substitute for hands-on experience, the
many reviews and data listings available should almost always give
the potential user a fair idea of when to expect benefits, and
roughly how much.

   To summarize: the factors that increase rates under pressure
include bond formation, charge development and crowding in the transition state, and those that improve the thermodynamics are the same
factors in the product. Conversely, bond cleavage, neutralization
and release of crowding operate in the opposite directions. The
chances for favorable shifts in the ratios of competing reactions can
obviously be assessed on the same basis.

FOOTNOTES AND REFERENCES

(1) W. J. le Noble, elsewhere in this volume.
(2) The author long ago ran into a startling example of the
     phenomenon of slow gas phase reactions. The decarboxylation

$$\underset{Me}{\overset{PhCH_2O}{\diagdown}}C=CHCO_2H \quad \xrightarrow{\Delta} \quad CO_2 \; + \; Ph\diagup\!\!\diagdown\!\!O\!\!=$$

     to give the liquid ether had been observed to occur readily upon
     melting the acid at 110°, and very rapidly upon further warming of
     the melt. On one occasion, a sample of the acid which had been
     sealed in an evacuated vessel was accidentally placed in an oven
     wrongly set at 210° rather than 110°, and left so overnight. The
     next morning, the error was discovered and the vessel removed from
     the oven. Within seconds, the acid began to crystallize on the
     cooling walls! The rapid heating on the previous day led the acid
     to vaporize before it could react.
(3) Matsumoto has written several extensive and thorough reviews
     specifically on the topic of synthesis at high pressure; thus, see
     K. Matsumoto, T. Uchida, and R. M. Acheson, Heterocycles 16,
     1367 (1981); K. Matsumoto, A. Sera, and T. Uchida, Synthesis, 1

(1985); K. Matsumoto and A. Sera, ibid., 999 (1985). References to the reviews on the mechanistic aspects may be found here and in Ref. 1.

(4) Y. Okamoto and Y. Shimagawa, Tetrahedron Lett., 317 (1966).

(5) W. J. le Noble and Y. Ogo, Tetrahedron., 26, 4119 (1970).

(6) E. W. Fawcett and R. O. Gibson, J. Chem. Soc., 387 (1934).

(7) M. G. Gonikberg and L. F. Vereschagin, Zh. Fiz. Khim., 23, 1447 (1949); C. Walling and J. Peisach, J. Am. Chem. Soc., 80, 5819 (1958); S. W. Benson and J. A. Berson, J. Am. Chem. Soc., 84, 152 (1962).

(8) R. A. Grieger and C. A. Eckert, J. Am. Chem. Soc., 92, 2918, 7149 (1970).

(9) C. A. Stewart, J. Am. Chem. Soc., 93, 4815 (1971) and 94, 635 (1972).

(10) C. Brun and G. Jenner, Tetrahedron, 28, 3113 (1972).

(11) W. J. le Noble, Rev. Phys. Chem. Japan, 50, 207 (1980).

(12) H. Takeshita, S. Sugiyama, and T. Hatsui, Bull. Chem. Soc. Jpn., 58, 2490 (1985).

(13) Y. E. Raifel'd, B. S. El'yanow, and S. M. Makin, Izves. Akad. Nauk SSSR, 1090 (1976).

(14) J. van Jouanne, H. Kelm, and R. Huisgen, J. Am. Chem. Soc. 101, 151 (1979).

(15) W. J. le Noble and R. Mukhtar, J. Am Chem. Soc., 97, 5938 (1975).

(16) W. G. Dauben, C. R. Kessel, and K. H. Takemura, J. Am. Chem. Soc., 102, 6893 (1980).

(17) J. Jurczak, T. Kozluk, S. Filipek, and C. H. Eugster, Helv. Chim. Acta, 661, 222 (1983).

(18) S. Srivastava, A. P. Marchand, and W. J. le Noble, unpublished observation.

(19) W. H. Jones, E. W. Tristram, and W. F. Benning, J. Am. Chem. Soc., 81, 2151 (1959).

(20) W. G. Dauben and R. A. Bunce, J. Org. Chem., 48, 4642 (1983).

(21) K. Matsumoto, Angew. Chem., Int. Ed. Engl., 20, 770 (1981).

(22) K. Matsumoto, Angew. Chem., Int. Ed. Engl., 21, 922 (1982).

(23) J. E. Rice and Y. Okamoto, J. Org. Chem., 47, 4189 (1982).

(24) W. J. le Noble, J. Am. Chem. Soc., 85, 1470 (1963).

(25) H. P. Kraemer and H. Plieninger, Tetrahedron, 34, 891 (1986).

(26) W. G. Dauben and R. A. Bunce, Tetrahedron Lett., 23, 4875 (1982). See also M. M. Midland and J. A. McLoughlin, J. Org. Chem., 49, 1316 (1984).

(27) W. H. Pirkle, J. R. Hauske, C. A. Eckert, and B. A. Scott, J. Org. Chem., 42, 3101 (1977).

(28) G. Jenner and M. Kellou, Chem. Commun., 851 (1979).

(29) T. Yamada, Y. Manabe, T. Miyazawa, S. Kuwata, and A. Sera, Chem. Commun., 1500 (1984).

(30) M. Pietraszkiewicz, P. Salanski, and J. Jurczak, Chem. Commun., 1184 (1983).

(31) W. G. Dauben, R. A. Bunce, J. M. Gerdes, K. E. Henegar, A. E. Cunningham, and T. B. Ottoboni, Tetrahedron Lett., 24, 5709 (1983).

(32) J. E. Rice and Y. Okamoto, J. Org. Chem., 46, 446 (1981).
(33) W. P. Roberts and G. Shoham, Tetrahedron Lett., 22, 4895 (1981).
(34) W. P. Roberts, S. Srivastava and W. J. le Noble, Industr. Chem. News, 5, 24 (1984).

KINETICS OF SOLVENT EXCHANGE REACTIONS AT HIGH PRESSURE

André E. Merbach
Institute of Inorganic and Analytical Chemistry
3, Place du Château
CH-1005 Lausanne
Switzerland

ABSTRACT. High pressure kinetic studies are particularly useful for the determination of the activation mode of simple inorganic substitution processes. Applications of high pressure multinuclear magnetic resonance to the elucidation of the mechanisms of solvent exchange reactions on metal ions in aqueous and non-aqueous media are presented. The effect of steric and electronic changes of the solvents and the metal ions on the mechanisms are discussed. The most striking results were obtained for divalent and trivalent high spin first row hexasolvated transition metal ions : for both series a gradual change-over in substitution mechanism occurs, with the early members showing associative activation mode and the later ones dissociative behaviour, the change in activation mode occurring after the d^5 configuration. Some methodological and technical aspects of high pressure NMR applied to solvent exchange studies are treated.

1. INTRODUCTION

The elucidation of the mechanisms of substitution reactions (Eqn. 1) involves the study of the dependence of the reaction rate on reactants

$$L_nMX + Y \longrightarrow L_nMY + X \qquad (1)$$

concentration, pH, ionic strength and solvent composition. The deduced empirical rate law, the electronic and steric effects induced by variations of leaving X, entering Y, and non-reacting L ligands on the rate constant, all the available experimental and theoretical chemical information on the system, and the variable temperature activation parameters, $\Delta H^{\#}$ and $\Delta S^{\#}$, are then used to assign a mechanism. Even so, additional experiments may be needed to further strengthen the assignment or to distinguish between alternatives.

Pressure is one of the fundamental physical variables influencing the rate of a chemical reaction in solution and is now widely used to supplement mechanistic assignments [1]. However, even if the first high pressure kinetic study in inorganic chemistry is more than thirty

R. van Eldik and J. Jonas (eds.), High Pressure Chemistry and Biochemistry, 311–331.
© *1987 by D. Reidel Publishing Company.*

years old, it is only during the last decade that a tremendous activity is this field has developped. This is related to the adaptation of most fast reaction techniques for use in high pressure kinetics : stopped-flow [2], nuclear magnetic resonance [3], temperature jump [4] and pressure jump [5]. The most commonly used are : stopped flow for the study of the wide class of unsymmetrical reactions, with reactants different from products (see, for example Eqn. 1), and nuclear magnetic resonance which applies to symmetrical reactions, like solvent exchange processes (Eqn. 2).

$$MS_n^{z+} + n\ast S \underset{}{\overset{k_{ex}}{\rightleftharpoons}} M\ast S_n^{z+} + nS \qquad (2)$$

In the present lecture I will focus on exchange reactions of solvent molecules between the first coordination sphere of tetrahedrally and octahedrally coordinated metal ions, and bulk solvent. But I will first discuss some methodological and technical aspects of these studies.

2. HIGH PRESSURE KINETICS DATA TREATMENT

The effect of pressure on the rate of a chemical reaction is usually interpreted in the framework of the transition state theory, which assumes that the transition state M is in true equilibrium with the reactants A and B (Eqn. 3). The volume of activation $\Delta V^{\#}$, the diffe-

$$A + B \rightleftharpoons \{M\}^{\#} \longrightarrow products \qquad (3)$$

rence between the partial molar volume of the transition state and the reactants, is related to the pressure derivative of lnk by Eqn. 4. Another parameter, the compressibility coefficient of activation $\Delta\beta^{\#}$,

$$(\partial lnk/\partial P)_T = -\Delta V^{\#}/RT \qquad (4)$$

defined by Eqn. 5, is also often introduced. It describes the pressure

$$\Delta\beta^{\#} = - (\partial\Delta V^{\#}/\partial P)_T \qquad (5)$$

dependence of $\Delta V^{\#}$. The problem of finding a suitable function to fit the lnk versus pressure data has been treated by various researchers [6]. The most widely used is the quadratic function (Eqn. 6), where $\Delta V_0^{\#} = -bRT$ and $\Delta\beta^{\#} = 2cRT$. But it must be added that there is no

$$lnk = a + bP + cP^2 \qquad (6)$$

physical justification for the use of a quadratic function.

$\Delta V^{\#}$ may be positive or negative. Bond stretching in a simple dissociative step gives rise to an increase in volume, which is manifested in a decrease in the rate of reaction with increasing pressure. Conversely, bond formation occurring in a simple associative process will lead to an increase in the rate constant with increasing pressure. For substitution reactions, the solvent electrostrictive effect when ions or dipoles are formed or neutralized at the transition state represents another important contribution in determining the sign and magnitude of $\Delta V^{\#}$. Therefore, the measured volume of activation $\Delta V^{\#}$ is usually considered the combination of an intrinsic contribution $\Delta V^{\#}_{int}$, resulting from changes in internuclear distances within the reactants during the formation of the transition state and an electrostrictive contribution $\Delta V^{\#}_{elec}$. Unfortunately, for substitution reactions involving charged species, the observed $\Delta V^{\#}$ may be dominated by the effect of electrostriction $\Delta V^{\#}_{elec}$, to the extent that the sign of $\Delta V^{\#}$ may differ from $\Delta V^{\#}_{int}$.

For solvent exchange reactions, the interpretation of $\Delta V^{\#}$ is fortunately simplified : there is no significant development of charge or dipole on going to the transition state ($\Delta V^{\#}_{elec} \simeq 0$). Therefore, the sign of $\Delta V^{\#}$ is immediately a diagnostic of the activation step : positive when bond breaking is favoured or negative when bond making is favoured [7].

3. METHODOLOGY

The first solvation sphere of a metal ion is composed of solvent molecules directly bound to the metal center. These have lost their freedom and move as one entity with the metal ion in solution. First order rate constants for the exchange of these solvent molecules with the bulk are spread over a wide scale. In water at 298 K, for instance, the k_{ex} values range from 10^{-8} s^{-1} for Rh^{3+} to little less than 10^{10} s^{-1} for the most labile ions [8]. For slow reactions ($k_{ex} < 10$ s^{-1}), the depletion or enrichment of a marked exchanging solvent molecule can be followed by NMR or mass spectrometry. For faster exchanges, NMR line-broadening is the dedicated technique which has allowed the study of reactions with k_{ex} as large as 10^9 s^{-1} for paramagnetic systems [9], and it has been used for many years in kinetic studies in solution [10].

Fig. 1 shows the effect of pressure on ^1H-NMR spectra for dimethylsulfoxide (DMSO) and tetramethylurea (TMU) exchanges on the tetrahedral BeS_4^{2+} in deuterated nitromethane at constant temperature. The signal at high field is due to free solvent and the other at low field to the four solvent molecules coordinated to Be^{2+}. In the DMSO case, with increasing pressure, the signals coalesce, meaning that the exchange becomes faster. The acceleration of the exchange process with pressure can be related to a bond making controlled process, whereas for TMU the reverse sequence of spectra shows a bond breaking process. Notice that the accuracy in the rate constant measurements will obviously be greater if the free and coordinated signals have similar intensity. This explains why most of the NMR relaxation rate constants for diamagnetic ions are determined in an inert diluent.

In water, though, it is much more difficult to find a suitable inert diluent, and it was thought easier to use ^{17}O line-broadening NMR in presence of Mn^{2+} (notice that ^1H cannot be used, the rate of

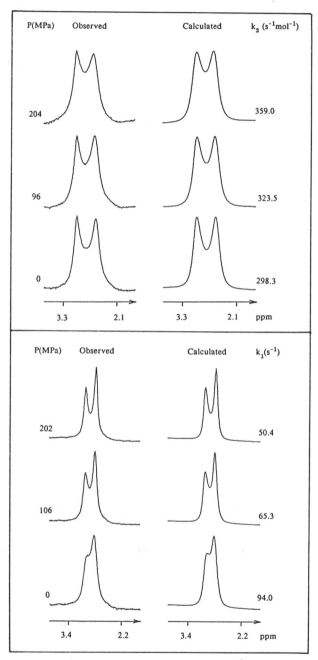

Fig. 1. Experimental and calculated 200 MHz ^{1}H–NMR spectra for DMSO, at 300 K, (top) and TMU, at 371 K, (bottom) exchanges on Be^{2+} at different pressures.

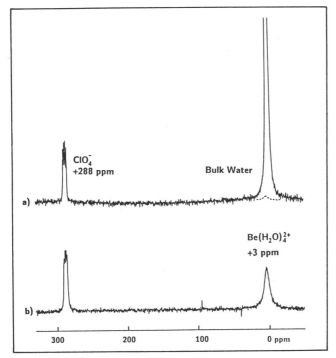

Fig. 2. 54.24 MHz ^{17}O-NMR spectra of 0.10 m Be(ClO$_4$)$_2$ and 0.40 m HClO$_4$ solutions, at 298 K. Top : in normal water (0.037 % ^{17}O). Bottom : with 0.10 m Mn(ClO$_4$)$_2$ in enriched water (0.40 % ^{17}O).

proton exchange being usually faster than the rate of exchange of a whole water molecule). If chemical exchange is slow, the ^{17}O-NMR spectrum of a dilute aqueous solution containing an aquated metal ion consists of two resonances : a large, intense peak due to bulk H$_2$O, and a smaller peak due to the M(H$_2$O)$_n^{2+}$ resonance. A natural-abundance ^{17}O spectrum of acidified Be(ClO$_4$)$_2$ consists of a narrow, intense signal due to bulk H$_2$O at 0 ppm, and a quadruplet due to ClO$_4^-$ at +288 ppm (Fig. 2, top). The bound-H$_2$O signal (dashed line), hidden under the large bulk water peak is invisible on this spectrum. The added Mn^{2+} ion is a very efficient relaxation agent for the bulk-H$_2$O signal due to its long electron relaxation time and its very fast coordinated/ bulk H$_2$O exchange rate. Its addition results in an extremely broad bulk water peak with negligible amplitude, revealing the underlying kinetically interesting bound water signal (Fig. 2, bottom). It is clear from Fig. 3 that the bound-H$_2$O transverse relaxation rate, $1/T_2^b$ (= π x full width at half height), is dominated by quadrupolar relaxation, $1/T_{2Q}^b$, at low temperatures and by solvent exchange, k_{ex}, at high temperatures. A double exponential analysis of the data according to Eqn. 7 leads to the temperature activation parameters for both processes (see Table 2). Experiments performed at high temperature (Fig. 4) show a broadening of the bound-H$_2$O signal when the pressu-

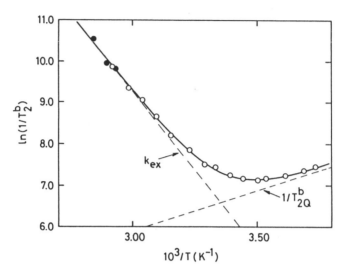

Fig. 3. Temperature dependence of the relaxation rates, $1/T_2^b$, of the bound-H_2O ^{17}O-NMR signal of 0.10 m $Be(H_2O)_4^{2+}$ solutions with added $Mn(ClO_4)_2$ and various $HClO_4$ concentrations (0.10-0.38 m), measured at 54.24 MHz.

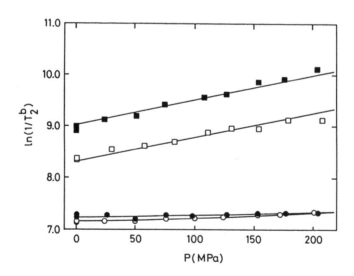

Fig. 4. Pressure dependence of $1/T_2^b$ in an acidified 0.10 m $Be(H_2O)_4^{2+}$ solution with added $Mn(ClO_4)_2$ in the exchange (■: 330 K, □: 321 K) and the quadrupolar (●: 274 K, ○: 278 K) regions, measured at 27.11 MHz.

$$1/T_2{}^b = k_{ex} + 1/T_{2Q}{}^b = k_B T/h \ \exp(\Delta S^{\#}/R - \Delta H^{\#}/RT)$$

$$+ \ (1/T_{2Q}{}^b)^{298} \ \exp[(E_Q{}^b/R)(1/T - 1/298.2)] \qquad (7)$$

re is increased, indicative of a bond-making process for water exchange on $Be(H_2O)_4{}^{2+}$.

For paramagnetic ions, the situation is different as shown in Fig. 5 for a dilute solution of Ni^{2+} in neat acetonitrile [11]. At low temperatures, the large sharp signal corresponds to free solvent, and the very small broad peak to the resonance of the six bound acetonitrile molecules, broadened by the paramagnetic influence of the proximate Ni^{2+} center. As the temperature increases, the peaks broaden and move together, eventually coalescing into a single broad peak. The reduced linewidth $1/T_{2r}$ of the free solvent is related to the rate constant $k_{ex} = 1/\tau_m$ and to the NMR parameters T_{2m}, T_{2os} and $\Delta\omega_m$ through a complex equation (Eqn. 8) proposed by Swift and Con-

$$\frac{1}{T_{2r}} = \frac{1}{\tau_m} \left[\frac{T_{2m}{}^{-2} + (T_{2m} \ \tau_m)^{-1} + \Delta\omega_m{}^2}{(T_{2m}{}^{-1} + \tau_m{}^{-1})^2 + \Delta\omega_m{}^2} \right] + \frac{1}{T_{2os}} \qquad (8)$$

nick [10]. All show a marked temperature (and pressure to some extent) dependence which complicates the extraction of rate data. This function, illustrated in Fig. 6, can be separated into four regions characterized by different predominant relaxation processes. The region where $\ln(1/T_{2r})$ is roughly proportional to the exchange rate $(1/\tau_m)$ is called the slow exchange domain. The temperatures corresponding to this region are ideal for the study of the effect of pressure on the rate constants with small or negligible contributions from the other relaxation processes. For Ni^{2+}, the line-narrowing observed corresponds to a decrease in the exchange rate (Fig. 7) reflected by a $\Delta V^{\#}$ value of $+9.6 \text{cm}^3 \text{ mol}^{-1}$. It should be stressed that the sign of $\Delta V^{\#}$

TABLE 1. Comparison of kinetic results for the exchange of acetonitrile on $[Ni(MeCN)_6]^{2+}$ [11].

Year	$10^{-3}k_{298}$, s^{-1}	$\Delta H^{\#}$, kJ mol^{-1}	$\Delta S^{\#}$, JK^{-1}mol^{-1}	Nucleus
1967	2.8	49	-15	^{1}H
1967	3.9	46	-37	^{1}H
1967	2.1	49	-16	^{1}H
1971	2.9	67	+43	^{1}H
1971	3.0	63.2	+41.8	^{14}N
1973	2.0	68	+50	^{14}N
1973	3.6	60	+23	^{1}H
1973	14.5	39.5	-32.6	^{14}N
1978	2.9	64.6	+37.9	^{1}H

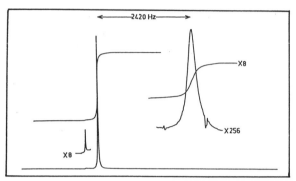

Fig. 5. 60 MHz ^{1}H-NMR spectrum of 0.1 m Ni(MeCN)$_6$$^{2+}$ in MeCN at 251 K with signal integrals. The resonances are (left to right) : 1 % internal benzene reference, free solvent and bound solvent.

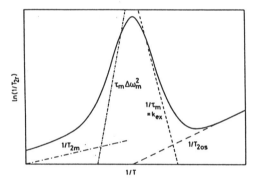

Fig. 6. Typical temperature dependence of $\ln(1/T_{2r})$. The solid-curve is obtained using Eqn. 8, and the dashed lines represent the four regions where predominant relaxation contributions lead to simplification of this equation.

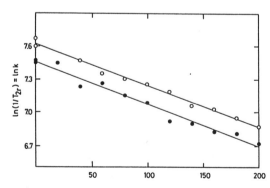

Fig. 7. Deceleration of the exchange rate with increasing pressure for MeCN exchange on Ni(MeCN)$_6$$^{2+}$ at two temperatures in the slow exchange domain.

is given unambiguously by the data. The experimental errors on $\Delta V^{\#}$ are of the order of ± 1 cm^3 mol^{-1} or $\pm$ 10 %, whatever the greatest, under normal experimental conditions. This is not the case for $\Delta S^{\#}$, obtained by extrapolation of the Eyring plot at 1/T=0, which may lead to large discrepancies for paramagnetic systems as illustrated in Table 1.

Although $\Delta S^{\#}$ values usually mimic the trends in $\Delta V^{\#}$, it has often been stressed that the large uncertainties associated with its determination by NMR line-broadening, especially in paramagnetic systems, are such as to restrict confidence in the use of this parameter to draw mechanistic conclusions.

4. HIGH PRESSURE HIGH RESOLUTION MULTINUCLEAR MAGNETIC RESONANCE

In 1954 already, Purcell [12], one of the founders of NMR spectroscopy, performed the first high pressure wide-line experiments on solids. In 1976 the first high pressure Fourier transform NMR probe with the high spectral resolution and sensitivity required to perform kinetic studies was built in our laboratory [3]. Fig. 8 is a schematic drawing of a probe-head recently designed for wide-bore (87 mm diameter) superconducting magnets (used up to 9.4 Tesla) [13]. It is made of two aluminium supports. The lower one, on the left, contains the pressure bomb itself, which can be used up to 250 MPa and is made of a non-magnetic beryllium-copper alloy. Three fluid connectors are located at the bottom : one for the pressure transmitting liquid, and two for the inlet and outlet of the thermostating liquid. The upper one, with the frequency adapter box at the bottom, is shown on the right. The adapter box contains a capacitive tuning network specific for each frequency range and which must therefore be changed to observe different nuclei or to work at different magnetic fields. This support contains two screwdrivers for access to the trimmer capacitors in the tuning network; it supports the radiofrequency connector to the spectrometer transmitter and receiver, and a platinum resistance connector to an ohm-meter for temperature measurements.

The internal components of the bomb and the sample tube are shown in Fig. 9. The radiofrequency coil is wound in saddle shape on a 7 mm o.d. glass tubing to produce a radiofrequency field perpendicular to the vertical static magnetic field. The temperature is measured inside the bomb by a platinum resistor placed just above the sample tube. The temperature stability is excellent ($\pm$ 0.1 $^\circ$C) and is not a limiting factor in the accuracy of measurement of $\Delta V^{\#}$. The glass tube containing the sample is a commercial 5 mm o.d. NMR tube cut to the right length. A Teflon separator transmits the hydrostatic pressure to the sample and avoids contact between the sample and the surrounding liquid. The surrounding pressure transmitting liquid is chosen as not to contain the observed nucleus. The spectral resolution and magnetic field stability observed with this non-spinning high pressure probe is as good as 1 Hz at 200 MHz, owing to the excellent homogeneity of todays superconducting magnets.

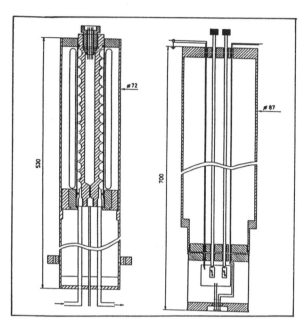

Fig. 8. Schematic drawing of the wide bore supraconducting magnet
multinuclear high pressure probe.

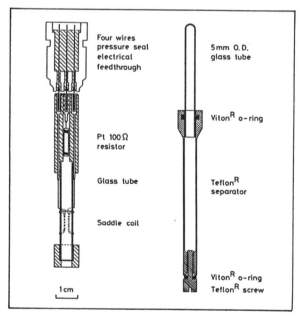

Fig. 9. Pressure seal and internal components of the bomb (left) ;
sample tube and separator assembly (right).

5. MECHANISMS

The classification of substitution reaction mechanisms originally proposed by Langford and Gray [14] is based on operational criteria. They
introduce, first of all, the concept of stoichiometric mechanism related to the sequence of bond breaking and bond making. If kinetic tests
show the presence of an intermediate of increased coordination number,
the mechanism is called associative and labelled A (Fig. 10). If an
intermediate of reduced coordination number can be detected, the mechanism is called dissociative and labelled D. When no intermediate
can be found, the mechanism is declared an interchange process, with
synchronous bond breaking and bond making, and labelled I. Within this
last category, a further distinction is made, depending on whether
evidence can be found for important incoming group influences (I_a)
or not (I_d). This second operational criterion is related to the
intimate mechanism which corresponds to associative or dissociative
activation mode. As discussed by Swaddle [15], this last operational
definition is too restrictive, and so it is useful to extend it.

For solvent exchange, where very few, if any, kinetic tests are
applicable, this extension is critical, and opens up to the use of
activation parameters as mechanistic criteria. In this context, volumes of activation are of primary importance and there has been indeed
considerable enthusiasm in recent years in collecting variable pressure data on solvent exchange kinetics, acknowledging in this way that
the volume of activation is almost the only criterion to get a better
understanding of their mechanisms. In a solvent exchange reaction, the
electrostriction can be assumed constant throughout the reaction, and
the measured volume of activation is solely constituted of the intrinsic contribution $\Delta V^{\#}_{intr}$ (see ch. 2.). In this fortunate case, the
sign of $\Delta V^{\#}$ therefore gives an immediate diagnostic of the activation mode (Fig. 11). The symmetrical nature of this exchange offers
other simplifications. Microscopic reversibility imposes that the forward reaction path be symmetrical to the reverse one. Hence, for limiting A or D mechanisms, with a reactive intermediate, the two transition states present along the reaction coordinate must necessarily
have identical energies. Along the same line, in an interchange process, with no true energy minimum in the reaction coordinate, the bond
lengths to the entering and leaving solvent molecules must be equivalent. It follows that for an I_a mechanism, the incoming and outgoing
entities will both have the same and considerable bonding at the transition state, whereas for an I_d mechanism, both bonds will be as
weak. From a structural point of view, the only difference between the
two mechanisms resides in the degree of expansion or contraction of
the transition state. Therefore the volume of activation is a direct
measure of the degree of associativity or dissociativity of the exchange reaction considered. A continuous spectrum of transition state
configurations can be envisaged, ranging from a very compact, highly
associative one, with a large negative $\Delta V^{\#}$ value, to a very expanded, highly dissociative one with a large positive $\Delta V^{\#}$ value. At each
extremity of this spectrum, limiting activation volumes $|\Delta V^{\#}_{lim}|$ will
be found for A and D mechanisms. For interchange mechanisms characte-

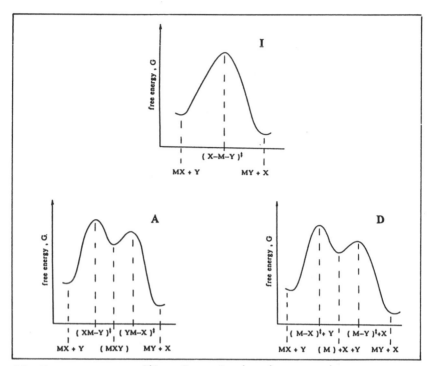

Fig. 10. Free energy profiles for substitution reactions.

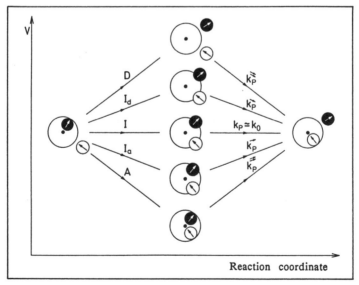

Fig. 11. Volume profiles for the spectrum of solvent exchange reactions mechanisms.

rized by $\Delta V^{\#}$ equal to zero or very small, the activation mode cannot be specified, and no subscript is associated to I.

6. SOLVENT EXCHANGE ON TETRAHEDRAL SOLVATES OF Be^{2+}

Lincoln and coworkers [16] have studied a series of non-aqueous solvent exchange reactions on the small tetrahedrally coordinated Be^{2+} ion in the diluent nitromethane. They have shown that the steric character of the solvent is an important factor determining the solvent exchange mechanism on that ion. We have complemented their work by a study as a function of pressure [17] (Table 2). For the solvents tetramethylurea (TMU) and dimethylpropyleneurea (DMPU), the rate law is first order, and the $\Delta S^{\#}$ and $\Delta V^{\#}$ are clearly positive. The conjunction of these facts indicates the occurrence of a limiting dissociative D mechanism, whereas for trimethylphosphate (TMP) and dimethylsulfoxide (DMSO) the second order rate law and the negative $\Delta S^{\#}$ and $\Delta V^{\#}$ suggest an associative activation mode for solvent exchange. The small absolute value for $\Delta V^{\#}$ does not allow the distinction between an associative interchange I_a and a limiting associative A mechanism. In the latter case, one may well imagine a large negative value due to bond formation, partially compensated by a positive contribution due to bond lengthening of the non-exchanging solvent molecules in a sterically crowded five coordinate intermediate. For DMF, both dissociative and associative pathways compete as indicated by the two terms rate law.

Our study of water exchange on Be^{2+}, performed in neat solvent, shows no acid dependence between 0.1 and 0.4 m $HClO_4$. The variable temperature data yield a value of $\Delta S^{\#}$ close to zero, therefore mechanistically useless. On the other hand, the $\Delta V^{\#}$ value of -13.6 cm^3 mol^{-1}, the most negative ever obtained for water exchange at a metal

TABLE 2. Rate constants and activation parameters for solvent S exchanges on BeS_4^{2+} (from [8]).^{a}

Solvent	k^{298}_{ex}	$\Delta H^{\#}$ kJ mol^{-1}	$\Delta S^{\#}$ JK^{-1}mol^{-1}	$\Delta V^{\#}$ cm^3 mol^{-1}	Mech.
H_2O	730^b	59.2	+8.4	−13.6	A
DMSO	213^c	35.0	−83.0	−2.5	A, I_a
TMP	4.2^c	43.5	−87.1	−4.1	A, I_a
DMF^d	16^c	52.0	−47.5	−3.1	A, I_a
	0.2^b	74.9	−7.3	−	D
TMU	1.0^b	79.6	+22.3	+10.5	D
DMPU	0.1^b	92.6	+47.5	+10.3	D

^{a} In CD_3NO_2 as diluent, except for H_2O in neat water. ^{b} First order rate constant (s^{-1}). ^{c} Second order rate constant (m^{-1} s^{-1}). ^{d} Two terms rate law.

ion, is clearly indicating an associative activation mode. The partial
molar volume of a water molecule electrostricted in the second coordination sphere can be estimated of about -15 cm^3 mol^{-1} [18]. The measured $\Delta V^{\#}$ approaches this value which can reasonably be estimated as a
maximum for an A mechanism, assuming that the incoming water molecule
is accomodated into an intermediate without any increase in volume accompanying the conversion of the tetrahedral initial state into a
pentacoordinated intermediate. The assignment of an A mechanism for
the small water molecule confirms that a sterically governed mechanistic changeover is taking place for solvent exchange on the small Be^{2+}
ion.

7. SOLVENT EXCHANGE ON OCTAHEDRAL METAL IONS

7.1 Trivalent cations

Solvent exchange around some diamagnetic trivalent metal ions has been
studied by NMR (Table 3). In non-aqueous solvents, the studies are
performed in the inert diluent, nitromethane, permitting the determination of the rate law. In the case of Sc(TMP)$_6^{3+}$, the similarity of
the results obtained for TMP exchange in neat solvent and in nitromethane gives confidence that this diluent does not significantly affect the reaction mechanism. The results for Al^{3+} and Ga^{3+} are very
convincingly conclusive of a dissociative activation mode, supported

TABLE 3. Kinetic parameters for solvent exchange on diamagnetic trivalent cations in CD$_3$NO$_2$ as diluent (from [8]).

	r_i pm	k^{298}	$\Delta H^{\#}$ kJ/mol	$\Delta S^{\#}$ J/K mol	$\Delta V^{\#}$ cm^3/mol	Mech.
Al(H$_2$O)$_6^{3+a}$	54	1.29^b	84.7	+41.6	+5.7	I_d
Al(DMSO)$_6^{3+}$		0.30^c	82.6	+22.3	+15.6	D
Al(DMF)$_6^{3+}$		5x10^{-2c}	88.3	+28.4	+13.7	D
Al(TMP)$_6^{3+}$		0.78^c	85.1	+38.2	+22.5	D
Ga(H$_2$O)$_6^{3+a}$	62	4.0x10^{2b}	67.1	+30.1	+5.0	I_d
Ga(DMSO)$_6^{3+}$		1.87^c	72.5	+3.5	+13.1	D
Ga(DMF)$_6^{3+}$		1.72^c	85.1	+45.1	+7.9	D
Ga(TMP)$_6^{3+}$		6.4^c	76.5	+27.0	+20.7	D
Sc(TMP)$_6^{3+a}$	75	736^b	34.1	−75.6	−23.8	A,I_a
Sc(TMP)$_6^{3+}$		39^c	21.2	−143.5	−18.7	
In(TMP)$_6^{3+}$	80	7.6^c	32.8	−118	−21.4	A,I_a

^{a} In neat solvent. ^{b} First order rate constant (s^{-1}).
^{c} Second order rate constant (m^{-1} s^{-1}).

by the first order rate laws and the large positive entropies and volumes of activation. For Sc^{3+} and In^{3+}, they are as convincing of an
associative activation mode, in view of the second order rate laws and
the negative sign of the entropies and volumes of activation. The
water-exchange results obtained in neat solvent are very welcome, since earlier ambient pressure results [19] suggested an associative behaviour for Ga^{3+} substitution in water solutions, in unexpected contrast to the situation in non-aqueous solutions. The mechanistic picture now seems more consistent in all solvents studied. The smaller
Al^{3+} and Ga^{3+} cations, being rather crowded, are easily willing to ex-
pell a solvent molecule from their first coordination sphere at the
transition state, whereas Sc^{3+} and In^{3+}, much larger, will more readily host a seventh molecule around them, at least partially, in forming
the activated complex.

A more difficult task nonetheless remains, which is to distinguish between interchange and limiting mechanisms. For water exchange
on octahedral first row transition metal ions, Swaddle used a semi-
empirical approach to predict average $\Delta V^{\#}_{lim}$ values of 13.5 cm^3
mol^{-1} for trivalent ions and 13.1 cm^3 mol^{-1} for divalent ions, almost
independent of the limiting mechanism (A or D) envisaged [15]. For
water exchange on Al^{3+} and Ga^{3+}, the volumes of activation are well
below the proposed limit and below the highest positive value so far
obtained on an hexaaqua metal ion, +7.2 cm^3 mol^{-1} for Ni^{2+} : these reactions can therefore be considered to proceed via dissociative interchanges (I_d). In non-aqueous solvents, assignment of a D mechanism
can be proposed for both ions on the basis of the first-order rate
laws which could be revealed and is consistent with the larger activation volumes. For Sc^{3+} and In^{3+}, it is more difficult to decide whether A or I_a mechanisms are operative, since both are compatible
with observed second order rate laws. Nevertheless, the similarity of
the $\Delta V^{\#}$ values with the volume of reaction for addition of TMP on
$Nd(TMP)_6^{3+}$ (r_i = 98 pm), $\Delta \bar{V}$ = -23.8 cm^3 mol^{-1} [20], suggests a common A mechanism for both ions. These data illustrate the relationship
between the increasing size of central ions and the mechanism : dissociative, associative and finally even a solvation equilibrium between
a six- and a seven-coordinate solvated ion.

A general view of the volumes of activation for first row high
spin transition metal ions is given in Table 4. A definite trend is
observable across the series. The $\Delta V^{\#}$ values are decreasingly negative on going along to Fe^{3+}, with a positive value for Ga^{3+}. This
changeover in mechanism from associative to dissociative activation
mode is confirmed by the few results in non-aqueous solvents. The value for Ti^{3+} is markedly more negative than those for the following
three members of the series. It is also rather close to the limiting
value of -13.5 cm^3 mol^{-1} yielded by the semi-empirical model of Swaddle. Therefore, although a limiting associative mechanism cannot be
attributed on the sole basis of this value, the character of water exchange on this center can be asserted to be strongly associative. The
smaller value for V^{3+}, Cr^{3+} and especially Fe^{3+} tend, on the other
hand, to favor the conclusion that an associative interchange (I_a)
is occurring for these three cations.

For Ga^{3+}, Fe^{3+}, Cr^{3+} and the low spin Ru^{3+} in water, hydrolysis

TABLE 4. Volumes of activation (cm^3/mol) for solvent S exchange on MS_6^{3+} of the first row transition metal series[a] [1].

M^{3+}	Sc	Ti	V	Cr	Fe	Ga
r_i(pm)	75	67	64	61	64	62
e_d^-	t_{2g}^0	t_{2g}^1	t_{2g}^2	t_{2g}^3	$t_{2g}^3 e_g^2$	$t_{2g}^6 e_g^4$
S=						
H_2O		−12.1	−8.9	−9.6	−5.4	+5.0
DMSO				−11.3	−3.1	+13.1[b]
DMF				−6.3	−0.9	+7.9[b]
TMP	−21.3					+20.7[b]

[a] By NMR exept for Cr^{3+} by isotopic labelling.
[b] In CD_3NO_2 as diluent.

is kinetically important. The conjugated base is in equilibrium with the hexaaquaion (Eqn. 9), offering an alternative pathway for water

$$M(H_2O)_6^{3+} \overset{K_a}{\rightleftharpoons} M(H_2O)_5OH^{2+} + H^+ \tag{9}$$

exchange, which must be accounted for. The overall observed rate constant will therefore be the sum of contributions from the two reaction paths, the water exchange on the hexaaqua species with rate constant k_{ex} and on its hydrolysed form with rate constant k_{OH} (Eqn. 10). Two main features are apparent from the rate constants and activation pa-

$$k = k_{ex} + k_{OH} \cdot K_a / [H^+] \tag{10}$$

TABLE 5. Rate constants and activation parameters for water exchange on some hexaaqua and monohydroxypentaqua metal ions (from [8]).

	k^{298} 1/s	k_{OH}/k_{ex}	$\Delta H^{\#}$ kJ/mol	$\Delta S^{\#}$ J/K mol	$\Delta V^{\#}$ cm^3/mol	pK_a
Ga^{3+}	4.0×10^2	275	67.1	+30.1	+5.0	~3.9
$Ga(OH)^{2+}$	1.1×10^5		58.9	−	+6.2	
Fe^{3+}	1.6×10^2	750	64.0	+12.1	−5.4	2.9
$Fe(OH)^{2+}$	1.2×10^5		42.4	+5.3	+7.0	
Cr^{3+}	2.4×10^{-6}	75	108.6	+11.6	−9.6	4.1
$Cr(OH)^{2+}$	1.8×10^{-4}		110.0	+55.6	+2.7	
Ru^{3+}	3.5×10^{-6}	100	89.8	−48.2	−8.3	2.5
$Ru(OH)^{2+}$	3.5×10^{-4}		91.0	−5.9	−3.5	

rameters reported in Table 5 : a higher reactivity (by a factor 75 to 750) and a larger dissociative character (more positive activation volumes) for the water exchange on the monohydroxypentaaqua metal ions. This drastic mechanistic difference between both exchange paths may be due to the strong electron donating capability of HO^-. The strong bonding between the metal center and this group will weaken the remaining metal-water bonds, most probably the trans one. The complex thus becomes more labile and dissociative activation is favoured.

7.2 Divalent cations

Some information from variable pressure data is available on almost every octahedral divalent member of the first row transition metal series, as apparent from Table 6. Ni^{2+} and Co^{2+} were the first cations studied, and the relatively large positive volumes of activation encountered, both in water and non aqueous solutions, supported the idea of a dissociative interchange mechanism operative for reactions involving the ions of this series. This idea proposed by Eigen and Wilkins

TABLE 6. Volumes of activation (cm^3/mol) for solvent S exchange on MS_6^{2+} of the first row transition metal series by NMR [1].

M^{2+}	V	Mn	Fe	Co	Ni	Cu
r_i(pm)	79	83	78	74	69	$(73)^a$
e_d^-	t_{2g}^3	$t_{2g}^3 e_g^2$	$t_{2g}^4 e_g^2$	$t_{2g}^5 e_g^2$	$t_{2g}^6 e_g^2$	$(t_{2g}^6 e_g^3)$
S						
H_2O	−4.1	−5.4	+3.8	+6.1	+7.2	
MeOH		−5.0	+0.4	+8.9	+11.4	+8.3
MeCN		−7.0	+3.0	+8.1	+8.5	
DMF			+8.5	+6.7	+9.1	
NH_3					+5.9	

a Effective ionic radius.

in the sixties [21, 22], was grounded on a comparison of ultrasonic absorption data of aqueous metal sulfates [23] with water exchange rates on the same series, Mn^{2+}, Fe^{2+}, Co^{2+} and Ni^{2+} [10]. The ultrasonic relaxation spectra could be separated into the first fast formation of an outer-sphere encounter complex, followed by the rate-determining slow interchange of a water molecule for a sulfate ligand in the first hydration sphere, as shown in Eqn. 11, neglecting charges. The close similarity between the rate of interchange in Eqn. 11 and

$$M(H_2O)_6 + L \underset{}{\overset{K_{os}}{\rightleftharpoons}} M(H_2O)_6 \cdots L \overset{k_I}{\longrightarrow} M(H_2O)_5 L + H_2O \quad (11)$$

the water exchange rate for a given cation led Eigen and Wilkins, although the latter author slightly revised his views for Mn^{2+} later on

[24], to conclude that k_I was independent of the nature of the entering ligand, and hence, that a dissociative interchange was the rule throughout the series. The variable pressure NMR results obtained subsequently for V^{2+}, Mn^{2+}, and even Fe^{2+}, forced though to depart from this, at the time commonly accepted, concept [25]. It is now generally accepted that a gradual changeover is occurring across the series, from the associatively activated V^{2+} to the dissociatively activated Ni^{2+}, at least in water, methanol and acetonitrile (Table 6). The small $\Delta V^{\#}$ values for Fe^{2+} are indicative of an interchange mechanism with almost as much contribution from bond making as from bond breaking. Considering the mean limiting values of $\Delta V^{\#}_{lim} = 13.1$ cm^3 mol^{-1} obtained from Swaddle's model in the extreme cases of A and D mechanisms for water exchange on divalent cations, one is tempted to attribute interchange mechanisms to every cation in Table 6, in view of the lower values of the volumes of activation reported. In water, methanol and acetonitrile, the changeover can therefore be assumed to proceed from an associative interchange for the early members of the series to a dissociative interchange for the late members, with the change in activation mode from the d^5 to the d^6 configuration (the value of $\Delta V^{\#} = +8.5$ cm^3 mol^{-1} recently obtained for Fe^{2+} in DMF tends to suggest that bond breaking is more important in this bulky solvent than in the other three; steric effects in DMF were already noticed for solvent exchange on the tetrahedral Be^{2+} ion).

7.3 Mechanistic trends

The progressive substitution mechanism changeover observed along the first row transition metal series for the divalent as well as for the trivalent octahedral solvates cannot be explained in terms of cationic size only, as was done for the limited group of the diamagnetic trivalent ions discussed earlier (Table 3). The electronic configuration in the valence shell also bears an important role in the mechanistic behaviour of these d elements. For a σ bonded octahedral complex, the t_{2g} orbitals are non-bonding whereas the e_g orbitals are antibonding (Fig. 11). Certainly, the gradual filling of the t_{2g} orbitals, spread out between ligands, will electrostatically disfavour the approach of a seventh molecule towards a face of the octahedron, and therefore decrease the ease of bond making. Similarly, an increased occupancy of the e_g orbitals, pointed to the ligands, will enhance the bond breaking tendency. These effects, combined with the steric effects outlined above, can explain the sequences of $\Delta V^{\#}$ values in Tables 4 and 6, with the exception of the values reported for V^{3+} or Cr^{3+}, which remain too low or too high, respectively. But this small discrepancy is not necessarily real since the value of $\Delta V^{\#}$ cannot be determined better than within ± 1 cm^3 mol^{-1} as discussed before. In any case, we are unable to distinguish which effect predominates, although the smaller value reported for V^{2+} ($t_{2g}^3 e_g^0$, $r_i = 79$ pm) compared to that for Mn^{2+} ($t_{2g}^3 e_g^2$, $r_i = 83$ pm) seems to indicate that the electronic effects more than compensates the steric effect.

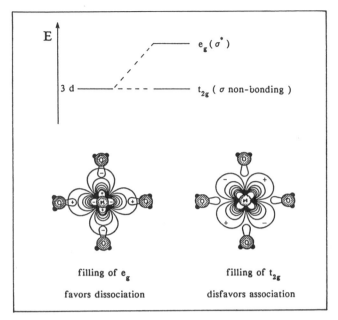

Fig. 11. Schematic representation of the 3d metal orbital splitting in an octahedral first row transition cation, and of the interactions of these orbitals with coordinated water.

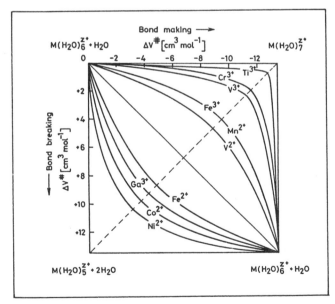

Fig. 12. Bond making and bond breaking contributions to the volumes of activation for water exchange on $M(H_2O)_6^{z+}$.

7.4 Visualization of exchange mechanisms

The systematic mechanistic trend for water exchange on the divalent
and trivalent 3d aquoions can be visualized, following Swaddle [15],
with the aid of a two dimensional More O'Ferral type plot (Fig. 12).
The sum of the coordinates of each point along a trajectory represents
the change in volume to reach that point. The volume of activation is
obtained at the intersection with the dashed diagonal. It should be
noted, however, that this square representation imposes that the activation volumes for the two limiting mechanisms be represented by the
same absolute value at the corners. In this diagram, we have chosen as
limiting values for A and D mechanisms the $\Delta V^{\#}_{lim}$ yielded by Swaddle's semi-empirical treatment, which fortuitously come out as approximately equal. Looking closer at Fig. 12, though, the possibility
cannot be excluded that the extreme value for a dissociative process
might be less than proposed by Swaddle's model.

8. CONCLUSION

High pressure multinuclear magnetic resonance has proven an invaluable
technique in the study of solvent exchange on metal ions, reactions
which are fundamental to the understanding of substitution and redox
reactions in inorganic chemistry. For these reactions, characterized
by symmetrical pathways and no major solvationnal changes, the acceleration or the retardation of the exchange rate constant with pressure
is unambiguously related to the activation mode. The most striking and
at the time unexpected results were obtained for divalent and trivalent high spin first row hexasolvated transition metal ions : for both
series a gruadual changeover in substitution mechanism occurs, with
the early members showing associative activation mode and the later
ones dissociative behaviour, the change in activation mode occurring
after the d^5 configuration.
  Finally, it should be stressed that the usefulness of $\Delta V^{\#}$ for
mechanistic discrimination, like for other activation parameters is
limited by the complexity of the reactions studied. For complex formation reactions, especially those preceeded by charge neutralization in
the formation of an outer sphere complex, a more cautious interpretation is required. But nevertheless, the results obtained so far produce a consistent picture [1] with those of solvent exchange reactions.
It is also not useless to stress that for any reaction with products
different from reactants, that is a non symmetrical-reaction, it is
imperative to have some knowledge of the volume profile before reaching a safe mechanistic assignment.

ACKNOWLEDGMENT

It is a great pleasure to acknowledge very stimulating cooperation and
discussions with Dr. Yves Ducommun and Dr. Lothar Helm, and to thank
the Swiss National Science Foundation for his continuing support of
this work. Portions of this text are taken from a review chapter on

Solvent exchange reactions written with Dr. Yves Ducommun for [1], and from a lecture presented at the XXIV International Conference on Coordination Chemistry in Athens, 1986 [8].

REFERENCES

1. Inorganic High Pressure Chemistry : Kinetics and Mechanisms, R. van Eldik, Ed., Elsevier, Amsterdam, 1986.
2. K. Heremans, J. Snauwaert and J. Rijkenberg, High Pressure Science and Technology, Vol. 1, K.D. Timmerhaus and M.S. Barber, Eds., Plenum, New York, p. 646, 1979.
3. H. Vanni, W.L. Earl, and A.E. Merbach, J. Magn. Reson., 29, 11 (1978).
4. E.F. Caldin, M.W. Grant, and B.B. Hasinoff, J. Chem. Soc., Faraday I, 68, 2247 (1972).
5. K.R. Brower, J. Am. Chem. Soc., 90, 5401 (1968).
6. H. Kelm and D.A. Palmer, High Pressure Chemistry, H. Kelm, Ed., Reidel, Dordrecht, p. 281, 1978.
7. T.W. Swaddle, Coord. Chem. Rev., 14, 217 (1974).
8. A.E. Merbach, Pure Appl. Chem., in press.
9. R.V. Southwood-Jones, W.L. Earl, K.E. Newman, and A.E. Merbach, J. Chem. Phys., 73, 5909 (1980).
10. T.J. Swift and R.E. Connick, J. Chem. Phys., 37, 307 (1962) and 41, 2553 (1964).
11. K.E. Newman, F.K. Meyer, and A.E. Merbach, J. Am. Chem. Soc., 101, 1470 (1979).
12. G.B. Benedek and E.M. Purcell, J. Chem. Phys., 22, 2208 (1954).
13. D.L. Pisaniello, L. Helm, P. Meier, and A.E. Merbach, J. Am. Chem. Soc., 105, 4528 (1983).
14. C.H. Langford and H.B. Gray, Ligand Substitution Processes, Ch. 1, W.A. Benjamin, New York, 1965.
15. T.W. Swaddle, Adv. Inorg. Bioinorg. Mechanisms, A.G. Sykes, Ed., 2, 95 (1983).
16. S.F. Lincoln and M.N. Tkcazuk, Ber. Bunseng. Phys. Chem., 85, 433 (1981) and 86, 221 (1982).
17. G. Elbaze, P.A. Pittet, L. Helm, and A.E. Merbach, to be submitted.
18. D.R. Stranks, Pure Appl. Chem., 38, 303 (1974).
19. J. Miceli and J. Stuehr, J. Am. Chem. Soc., 90, 6967 (1968).
20. D.L. Pisaniello, P.J. Nichols, Y. Ducommun, and A.E. Merbach, Helv. Chim. Acta, 65, 1025 (1982).
21. M. Eigen, Z. Elektrochem., 64, 115 (1960).
22. R.G. Wilkins and M. Eigen, Adv. Chem. Ser., 49, 55 (1965).
23. M. Eigen and K. Tamm, Z. Elektrochem., 66, 93 (1962).
24. R.G. Wilkins, Acc. Chem. Res., 3, 408 (1970).
25. J. Burgess, Metal Ions in Solution, Ellis Horwood, Chichester, 1978.

HIGH PRESSURE STUDIES OF INORGANIC REACTION MECHANISMS

Rudi van Eldik
Institute for Physical and Theoretical Chemistry
University of Frankfurt
Niederurseler Hang
6000 Frankfurt am Main 50
Federal Republic of Germany

ABSTRACT. The effect of pressure on a wide range of inorganic and organo-metallic reactions is discussed to illustrate the general applicability of high pressure kinetics in the elucidation of reaction mechanisms. The selected reaction types include substitution and isomerization reactions of square-planar and octahedral complexes, as well as electron transfer and addition/elimination reactions of octahedral systems. A critical evaluation is given, general trends are discussed, and recent advances are emphasized.

1. INTRODUCTION

The application of high pressure kinetic techniques in the elucidation of the reaction mechanisms of inorganic and organometallic complexes in solution has received considerable attention in recent years. The notable explosion in the available data over the last decade[1-3] is due to the fundamental insight obtained with this method and the more general availability of equipment that can handle such measurements at pressures up to 300 MPa on a milli-, micro- and nanosecond time scale. Since pressure is a fundamental physical variable, the obtained information not only assists in the elucidation of the reaction mechanism, but also adds a further dimension to the kinetic parameters that the suggested reaction mechanism must account for.

   The fundamental aspects of this method have been reviewed in detail elsewhere[1-11], and also in the contributions by le Noble and Merbach in this monograph. According to the transition state theory, the volume of activation can be determined from the pressure dependence of the rate constant, viz. $\Delta V^{\neq} = -RT(\partial \ln k/\partial P)_T$, and combined with the partial molar volumes of reactant and product species (obtainable from density measurements) can construct a volume profile for the reaction under investigation. For a reaction of the type $A + B \xrightarrow{k} AB$, the simplest conceivable mechanism according to the transition state theory can be formulated as in (1), and a possible volume profile is given in Figure 1. The magnitude and sign of the reaction volume, $\Delta \bar{V} = \bar{V}_{AB} - \bar{V}_A - \bar{V}_B$, and of the

R. van Eldik and J. Jonas (eds.), High Pressure Chemistry and Biochemistry, 333–356.
© *1987 by D. Reidel Publishing Company.*

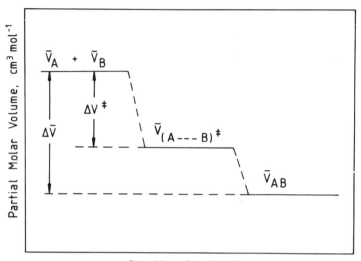

Figure 1. Volume profile for the reaction A + B $\rightleftharpoons$ (A---B)$^{\#}$ $\longrightarrow$ AB.

$$A + B \rightleftharpoons [A\text{---}B]^{\#} \longrightarrow AB \qquad (1)$$

activation volume, $\Delta V^{\#} = \bar{V}_{\#} - \bar{V}_A - \bar{V}_B$, depend on the nature of the chemical species involved and their environment. It follows that such a profile is a pictorial view of the chemical process on the basis of volume changes.

In general, $\Delta V^{\#}$ is thought of as the sum of an intrinsic part ($\Delta V^{\#}_{intr}$), which represents the change in volume due to changes in bond lengths and angles, and a solvational part ($\Delta V^{\#}_{solv}$), which represents the change in volume due to changes in the surrounding medium (electrostriction) during the activation step. Schematic representations of these components for typical bond formation and bond cleavage processes, are given in Figure 2. It is of course principally the $\Delta V^{\#}_{intr}$ part that is the mechanistic indicator in terms of changes in bond lengths and angles. In reactions with large polarity changes, $\Delta V^{\#}_{solv}$ may be large and even larger than $\Delta V^{\#}_{intr}$ so that it can in fact counteract and swamp the $\Delta V^{\#}_{intr}$ contribution. Details on instrumentation used for measurements in a batch reactor (autoclave), stopped-flow, temperature-jump, pressure-jump and NMR instruments, have recently been treated in detail[12]. Most of the high pressure kinetic equipment is now commercially available or described in detail and easy to build.

In this paper we want to report the advances achieved in this area during the nine years since the previous Nato Advanced Study Institute on High Pressure Chemistry[13]. It is not our intention to report a complete review, but rather to highlight some typical examples that will demonstrate the type of results obtained for a wide range of different reactions involving a variety of inorganic and organometallic complexes. Readers are advised to consult other reviews[1-3] for more detail information. In

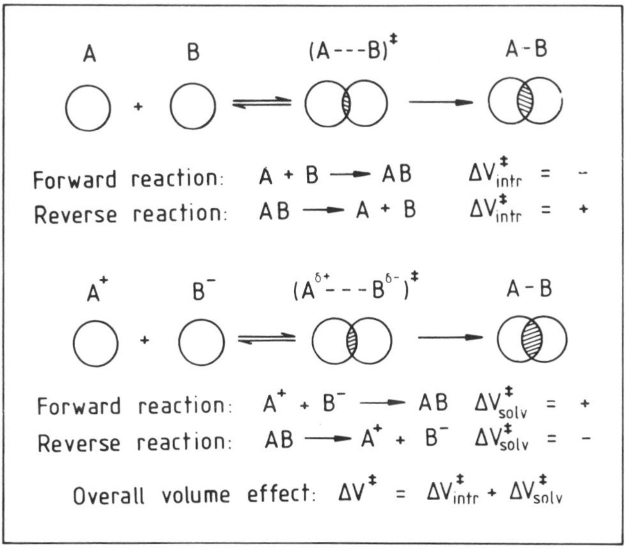

Figure 2. Schematic presentation to illustrate the sign and components of $\Delta V^{\neq}$.

addition, no detail information on solvent exchange processes will be treated here since this is done in a separate paper by Merbach in this monograph and elsewhere[5,14,15]. A critical evaluation of this method will be given in the concluding section following the presentation of the various reaction types.

2. SUBSTITUTION REACTIONS OF TRANSITION METAL COMPLEXES

Consider a general substitution reaction of the type outlined in (2) in which M is the metal center, L the non-participating ligand, and X and Y the leaving and entering groups, respectively. Such reactions are usually

$$ML_nX + Y \longrightarrow ML_nY + X \qquad (2)$$

discussed in terms of the classification proposed by Langford and Gray[15], according to which the substitution mechanism can be of the dissociative (D), interchange (I) or associative (A) type. In the D and A mechanisms the reaction involves the formation of an intermediate of lower and higher coordination number, respectively. In the I process bond formation and bond cleavage occur simultaneously (synchronous or concerted), which can either be more associative (I_a) or more dissociative (I_d) in nature. The pictorial view presented in Figure 3 is useful to visualize the intimate nature of these mechanisms, and the corresponding energy profiles (in which ML_n is represented by M) are given in Figure 4. If bond making is predominant, a significant decrease in molar volume of the reactants is expected (negative $\Delta V^{\#}$); in contrast, a significant increase in volume will occur in a dissociatively activated process (positive $\Delta V^{\#}$). In the case of an interchange process, bond breakage and bond formation occur to varying degrees within the precursor (ion-pair) species and small effects are expected, i.e. a slightly negative $\Delta V^{\#}$ for I_a and a slightly positive $\Delta V^{\#}$ for I_d. This interpretation is solely based on considering intrinsic volume contributions originating from changes in bond lengths and bond angles, and applies excellently to symmetrical chemical reactions such as solvent exchange. Many unsymmetrical substitution reactions are accompanied by major changes in charge distribution, dipole moment and dipole-dipole interactions, so that volume changes due to electrostriction effects, i.e. $\Delta V^{\#}_{solv}$, must be taken into consideration. For this reason it is important to know the overall reaction volume and to construct a volume profile before a safe mechanistic assignment can be reached. Examples in the forthcoming sections will demonstrate this point.

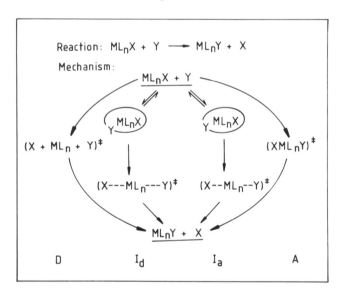

Figure 3. Pictorial view of the different types of substitution mechanisms for reaction (2).

2.1. Substitution at square-planar metal centers

In general, square-planar complexes of d^8 metal ions, viz. Rh(I), Pd(II) and Pt(II), undergo substitution according to an associative (A) mechanism, which is characterized by the well-known two-term rate law as indicated in Figure 5. This is true for all three types of reactions included, viz. solvolysis (k_1), anation when S = H_2O (k_{an}) and direct ligand substitution (k_2). The k_1 path strongly depends on the nature (especially nucleophilicity) of the solvent and is a common constant for various Y when studied in the same solvent. The k_2 path depends markedly on the nature (nucleophilicity and size) of Y[16]. The solvento intermediate ML_3S usually undergoes rapid substitution, i.e. anation when $S = H_2O$, due to the lability of the solvento species. k_1 cannot always be determined exactly, especially when $k_1 \ll k_2[Y]$, but in some cases it represents the major reaction route.

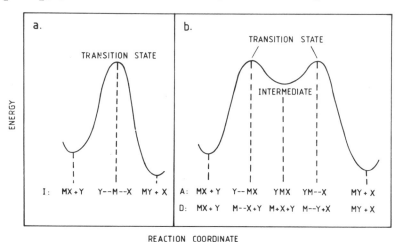

REACTION COORDINATE

Figure 4. Energy profiles for subsitution mechanisms for reaction (2), where ML_n is represented by M.
a - I; b - A and D.

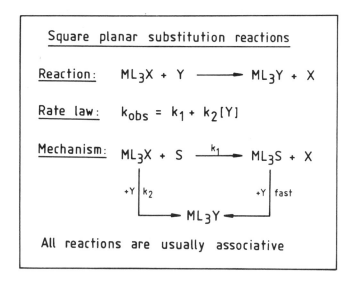

Figure 5. A general square-planar substitution process.

Investigators in this area have raised the question whether an increase in steric hindrance around the metal center could hinder the direct attack of the entering ligand (Y or S) and so force the substitution process into a dissociative reaction mode. Such a changeover in mechanism on increasing the steric hindrance should clearly show up in the $\Delta V^{\#}$ data. We have good reasons to believe that steric hindrance alone cannot induce a changeover from A to D. This is demonstrated by the data in Table I[17], where an increase in steric hindrance on the dien (diethylenetriamine) ligand is accompanied by a drastic decrease in the value of k_1. However, the mechanism remains associative throughout the series as seen from the fairly constant and significantly negative values of $\Delta S^{\#}$ and $\Delta V^{\#}$, and the entire effect only shows up in $\Delta H^{\#}$. The average volume of activation of -12 ± 2 cm^3 mol^{-1} is close to the maximum value expected for the associative entrance of a water molecule into the coordination sphere of an octahedral complex ion[18], but significantly more negative than the values reported for typical solvent exchange reactions of square-planar complexes, viz. -2.2 cm^3 mol^{-1} for Pd(H$_2$O)$_4^{2+}$ [19], -4.6 cm^3 mol^{-1} for Pt(H$_2$O)$_4^{2+}$ [20], and -2.8 cm^3 mol^{-1} for Pd(dien)H$_2$O^{2+} [21]. This means that the quoted solvolysis reactions may be accompanied by significant changes in electrostriction since the leaving group is a charged species. Recent measurements on similar solvolysis reactions with neutral leaving groups[22,23], viz. NH$_3$ and py, supported this suggestion and $\Delta V^{\#}$ values of between -3 and -4 cm^3 mol^{-1} were found. The relatively small absolute value of $\Delta V^{\#}$ for solvent exchange and solvolysis reactions involving neutral leaving groups could be an indication for an I_a rather than an A mechanism. However, this does not necessarily have to be the case since a substantial volume increase could occur when going from a tetragonal-pyramidal to a tri-

TABLE I. Kinetic parameters for the reaction

$$Pd(L)Cl^+ + H_2O \xrightarrow[\text{slow}]{k_1} Pd(L)H_2O^{2+} + Cl^-$$

$$+I^- \downarrow \text{ fast}$$

$$Pd(L)I^+ + H_2O$$

L	k_1 at 25°C s^{-1}	$\Delta H^{\#}$ kJ mol^{-1}	$\Delta S^{\#}$ J K^{-1} mol^{-1}	$\Delta V^{\#}$ at 25°C cm^3 mol^{-1}
dien	43.8 $\pm$ 0.5	43 $\pm$ 3	−69 $\pm$ 12	−10.0 $\pm$ 0.6
1,4,7-Me$_3$dien	25.0 $\pm$ 4.2	38 $\pm$ 4	−87 $\pm$ 15	−9.2 $\pm$ 0.6
1,4,7-Et$_3$dien	10.0 $\pm$ 0.1	41 $\pm$ 5	−86 $\pm$ 18	−10.8 $\pm$ 1.0
1,1,7,7-Me$_4$dien	0.99 $\pm$ 0.02	49 $\pm$ 1	−79 $\pm$ 3	−13.4 $\pm$ 1.9
1,1,4-Et$_3$dien	0.77 $\pm$ 0.01	51 $\pm$ 1	−76 $\pm$ 3	−14.5 $\pm$ 1.2
1,1,4,7,7-Me$_5$dien	(2.76 $\pm$ 0.04)x10^{-1}	50 $\pm$ 1	−88 $\pm$ 3	−10.9 $\pm$ 0.3
1,1,7,7-Et$_4$dien	(2.1 $\pm$ 0.4)x10^{-3}	69 $\pm$ 2	−67 $\pm$ 8	−14.9 $\pm$ 0.2
4-Me-1,1,7,7-Et$_4$dien	(6.8 $\pm$ 0.1)x10^{-4}	66 $\pm$ 7	−84 $\pm$ 25	−14.3 $\pm$ 0.6
1,1,4,7,7-Et$_5$dien	(6.7 $\pm$ 0.1)x10^{-4}	59 $\pm$ 3	−106 $\pm$ 9	−12.8 $\pm$ 0.8

TABLE II. Kinetic parameters for the reaction

$$Pd(L)H_2O^{2+} + Y \longrightarrow Pd(L)Y^{2+} + H_2O$$

L	Y	$k_{anation}$[a]	$\Delta H^{\#}$	$\Delta S^{\#}$	$\Delta V^{\#a}$	Ref.
		$M^{-1} s^{-1}$	$kJ\ mol^{-1}$	$J\ K^{-1}\ mol^{-1}$	$cm^3\ mol^{-1}$	
$(H_2O)_3$	Me_2SO	2.45	58 ± 1	-44 ± 3	-10.4 ± 0.5	24
Me_5dien	$SC(NH_2)_2$	1580	30 ± 1	-85 ± 4	-9.3 ± 0.4	25
	$SC(NHMe)_2$	720	34 ± 1	-78 ± 3	-9.1 ± 0.6	
	$SC(NMe_2)_2$	195	37 ± 1	-81 ± 3	-13.4 ± 0.7	
Et_5dien	$SC(NH_2)_2$	1.12	54 ± 2	-69 ± 7	-8.3 ± 0.3	25
	$SC(NHMe)_2$	0.54	59 ± 2	-58 ± 3	-10.2 ± 0.6	
	$SC(NMe_2)_2$	0.12	63 ± 1	-56 ± 4	-12.7 ± 0.6	

[a]Data at 25°C

TABLE III. Kinetic parameters for the reaction[a]

$$Pd(L)H_2O^{2+} + Y^- \longrightarrow Pd(L)Y^+ + H_2O$$

L	Y^-	$k_{anation}$[b]	$\Delta H^{\#}$	$\Delta S^{\#}$	$\Delta V^{\#}$[b]
		$M^{-1} s^{-1}$	$kJ\ mol^{-1}$	$J\ K^{-1}\ mol^{-1}$	$cm^3\ mol^{-1}$
$1,1,7,7-Me_4dien$	Cl^-	1908 ± 50	40 ± 2	-49 ± 6	-7.2 ± 0.2
	Br^-	3126 ± 120	39 ± 1	-47 ± 5	-7.6 ± 0.3
	I^-	8081 ± 890	34 ± 2	-55 ± 6	-9.3 ± 0.8
$1,1,4-Et_3dien$	Cl^-	1558 ± 7	44 ± 3	-36 ± 10	-2.7 ± 0.2
$1,1,4,7,7-Me_5dien$	Cl^-	629 ± 19	44 ± 2	-42 ± 8	-4.9 ± 0.4
	Br^-	1087 ± 20	40 ± 2	-54 ± 7	-7.3 ± 0.4
	I^-	4162 ± 140	33 ± 5	-63 ± 17	-9.9 ± 1.2
	N_3^-	6678 ± 440	31 ± 2	-66 ± 9	-11.7 ± 1.2
$1,1,7,7-Et_4dien$	Cl^-	4.19 ± 0.06	56 ± 2	-45 ± 5	-3.0 ± 0.2
$1,1,4,7,7-Et_5dien$	Cl^-	2.27 ± 0.01	70 ± 2	-1 ± 5	-7.7 ± 0.5

[a]Data taken from ref. 26 - 28
[b]Data at 25°C

gonal-bipyramidal transition state during the associative (A) reaction mode. Furthermore, this will also account for the significantly more negative values reported for the reactions involving charged leaving groups[17,22,23] since major solvational effects are expected to occur during such an internal rearrangement. There are many solvolysis reactions of a variety of Pt(II) and Pd(II) complexes that all exhibit $\Delta V^{\#}$ values between -9 and -18 $cm^3\ mol^{-1}$ when the leaving group is a singly charged anionic species[23].

Kinetic investigations of various anation reactions of the type outlined in (3) show that the pseudo-first-order rate constant depends

$$ML_3H_2O^{2+} + Y^{n-} \longrightarrow ML_3Y^{(2-n)+} + H_2O \qquad (3)$$

linearly on the concentration of the entering group Y^{n-}. The volumes of activation for such reactions should exhibit a close correlation with the corresponding solvent exchange reactions especially in those cases where Y^{n-} is a neutral species and no major changes in electrostriction are expected. The available results are summarized in Tables II and III for neutral and charged anating ligands, respectively. The values of $\Delta V^{\#}$ in Table II are significantly more negative than the corresponding solvent exchange data mentioned above, which can be accounted for in terms of the increasing size of the entering ligand, i.e. the more effective overlap of the van der Waals radii. Both series of data in Tables II and III clearly demonstrate this trend, i.e. more negative $\Delta V^{\#}$ for larger entering groups. Partial charge neutralization during bond formation is expected to affect the values of $\Delta V^{\#}$ reported for anation by the singly charged anions in Table III, but the overall trends are clearly in line with an associative (A) mechanism. The earlier remarks concerning the effects of steric hindrance

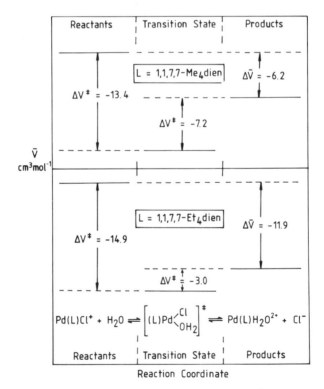

Figure 6. Volume profiles for the reaction $Pd(L)Cl^+ + H_2O \rightleftharpoons Pd(L)H_2O^{2+} + Cl^-$.

also account for such trends observed in the data of Tables II and III.
    The $\Delta V^{\neq}$ data in Tables I and III enable us to construct a relative
reaction volume profile for reaction (4) although the partial molar volumes

$$Pd(L)Cl^+ + H_2O \underset{k_{anation}}{\overset{k_1}{\rightleftharpoons}} Pd(L)H_2O^{2+} + Cl^- \tag{4}$$

of the reactant and product complexes are not available. This is due to
spontaneous solvolysis[29] that interfers with such measurements for the
reactant complex, and our inability to isolate pure and stable samples of
the product complex. For reaction (4), $\Delta \bar{V} = \Delta V^{\neq}(k_1) - \Delta V^{\neq}(k_{anation})$, and
two representative examples are given in Figure 6. Throughout the series of
chloro complexes $\Delta \bar{V}$ is significantly negative, varying between −6 and −12
$cm^3 \, mol^{-1}$, which is in line with the overall effect of charge creation, i.e.
increasing electrostriction. Furthermore, the transition state has a
significantly lower partial molar volume than either the reactant or
product species, demonstrating the associative character of the
substitution process.
    The intimate nature of the direct nucleophilic substitution route,
denoted by k_2 in Figure 5, has been investigated in detail in numerous
studies during the last decade[23]. Special emphasis was placed on solvent
effects and the possible role of conjugate base species during substitution
by OH^-. In the majority of cases the reported activation volumes are
significantly negative and can be interpreted in terms of an associative (A)
process as for the solvolysis and anation reactions described above. A
complete treatment of these systems has been published[23] and only a few

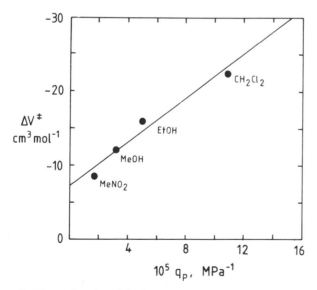

Figure 7. The relationship between the volume of activation and q_p parameter
for the substitution of trans-Pt(py)$_2$(NO$_2$)Cl by pyridine (ref. 23, 30).

interesting findings will be highlighted here. Substitution reactions involving charged entering and/or leaving groups exhibit $\Delta V^{\#}$ values typical for processes in which a significant contribution from $\Delta V^{\#}_{solv}$ can be expected. In one study[30] this contribution was represented by the solvent parameter q_p, which is obtained from the pressure derivative of the Kirkwood equation[31]. A plot of $\Delta V^{\#}$ versus q_p for reaction (5) in a range of solvents resulted in an intercept ($\Delta V^{\#}_{intr}$) of $- 7 \pm 1$ cm^3 mol^{-1} (see

$$\text{trans-Pt(py)}_2(\text{NO}_2)\text{Cl} + \text{py} \longrightarrow \text{trans-Pt(py)}_3\text{NO}_2^+ + \text{Cl}^- \qquad (5)$$

Figure 7). This intrinsic value is consistent with an associative process in which the Pt-py bond appears to be partially formed in the transition state. The values of $\Delta V^{\#}_{solv}$, i.e. q_p, show the expected solvent dependence. In a few cases only, positive $\Delta V^{\#}$ values were reported for the substitution of Pd(II) dien complexes by OH$^-$ [17]. This is presumably due to the participation of a conjugate base species that will involve neutralization of charge, the release of a water molecule and the possibility of a dissociative reaction mode.

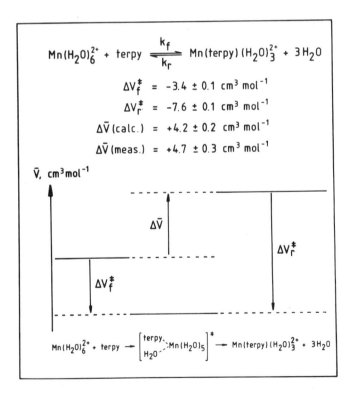

Figure 8. Typical example of a volume profile for a complex formation reaction of Mn(II) (ref. 10,37).

2.2. Substitution at octahedral metal centers

High pressure techniques have been used to investigate numerous substitution reactions of octahedral complexes, and this topic has been reviewed in detail[1,2,5-7,32]. It can in general be concluded from the $\Delta V^{\neq}$ data for solvent exchange reactions of octahedral complexes (see contribution by Merbach or ref. 5 and 14), that gradual mechanistic changeovers occur along specific series of the transition metals. This is of fundamental importance to the understanding of the general substitution behaviour of such species, and it is important to know whether complex formation and related processes exhibit similar trends. The available results for first-row transition-metal complexes are summarized in Table IV and exhibit a similar trend in the complex formation data as for the solvent exchange data. The volume profile for the complexation of $Mn(H_2O)_6^{2+}$ by terpy (Figure 8) clearly demonstrates that the transition state is of a significantly lower volume than either the reactant or product states, thus favouring an associative (probably I_a) reaction mode[37]. It follows that the reported mechanistic changeover also holds for complex formation reactions and thus contradicts the traditional "Eigen-Wilkins" (I_d) mechanism previously suggested for such labile metal species. Similar volume profiles have been reported for complex formation reactions of V(II)[35] and Ni(II)[38,39].

TABLE IV. A comparison of $\Delta V^{\neq}$ for solvent exchange and complex formation of first-row transition-metal elements in aqueous solution

Entering ligand	V^{2+}	Mn^{2+}	Fe^{2+}	Co^{2+}	Ni^{2+}	Ref.
H_2O	−4.1±0.1	−5.4±0.1	+3.8±0.2	+6.1±0.2	+7.2±0.3	5
NH_3				+4.8±0.7	+6.0±0.3	33
pada[a]				+7.2±0.2	+7.7±0.3	33
2,2´bipyridine		−1.2±0.2[b]		+4.3±1.0	+5.5±0.3	34[d]
				+7.5±1.4	+5.1±0.4	34[e]
2,2´:6´,2"-terpyridine		−3.4±0.7[c]	+3.4±0.6	+4.5±0.8	+6.7±0.2	34[d]
			+3.7±0.8	+3.7±1.3	+4.5±0.4	34[e]
SCN^-	−2.1±0.8					35

[a] pada = pyridine-2-azo(p-dimethylaniline)
[b] Ref. 36
[c] Ref. 37
[d] Experimental conditions such that [L] >> [M]
[e] Experimental conditions such that [M] >> [L]

Another category of substitution reactions of octahedral complexes that has received remarkable attention concerns the aquation reactions of various Cr(III) and Co(III) complexes for which more than 120 volumes of activation have been measured[32]. It is impossible to go into the details here and the discussion will be restricted to a few general remarks. The results for the aquation of Cr(III) and Co(III) complexes are such that in many cases an I_a reaction mode is favoured in the former and an I_d reaction mode in the latter case. However, the interpretation of the $\Delta V^{\#}$ data is in many cases complicated by large contributions arising from electro-striction effects when the leaving groups are charged species. So for instance, is $\Delta V^{\#}$ significantly negative for the aquation of a series of $Co(NH_3)_5X^{(3-n)+}$ complexes[40], where $X^{n-} = Cl^-$, Br^-, NO_3^-, SCN^- and SO_4^{2-}, although the results are discussed in terms of an I_d mechanism. In some more recent work, this complication was eliminated by studying the aquation reactions of a series of $Cr(NH_3)_5X^{3+}$ and $Co(NH_3)_5X^{3+}$ complexes, where X is a neutral leaving group[41,42]. The results in Table V show a remarkable agreement with the solvent exchange data for the corresponding aqua complexes and underline the I_a and I_d nature of the aquation reactions for Cr(III) and Co(III), respectively. As a final remark it should be mentioned that volume correlations (plots of $\Delta V^{\#}$ versus $\Delta \bar{V}$) and volume equation calculations have been adopted extensively to contribute towards the interpretation of $\Delta V^{\#}$ data for aquation reactions of Ni(II), Co(III) and Cr(III) complexes[32,45,46].

TABLE V. Volumes of activation for the aquation of a series of complexes according to the reaction[a]

$$M(NH_3)_5X^{3+} + H_2O \longrightarrow M(NH_3)_5H_2O^{3+} + X$$

X	$\Delta V^{\#}$, cm^3 mol^{-1}	
	M = Co(III)	M = Cr(III)
OH_2	$+1.2^b$	-5.8^c
OHMe	$+2.2$	
OHEt	$+2.9$	
$OHCHMe_2$	$+3.8$	
$OCH.NH_2$	$+1.1$	-4.8
OCH.NHMe	$+1.7$	
$OCH.NMe_2$	$+2.6$	-7.4
$OCMe.NMe_2$		-6.2
$OSMe_2$	$+2.0$	-3.2
$OC(NH_2)_2$	$+1.3$	-8.2
$OC(NH_2)(NHMe)$	$+0.3$	
$OC(NHMe)_2$	$+1.5$	-3.8
$OP(OMe)_3$		-8.7

[a]Data taken from ref. 41 and 42
[b]Data taken from ref. 43
[c]Data taken from ref. 44

Volumes of activation and reaction volume profiles not only reveal significant mechanistic detail for relatively simple substitution reactions as those described up to know, but also for more complicated processes. One of these concerns the base hydrolysis reactions of Co(III) complexes, which are generally accepted to proceed according to the S_N1CB mechanism as outlined in (6). The expected volume profile for this rather complex scheme[47] is given in Figure 9. In this mechanism, formation of the conjugate base species is followed by rate-determining dissociation of X^{n-} to produce the five-coordinate $Co(NH_3)_4NH_2^{2+}$ species, which rapidly reacts with the solvent to produce the hydrolysis product. Since $k_{obs} = kK[OH^-]$ for

$$Co(NH_3)_5X^{(3-n)+} + OH^- \xrightleftharpoons[]{K} Co(NH_3)_4(NH_2)X^{(2-n)+} + H_2O$$

$$Co(NH_3)_4(NH_2)X^{(2-n)+} \xrightarrow{k} Co(NH_3)_4NH_2^{2+} + X^{n-}$$

$$Co(NH_3)_4NH_2^{2+} + H_2O \xrightarrow{fast} Co(NH_3)_5OH^{2+} \qquad (6)$$

these reactions, it follows that $\Delta V^{\#} = \Delta\bar{V}(K) + \Delta V^{\#}(k)$. Solvational effects are expected to influence the values of $\Delta\bar{V}(K)$ and $\Delta V^{\#}(k)$ significantly, and also that of the overall volume change $\Delta\bar{V}$, since significant charge creation will be present when X^{n-} is a charged leaving group. This was indeed found to be the case[47,48] and typical $\Delta V^{\#}$ values are (X^{n-}): $+40.2 \pm 0.5$ (Me_2SO); $+31.0 \pm 0.8$ (NO_3^-); $+33.6 \pm 1.0$ (I^-); $+32.5 \pm 1.4$ (Br^-); $+33.0 \pm 1.4$ (Cl^-); $+26.4 \pm 1.0$ (F^-); and $+22.2 \pm 0.7$ (SO_4^{2-}) cm^3 mol^{-1} at $25^\circ C$. A similar trend is seen in $\Delta\bar{V}$: $+21.2 \pm 0.3$ (Me_2SO); $+13.2 \pm 0.4$ (NO_3^-); $+15.5 \pm 0.3$ (I^-); $+11.1 \pm 0.2$ (Br^-); $+9.9 \pm 0.2$ (Cl^-); $+7.4 \pm 1.0$ (F^-); and -3.9 ± 0.4 (SO_4^{2-}) cm^3 mol^{-1} at $25^\circ C$. More interesting, however, is the possibility to estimate the partial molar volume of the five-coordinate intermediate on the assumption that the rate-determining step in (6) has a late (product-like) transition state. Furthermore, this volume should be independent of the nature of X^{n-}, and can be estimated from the expression given in (7).

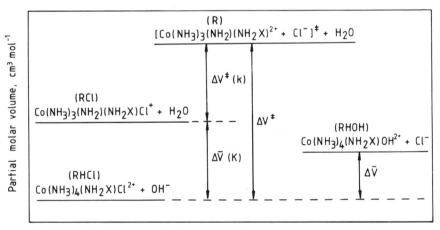

Reaction coordinate

Figure 9. Volume profile for the base hydrolysis of $Co(NH_3)_5X^{(3-n)+}$ according to an S_N1CB mechanism.

$$\Delta V^{\#} = \bar{V}_{\#} + \bar{V}_{H_2O} - \bar{V}_{RHX} - \bar{V}_{OH}$$

$$\sim \bar{V}_R + \bar{V}_X + \bar{V}_{H_2O} - \bar{V}_{RHX} - \bar{V}_{OH} \qquad (7)$$

The calculations[47] show that $\bar{V}_R$ is indeed constant, independent of X^{n-}, with an average value of $71 + 4$ cm^3 mol^{-1}, which is remarkably close to the partial molar volume of the six-coordinate $Co(NH_3)_5OH^{2+}$ species, viz. 68 cm^3 mol^{-1}.

   This approximate equality of the partial molar volume of five- and six-coordinate species encouraged us to introduce substituents on the ammine ligand to systematically increase the size of the five-coordinate species. Recent measurements[49] on systems of the type cis- and trans-$Co(NH_3)_4(NH_2X)Cl^{2+}$, where $X = Me$, Et, n-Pr, n-Bu and i-Bu, indicate that the partial molar volumes of the five-coordinate species $Co(NH_3)_3(NH_2X)NH_2^{2+}$ equal those for the corresponding hydroxo species and increase linearly with increasing size of NH_2X as shown by the correlation in Figure 10. These findings are of fundamental importance and contribute to a long standing discrepancy regarding the partial molar volumes of such intermediate species. A further interesting point is that $\Delta V^{\#} - \Delta\bar{V}$ for the volume profile in Figure 9 and the reaction scheme in (6) should be independent of the nature of X^{n-} since it only involves the reaction of H_2O with $Co(NH_3)_4NH_2^{2+}$ to produce $Co(NH_3)_5OH^{2+}$. This difference is indeed constant and has an average value of $20 + 2$ cm^3 mol^{-1} for the series of pentaammine complexes[47], and a value of $18 + \bar{2}$ cm^3 mol^{-1} for the monoalkyl substituted complexes[49]. These values almost equal the partial molar volume of water, demonstrating that the water molecule is completely absorbed by the five-coordinate

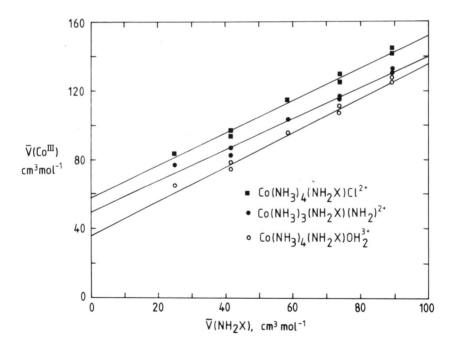

Figure 10. Plot of $\bar{V}(Co^{III})$ versus $\bar{V}(NH_2X)$ for a series of monoalkyl-substituted ammine complexes (ref. 49).

intermediate during the final step of the process and accounting for the volume equality of the five- and six- coordinate complexes mentioned above. It follows that the volume profile in Figure 9 presents an accurate description of the underlying reaction mechanism.

To conclude this section on substitution reactions of octahedral metal centers, we would like to draw attention to some work performed on metal carbonyls. The first contribution in this area came from Brower and Chen[50] for reactions of the type given in (8), where M = Cr, Mo and W. The reported

$$M(CO)_6 + PPh_3 \longrightarrow M(CO)_5PPh_3 + CO \tag{8}$$

$\Delta V^{\#}$ vaues are +15 (M = Cr), +10 (M = Mo) and -10 (M = W) cm^3 mol^{-1} and in line with a dissociative reaction mode for the first two complexes and an associative reaction mode for the last one. More recently it was reported[51-54] that substitution reactions of tetra- and pentacarbonyl complexes of Mo(0) exhibit $\Delta V^{\#}$ values ranging from significantly negative to slightly positive. Some of the data can be interpreted in terms of solvational effects, but the majority of data suggests that Mo(0) may be a borderline case and the operation of an interchange mechanism should be considered. On the contrary, such complexes of low oxidation state metal ions with neutral ligands should exhibit minor solvational effects and the observed $\Delta V^{\#}$ should mainly represent the intrinsic (mechanistic important) component. Along these lines it is important to note that some organometallic kineticists have run into problems with the interpretation of $\Delta S^{\#}$ data which presumably contradicts the observed kinetic trends. In these cases the obtained volumes of activation gave a clear cut answer and were in agreement with the overall kinetic trends. For instance, substitution of Cr(CO)$_4$(BTE) by P(OEt)$_3$ in 1,2-dichlorethane, where BTE = bis(tert-butylthio)ethylene, exhibits a $\Delta V^{\#}$ value of +14.0 $\pm$ 0.6 cm^3 mol^{-1} [55], which clearly supports the dissociative nature of the process although a negative $\Delta S^{\#}$ value was found before[56].

We recently determined the volumes of activation for the substitution reactions of the triruthenium clusters HRu$_3$(CO)$_{11}^-$ and Ru$_3$(CO)$_{11}$(CO$_2$CH$_3$)$^-$. Both these species have been suggested to undergo dissociative loss of CO prior to the reaction with the entering ligand as indicated in (9) and (10), respectively[57]. Under limiting conditions the loss of CO is the rate-

$$HRu_3(CO)_{11}^- \rightleftharpoons HRu_3(CO)_{10}^- + CO$$
$$HRu_3(CO)_{10}^- + PPh_3 \rightleftharpoons HRu_3(CO)_{10}(PPh_3)^- \tag{9}$$

$$Ru_3(CO)_{11}(CO_2CH_3)^- \rightleftharpoons Ru_3(CO)_{10}(CO_2CH_3)^- + CO$$
$$Ru_3(CO)_{10}(CO_2CH_3)^- + P(OMe)_3 \rightleftharpoons Ru_3(CO)_{10}(P(OMe)_3)(CO_2CH_3)^- \tag{10}$$

determining step. The corresponding volumes of activation were found to be +21.2 $\pm$ 1.4 and +16 $\pm$ 2 cm^3 mol^{-1} for reactions (9) and (10) in THF and 90/10 THF/MeOH, respectively. These values underline the dissociative nature of the process, which is in line with the kinetic observations although slightly negative $\Delta S^{\#}$ values were reported in both cases.

A similarly large positive $\Delta V^{\#}$ value of +20.6 $\pm$ 0.4 cm^3 mol^{-1} was recently found[58] for the release of CO in reaction (11), which was suggested to follow a dissociative mechanism. These large positive $\Delta V^{\#}$ values for the

$$Mn(CO)_5Cl + dab \longrightarrow Mn(CO)_3(Cl)(dab) + 2CO \qquad (11)$$
$$dab = biacetylbis(phenylimine)$$

dissociation of CO strongly support the dissociative nature of the process and underline the mechanistic discrimination ability of this parameter (see Concluding Remarks).

3. ISOMERIZATION REACTIONS OF TRANSITION METAL COMPLEXES

High pressure kinetic measurements have also assisted in the understanding of racemization, geometrical and linkage isomerization reactions of transition metal complexes[1,32,59]. Various mechanisms have been proposed for the racemization of octahedral complexes and these are presented schematically in Figure 11. Although it is in many cases not possible to distinguish between all of these mechanisms, some racemization reactions exhibit very distinct pressure effects. For instance, $\Delta V^{\#}$ for the racemization of $Fe(phen)_3^{2+}$ is $+15.6 \pm 0.3$ cm^3 mol^{-1} compared to a value of -12.3 ± 0.3 cm^3 mol^{-1} reported for the racemization of $Cr(C_2O_4)_2(phen)^-$ [59]. The former result indicates a dissociative mechanism, whereas the latter an intramolecular one-ended ring-opening of the oxalate ligand. The large negative $\Delta V^{\#}$ originates from the increase in electrostriction during ring-

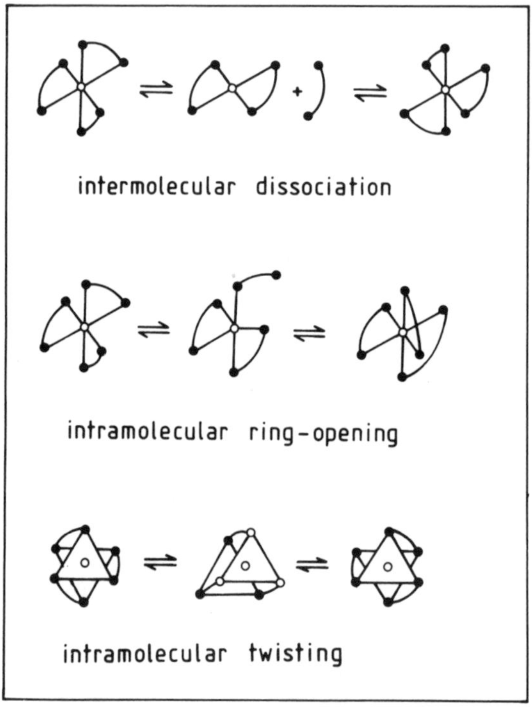

Figure 11. Schematic representation of different racemization mechanisms of octahedral species.

opening of the oxalate chelate. A similar conclusion was drawn from the $\Delta V^{\#}$ value of $-16.6 + 0.5$ cm^3 mol^{-1} reported[60] for the geometrical isomerization of trans-$Cr(C_2O_4)_2(H_2O)_2^-$. On the other hand, a $\Delta V^{\#}$ value of $+14.3 + 0.2$ cm^3 mol^{-1} for the trans to cis isomerization of $Co(en)_2(H_2O)_2^{3+}$ clearly supports the dissociative nature (release of H_2O) of the process[61]. Numerous sets of data are available for closely related systems[32], and meaningful conclusions have been reached for both racemization and geometrical isomerization processes.

Linkage isomerization reactions are usually concerted intramolecular processes, and in general exhibit small pressure effects[2,59]. This is understandable since such reactions, viz. $M-ONO \longrightarrow M-NO_2$, $M-OSO_2 \longrightarrow M-SO_3$, and $M-SCN \longrightarrow M-NCS$, are expected to proceed via a transition state in which both coordination sites of the ligand are weakly bonded to the metal center. A volume profile for the linkage isomerization of $Co(NH_3)_5ONO^{2+}$ demonstrates that the transition state lies approximately halfway between the reactant and product states on a volume basis, i.e. $\Delta V^{\#} \sim 0.5 \, \Delta \bar{V}$[62].

In contrast to octahedral systems for which geometrical isomerization usually occurs via a trigonal-bipyramidal intermediate, square-planar complexes isomerize via quite a number of different mechanisms. For instance, some of these reactions are catalyzed by uncoordinated ligands such as phosphines, arsines, stibines and amines, and the isomerization process is believed to involve either pseudorotation of a five-coordinate intermediate or consecutive displacement of ligands[63]. The mechanism of spontaneous cis-trans isomerization of sterically hindered complexes of the type cis-$Pt(PEt_3)_2(R)X$, where R = alkyl or substituted aryl groups and X = Cl^- or Br^-, in protic solvents is still a controversial issue. Romeo and co-workers[64,65] suggested that the reactions proceed through a dissociative asynchronous mechanism in which the rate-determining step is Pt-X bond cleavage to produce a three-coordinate "cis-like" intermediate. An alternative mechanism was proposed[66] in which the cis-$Pt(PEt_3)_2(R)X$ species exists in rapid equilibrium with the solvento species, cis-$Pt(PEt_3)_2(R)S^+$, which then undergoes rate-determining isomerization. The average volume of activation for the case R = Ph and X = Cl^-, Br^- and I^- is $+6.3 \pm 1.0$ cm^3 mol^{-1}, which clearly suggests a completely different mechanism for the isomerization step as compared to the associative solvolysis step. The magnitude of $\Delta V^{\#}$ suggests that either dissociation of the solvent molecule or pseudorotation of a five-coordinate species must present the rate-determining step of the mechanism.

4. ELECTRON TRANSFER REACTIONS OF TRANSITION METAL COMPLEXES

Electron transfer reactions are generally classified as outer sphere or inner sphere depending on the nature of the precursor intermediate species, Ox//Red, in (12), which can be an ion-pair or encounter complex, or a bridged intermediate, respectively. In many cases the electron transfer step is

$$Ox + Red \underset{}{\overset{K}{\rightleftharpoons}} Ox//Red$$

$$Ox//Red \xrightarrow{k_{ET}} Ox^-//Red^+$$

$$Ox^-//Red^+ \rightleftharpoons Ox^- + Red^+ \tag{12}$$

rate-determining, and in the presence of excess Red, $k_{obs} = k_{ET}K[Red]/\{1 + K[Red]\}$. In some cases where K is large it is possible to separate k_{ET} and K kinetically, but in the majority of cases K is small and the rate expression reduces to $k_{obs} = k_{ET}K[Red]$, such that the observed second-order rate constant ($k_{ET}K$) is a composite quantity. A detailed discussion of the effect of pressure on inner sphere and outer sphere electron transfer reactions was recently presented by Swaddle[67]. We will only focus on a few important aspects here.

   Stranks[68] reported $\Delta V^{\#}$ data for some self-exchange reactions, viz. $Fe(H_2O)_6^{2+/3+}$, $Co(en)_3^{2+/3+}$ and $Tl(H_2O)_6^{+/3+}$, and found a remarkably close agreement with the predicted values for an outer sphere process based on the Hush–Marcus theory. Failure of such predictions of $\Delta V^{\#}$ for the reactions involving $M(H_2O)_6^{2+}$ and $M(H_2O)_5OH^{2+}$ (M = Cr(III), Fe(III)) was taken as evidence for an inner sphere electron transfer process. Stranks′ procedure involved calculation of the internal reorganization (IR), solvent reorganization (SR), coulombic work (coul), and Debye Hueckel (DH) contributions to $\Delta V^{\#}$ according to the expression in (13). Unfortunately,

$$\Delta V^{\#} = \Delta V^{\#}_{IR} + \Delta V^{\#}_{SR} + \Delta V^{\#}_{coul} + \Delta V^{\#}_{DH} \tag{13}$$

Wherland[69] pointed out that the sign of the $\Delta V^{\#}_{DH}$ term had become inversed in Stranks′ paper[68], with the result that the apparent agreement for outer sphere processes is in error and the calculated values deviate significantly from the observed $\Delta V^{\#}$′s. In a recent study, Spiccia and Swaddle[70] reported a $\Delta V^{\#}$ of -21 cm^3 mol^{-1} for the outer sphere electron transfer between MnO_4^- and MnO_4^{2-}. The theoretical value according to the corrected Stranks equation is -6.6 cm^3 mol^{-1}, which also underlines the invalidity of this equation. However, we[71] recently found $\Delta V^{\#}$ for the reaction between $Co(terpy)_2^{2+}$ and $Co(bipy)_3^{3+}$ to be -9.4 ± 0.4 cm^3 mol^{-1}, which is remarkably close to the theoretically predicted value of -7.3 cm^3 mol^{-1}. This apparent agreement may be due to the fact that this is the only system that could be studied at low ionic strengths. Swaddle et al.[67,70,72] have in the meantime modified the theoretical prediction of $\Delta V^{\#}$ for outer sphere electron transfer reactions, and more experimental results to probe these equations are forthcoming.

   The pressure dependence of a series of outer sphere electron transfer reactions in which significant ion-pair formation occurs, was also investigated[73,74]. In this case K and k_{ET} could be separated kinetically for the reactions given in (14), and the pressure dependence of k_{ET} resulted in

$$Co(NH_3)_5X^{3+} + Fe(CN)_6^{4-} \underset{}{\overset{K}{\rightleftharpoons}} \{Co(NH_3)_5X^{3+} \cdot Fe(CN)_6^{4-}\}$$

X = H$_2$O, Me$_2$SO, py

$$\Big\downarrow k_{ET}$$

$$Co^{2+} + 5NH_3 + Fe(CN)_6^{3-} + X \tag{14}$$

the following $\Delta V^{\#}$ values: $+26.5 \pm 2.4$ (X = H$_2$O), $+29.8 \pm 1.4$ (X = py) and $+34.4 \pm 1.1$ (X = Me$_2$SO) cm^3 mol^{-1}. Partial molar volume considerations showed that these volume increases are due to the large volume increase in going from the precursor to the successor ion-pair, e.g. 65 cm^3 mol^{-1} when $Fe(CN)_6^{4-}$ is oxidized to $Fe(CN)_6^{3-}$ and $Co(NH_3)_5X^{3+}$ is reduced to "$Co(NH_3)_5X^{2+}$". This means that the transition state for the electron transfer step is approximately halfway between the two ion-pairs on a volume basis, and that the metal centers have a geometry halfway between the oxidation states involved.

Halpern et al.[75] studied the pressure dependence of some typical inner-sphere electron transfer reactions of the type shown in (15), and their

$$Co(NH_3)_5X^{2+} + Fe(H_2O)_6^{2+} \longrightarrow Co^{2+} + 5NH_3 + Fe(H_2O)_5X^{2+} + H_2O \qquad (15)$$

results are summarized along with data for these reactions in Me_2SO in Table VI. It is seen that these values are indeed significantly smaller than those reported for similar outer sphere type of reactions, which may be the basis for future mechanistic assignments. A particularly interesting aspect of the data in Table VI is our ability to rule out Watts´ hypothesis[78] that the reactive Fe(II) species is tetrahedral in the bridged intermediate for Me_2SO as solvent. If this were the case, one would expect $\Delta V^{\neq}$ to be significantly larger in Me_2SO than in H_2O. Finally, the internal redox decomposition of $Co(NH_3)_5OSO_2^+$ exhibits a remarkably high $\Delta V^{\neq}$ value (+3.5 cm^3 mol^{-1} [79]), which can either be due to the dissociation of NH_3 prior to electron transfer due to the strong trans labilization of sulfite, or due to the break up of the solvent cage around the newly-formed {$Co(NH_3)_5^{2+}$, ˙OSO_2^-} following Co-O homolysis[67].

TABLE VI. Volumes of activation for the reduction of $Co(NH_3)_5X^{2+}$ by Fe(II)

X	$\Delta V^{\neq}$ in H_2O cm^3 mol^{-1}	$\Delta V^{\neq}$ in Me_2SO cm^3 mol^{-1}
F	+10.7[a]	+10.3[b]
Cl	+8.7[a]	+3.8[b]
Br	+6.4[a]	0.0[b]
N_3	+12.1[c]	+6.5[c]

[a]Recalculated from ref. 75
[b]Ref. 76
[c]Ref. 77

5. MISCELLANEOUS REACTIONS OF TRANSITION METAL COMPLEXES

Processes that involve addition or elimination reactions are expected mainly to exhibit intrinsic volumes of activation when the process involves the addition or elimination of neutral species. For instance, CO_2 uptake by complexes of the type $M(NH_3)_5OH^{2+}$ to produce $M(NH_3)_5OCO_2^+$ is characterized by negative volumes of activation, viz. -10.1 (M = Co(III)), -4.7 (M = Rh(III)) and -4.0 (M = Ir(III)) cm^3 mol^{-1}, whereas the reverse de-carboxylation reactions are characterized by positive volumes of activation, viz. +6.8 (M = Co(III)), +5.2 (M = Rh(III)) and + 2.5 (M = Ir(III)) cm^3 mol^{-1} [80]. Much larger volume changes have been reported for oxidative addition reactions involving the addition of HCl, CH$_3$I and H$_2$ to Ir(I) centers with $\Delta V^{\#}$ ranging between −18 and −30 cm^3 mol^{-1} [81-83]. Extremely negative values were found[84] for the oxidative addition of O$_2$ to Ir(cod)(phen)$^+$ and Ir(cod)(phen)I (cod = cycloocta-1,5- diene, phen = 1,10-phenanthroline) to produce Ir(cod)(phen)O$_2^+$, viz. −31.1 $\pm$ 1.7 and −44.4 $\pm$ 1.6 cm^3 mol^{-1}, respectively. Such large negative volumes of activation demonstrate the importance of bond formation in these processes and the significant catalytic effect pressure would have in such reactions.

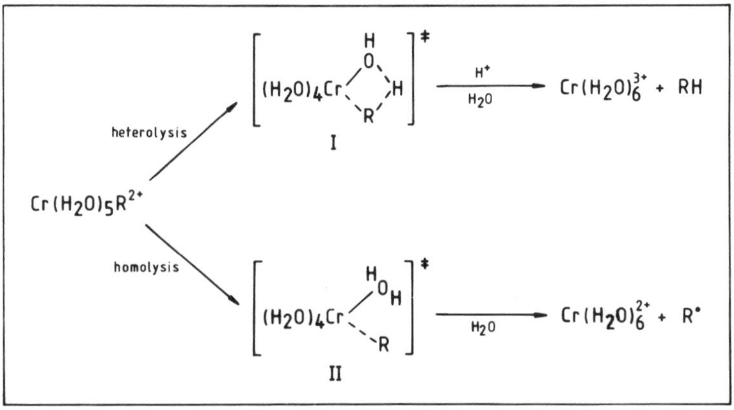

Figure 12. Transition states for the heterolysis and homolysis reactions of organochromium(III) species (ref. 84).

Homolytic and heterolytic decomposition reactions of coordination complexes are expected to exhibit characteristic pressure dependencies. Heterolysis of Cr(H$_2$O)$_5$R^{2+} exhibits volumes of activation of +0.3 $\pm$ 0.2 and −0.2 $\pm$ 0.2 cm^3 mol^{-1} for R = C(CH$_3$)$_2$OH and CH(CH$_3$)$_2$, respectively[84]. These small effects point to a transition state (I in Figure 12) that involves no net development of charge or major net changes in bond lengths. On the contrary, $\Delta V^{\#}$ for homolysis of these complexes is +15.1 $\pm$ 1.6 and +26 $\pm$ 2 cm^3 mol^{-1}, respectively. These large positive values are ascribed to massive desolvation, i.e. breakup of the solvent cage, as the organic radicals separate from Cr^{2+} in the transition state (II in Figure 12). Such mechanistic detail could not be concluded from the conventional activation parameters and demonstrates the useful application of high pressre measurements in such systems.

6. CONCLUDING REMARKS

We have described in this paper the advances that have been made in the understanding of inorganic reaction mechanisms through the application of high pressure kinetic techniques. Attention has often been called in the past to the apparent correlation between $\Delta S^{\#}$ and $\Delta V^{\#}$ data[3]. However, there is no thermodynamic equation that links $\Delta V^{\#}$ and $\Delta S^{\#}$ directly, and it could be rather misleading to use such a generalized correlation. There are numerous examples where these parameters have opposite signs, although some form of relation between the randomness at the transition state and its molar volume is expected especially for significantly positive or significantly negative $\Delta S^{\#}$ and $\Delta V^{\#}$. In this respect it is important to note that $\Delta S^{\#}$ data are usually subjected to large error limits since they are determined via an extrapolation of $1/T \rightarrow 0$. In contrast, $\Delta V^{\#}$ can usually be determined fairly accurately since it is estimated from the slope of a plot, similar to $\Delta H^{\#}$.

The correlations referred to above have led to the development of theoretical models to understand the overall trend. In this respect Phillips[85],[86] introduced the composite ionberg-iceberg model and suggested that there is a gas of ionized or dissociated water molecules between large solute molecules and ice casings. This model enables him to account for the slope of the $\Delta S^{\#}/\Delta V^{\#}$ correlations.

High pressure studies have also led to a modification of the transition state theory by introducing stochastic[87] and transport[88] models. In these models special emphasis is placed on the role of solvent friction and the transport aspect of the reaction. In other words, the role of solvent dynamics in the interpretation of pressure effects on rate constants is taken into account[3]. A major advantage of the volume of activation is the fact that it is the only activation parameter than can be correlated with a ground state property in an experimentally rather simple manner. Expecially in the case of non-symmetrical reactions, as treated in this paper, it is essential to construct a volume profile before a mechanistic assignment based on $\Delta V^{\#}$ can be made with some certainty.

As kineticists it is our goal to contribute to the understanding of the intimate mechanism of a chemical reaction. High pressure kinetic data adds a further dimension to the information obtained from conventional kinetic measurements at ambient pressure. It has almost become a necessity for the realistic assignment of the underlying reaction mechanism.

ACKNOWLEDGEMENTS
The author gratefully acknowledges financial support from the Deutsche Forschungsgemeinschaft, the Fonds der Chemischen Industrie, the Max-Buchner-Forschungstiftung and the Scientific Affairs Division of NATO. He sincerely appreciates the fine collaboration with and stimulating environment created by those cited on the papers from our laboratories.

7. REFERENCES

1. D.A. Palmer and H. Kelm, Coord. Chem. Rev., 36, 89 (1981).
2. R. van Eldik, in Mechanisms of Inorganic and Organometallic Reactions Vol. 3, M. Twigg (Ed.), Plenum, 399 (1985).
3. R. van Eldik (Ed.), Inorganic High Pressure Chemistry: Kinetics and Mechanisms, Elsevier, 448pp (1986).
4. W.J. le Noble and H. Kelm, Angew. Chem. Int. Ed. Engl., 19, 841 (1980).
5. A.E. Merbach, Pure Appl. Chem., 54, 1479 (1982).
6. T.W. Swaddle, in Mechanistic Aspects of Inorganic Reactions, D.B. Rorabacher and J.F. Endicott (Eds.), ACS Symposium Series no 198, 39 (1982).
7. T.W. Swaddle, in Inorganic and Bioinorganic Mechanisms Vol. 2, A.G. Sykes (Ed.), Academic Press, 95 (1983).
8. M.J. Blandamer and J. Burgess, Pure Appl. Chem., 55, 55 (1983).
9. P. Moore, Pure Appl. Chem., 57, 347 (1985).
10. R. van Eldik, Comments Inorg. Chem., 5, 135 (1986).
11. R. van Eldik, Angew. Chem. Int. Ed. Engl., 25 673 (1986)
12. R. van Eldik, in Inorganic High Pressure Chemistry: Kinetics and Mechanisms, R. van Eldik (Ed.), Elsevier, Chapter 1 (1986).
13. H. Kelm (Ed.), High Pressure Chemistry, D. Reidel Publishing Co., 600pp (1978).
14. Y. Ducommun and A.E. Merbach, in Inorganic High Pressure Chemistry: Kinetics and Mechanisms, R. van Eldik (Ed.), Elsevier, Chapter 2 (1986).
15. C.H. Langford and H.B. Gray, Ligand Substitution Processes, W.A. Benjamin Inc., 1965.
16. M.L. Tobe, Inorganic Reaction Mechanisms, Nelson, Chapter 5 (1972).
17. E.L.J. Breet and R. van Eldik, Inorg. Chem., 23, 1865 (1984).
18. T.W. Swaddle, Inorg. Chem., 22, 2663 (1983).
19. L. Helm, L.I. Elding and A.E. Merbach, Helv. Chim. Acta, 67, 1453 (1984).
20. L. Helm, L.I. Elding and A.E. Merbach, Inorg. Chem., 24, 1719 (1985).
21. L. Helm, M. Kotowski, A.E. Merbach and R. van Eldik, unpublished results.
22. M. Kotowski and R. van Eldik, Inorg. Chem., in press.
23. M. Kotowski and R. van Eldik, in Inorganic High Pressure Chemistry: Kinetics and Mechanisms, R. van Eldik (Ed.), Elsevier, Chapter 4 (1986).
24. Y. Ducommun, P.J. Nichols, L. Helm, L.I. Elding and A.E. Merbach, J. de Phys., 45, C8-221 (1984), see also ref. 19.
25. J. Berger, M. Kotowski and R. van Eldik, unpublished results, data taken from J. Berger, Diplomarbeit, University of Frankfurt (1986).
26. R. van Eldik, D.A. Palmer, R. Schmidt and H. Kelm, Inorg. Chim. Acta, 50, 131 (1981).
27. D.A. Palmer, R. Schmidt, R. van Eldik and H. Kelm, Inorg. Chim. Acta, 29, 261 (1978).

28. E.L.J. Breet, R. van Eldik and H. Kelm, Polyhedron, 2, 1181 (1983).
29. M. Kotowski and R. van Eldik, Inorg. Chem., 23, 310 (1984).
30. M. Kotowski, D.A. Palmer and H. Kelm, Inorg. Chem., 18, 2555 (1979).
31. J.G. Kirkwood, J. Chem. Phys., 2, 351 (1934).
32. R. van Eldik, in Inorganic High Pressure Chemistry: Kinetics and Mechanisms, R. van Eldik (Ed.), Elsevier, Chapter 3 (1986).
33. E.F. Caldin, M.W. Grant, B.B. Hasinoff and P.A. Tregloan, J. Phys. (E), 6, 349 (1973).
34. R. Mohr and R. van Eldik, Inorg. Chem., 24, 3396 (1985).
35. P.J. Nichols, Y. Ducommun and A.E. Merbach, Inorg. Chem., 22, 3993 (1983).
36. R. Doss and R. van Eldik, Inorg. Chem., 21, 4108 (1982).
37. R. Mohr, L.A. Mietta, Y. Ducommun and R. van Eldik, Inorg. Chem., 24, 757 (1985).
38. T. Inoue, K. Kojima and R. Shimozawa, Inorg. Chem., 22, 3972 (1983).
39. T. Inoue, K. Sugahara, K. Kojima and R. Shimozawa, Inorg. Chem., 22, 3977 (1983).
40. W.E. Jones, L.R. Carey and T.W. Swaddle, Can. J. Chem., 50, 2739 (1972).
41. G.A. Lawrance, Inorg. Chem., 21, 3687 (1982).
42. N. Curtis, G.A. Lawrance and R. van Eldik, prepared for publication.
43. H.R. Hunt and H. Taube, J. Am. Chem. Soc., 80, 2642 (1958).
44. T.W. Swaddle and D.R. Stranks, J. Am. Chem. Soc., 94, 8357 (1972).
45. G. Guastalla and T.W. Swaddle, Can. J. Chem., 51, 821 (1973).
46. D.A. Palmer and H. Kelm, Inorg. Chem., 16, 3139 (1977).
47. Y. Kitamura, R. van Eldk and H. Kelm, Inorg. Chem., 23, 2038 (1984).
48. Y. Kitamura and R. van Eldik, Ber. Bunsenges. Phys. Chem., 88, 418 (1984).
49. Y. Kitamura, R. van Eldik and C.R. Piriz Mac-Coll, Inorg. Chem., in press.
50. K.R. Brower and T.S. Chen, Inorg. Chem., 12, 2198 (1973).
51. J. Burgess, A.J. Duffield and R. Sherry, J. Chem. Soc., Chem. Commun., 350 (1980).
52. M.J. Blandamer, J. Burgess, J.G. Chambers and A.J. Duffield, Transition Met. Chem., 6, 156 (1981).
53. H.-T. Macholdt and H. Elias, Inorg. Chem., 23, 4315 (1984).
54. H.-T. Macholdt and R. van Eldik, Transition Met. Chem., 10, 323 (1985).
55. H.-T. Macholdt, R. van Eldik and G.R. Dobson, Inorg. Chem., 25, 1914 (1986).
56. G.R. Dobson, Z.Y. Al-Saigh and N.S. Binzet, J. Coord. Chem., 11, 159 (1981).
57. D. Taube, R. van Eldik and P.C. Ford, Organometallics, in press, and references cited therein.
58. G. Schmidt, H. Elias and R. van Eldik, prepared for publication.

59. G.A. Lawrance and D.R. Stranks, Acc. Chem. Res., 12, 403 (1979).
60. P.L. Kendall, G.A. Lawrance and D.R. Stranks, Inorg. Chem., 17, 1166 (1978).
61. D.R. Stranks and N. Vanderhoek, Inorg. Chem., 15, 2639 (1976).
62. M. Mares, D.A. Palmer and H. Kelm, Inorg. Chim. Acta, 27, 153 (1978).
63. W.J. Louw, Inorg. Chem., 16, 2147 (1977), and references cited therein.
64. R. Romeo, D. Minniti and M. Trozzi, Inorg. Chim. Acta, 14, L15 (1975).
65. R. Romeo, D. Minniti and M. Trozzi, Inorg. Chem., 15, 1134 (1976).
66. R. van Eldik, D.A. Palmer and H. Kelm, Inorg. Chem., 18, 572 (1979).
67. T.W. Swaddle, in Inorganic High Pressure Chemistry: Kinetics and Mechanisms, R. van Eldik (Ed.), Elsevier, Chapter 5 (1986).
68. D.R. Stranks, Pure Appl. Chem., 38, 303 (1974).
69. S. Wherland, Inorg. Chem., 22, 2349 (1983).
70. L. Spiccia and T.W. Swaddle, J. Chem. Soc., Chem. Commun., 67 (1985).
71. P. Braun and R. van Eldik, J. Chem. Soc., Chem. Commun., 1349 (1985).
72. T.W. Swaddle and L. Spiccia, Physica B, 139/140, 684.
73. R. van Eldik and H. Kelm, Inorg. Chim. Acta, 73, 91 (1983).
74. I. Krack and R. van Eldik, Inorg. Chem., 25, 1743 (1986).
75. J.P. Candlin and J. Halpern, Inorg. Chem., 4, 1086 (1965).
76. R. van Eldik, D.A. Palmer and H. Kelm, Inorg. Chim. Acta, 29, 253 (1978).
77. R. van Eldik, Inorg. Chem., 21, 2501 (1982).
78. B.A. Matthews and D.W. Watts, Inorg. Chim. Acta, 11, 127 (1974).
79. R. van Eldik, Inorg. Chem., 22, 353 (1983).
80. U. Spitzer, R. van Eldk and H. Kelm, Inorg. Chem., 21, 2821 (1982).
81. M. Walper and H. Kelm, Z. Phys. Chem. N.F., 113, 207 (1978).
82. H. Stieger and H. Kelm, J. Phys. Chem., 77, 290 (1973).
83. R. Schmidt, M. Geis and H. Kelm, Z. Phys. Chem. N.F., 92, 223 (1974).
84. M.J. Sisley, W. Rindermann, R. van Eldik and T.W. Swaddle, J. Am. Chem. Soc., 106, 7432 (1984).
85. J.C. Phillips, J. Chem. Phys., 81, 478 (1984).
86. J.C. Phillips, J. Phys. Chem., 89, 3060 (1985).
87. J. Jonas, Acc. Chem. Res., 17, 74 (1984).
88. J. Troe, J. Phys. Chem., 90, 357 (1986).

INORGANIC PHOTOCHEMISTRY AT HIGH PRESSURE

Rudi van Eldik
Institute for Physical and Theoretical Chemistry
University of Frankfurt
Niederurseler Hang
6000 Frankfurt am Main 50
Federal Republic of Germany

ABSTRACT. Pressure effects on the photochemical reactions of transition metal complexes in solution are discussed to demonstrate the information that can be obtained on the dynamics of the excited state species. Special attention is given to photochemical substitution, charge transfer, and isomerization reactions, and typical examples are discussed in each case. A critical evaluation is given to emphasize the limitations of this method at the present stage.

1. INTRODUCTION

In two earlier chapters (see contributions by Merbach and van Eldik) it was demonstrated how high pressure kinetic methods can be employed to assist in the elucidation of the mechanisms of thermal reactions in inorganic chemistry. In the case of symmetrical reactions where the reaction volume is zero, $\Delta V^{\#}$ alone can be used as a mechanistic indicator. However, in the case of unsymmetrical reactions ($\Delta \bar{V} \neq 0$), $\Delta V^{\#}$ must be seen in terms of the overall volume change, and a volume profile must be constructed in order to make a realistic mechanistic assignment. In this contribution we want to extend this method to the dynamics of excited state species and demonstrate how one can try to obtain insight into the mechanistic nature of photochemical reactions. In contrast to thermal processes where the application of high pressure techniques can reveal <u>additional</u> mechanistic information, high pressure studies is one of the few methods available to give mechanistic information concerning the reactions of excited state species[1]. In this paper we will focus on the effect of pressure on photochemical processes, since the pressure dependence of photophysical processes is addressed in the contribution by Drickamer. Furthermore, we will distinguish in our presentation between ligand field and charge transfer photochemical processes in photosubstitution and photoredox reactions, respectively.

R. van Eldik and J. Jonas (eds.), High Pressure Chemistry and Biochemistry, 357–378.
© *1987 by D. Reidel Publishing Company.*

2. PHOTOCHEMICAL SUBSTITUTION REACTIONS OF TRANSITION METAL COMPLEXES

In photochemical substitution reactions of transition metal complexes we are dealing with the substitution behaviour of ligand field excited state species. Since so little is known about the molecular dynamics of such species, the pressure dependence of this reaction may help reveal the intimate nature of the process.

Ligand field photolysis of Cr(III) ammine complexes occurs according to the reactions given in (1). The two photoreactions are thought to occur via different electronic excited states, their exact nature still being

$$Cr(NH_3)_5X^{(3-n)+} + H_2O \quad \xrightarrow[\text{LF}]{h\nu} \quad \begin{array}{l} Cr(NH_3)_5H_2O^{3+} + X^{n-} \\[1em] cis\text{-}Cr(NH_3)_4(H_2O)X^{(3-n)+} + NH_3 \end{array} \qquad (1)$$

$$(n = 1: X = Cl, Br, NCS; n = 0: X = NH_3)$$

debated in the literature. The pressure dependence of the photoaquation quantum yields resulted in significantly negative apparent volumes of activation[2], viz. an average value of -6 cm^3 mol^{-1} for aquation of NH$_3$, and between -10 and -13 cm^3 mol^{-1} for aquation of X^{n-}. Since little is known about the pressure dependence of the other deactivation processes, it was assumed that the apparent volumes of activation mainly represent that for the primary photochemical reaction. The mentioned results were interpreted in terms of an I$_a$ substitution mechanism, the more negative values for aquation of X^{n-} being ascribed to a solvational contribution originating from partial charge separation in the transition state.

Significantly more progress was made in studying the photosolvolysis reactions of Rh(III) ammine complexes[3-6] outlined in (2). Earlier studies by Ford and co-workers and others (see literature cited in ref. 3) indicated that ligand field photolysis can be presented by the Jablonski diagram given in Figure 1. The lowest energy excited state is a ligand field triplet state

$$Rh(NH3)_5X^{(3-n)+} + S \quad \xrightarrow[\text{LF}]{h\nu} \quad \begin{array}{l} Rh(NH_3)_5S^{3+} + X^{n-} \\[1em] trans\text{-}Rh(NH_3)_4(S)X^{(3-n)+} + NH_3 \end{array} \qquad (2)$$

$$(S = solvent)$$

from which the primary photoreaction (k$_p$), nonradiative (k$_n$) and radiative deactivation (k$_r$) occur. It can furthermore be assumed that $\phi_{ISC} \sim 1$ and k$_r$ $\ll$ (k$_n$ + k$_p$), so that the expression for the photochemical quantum yield,

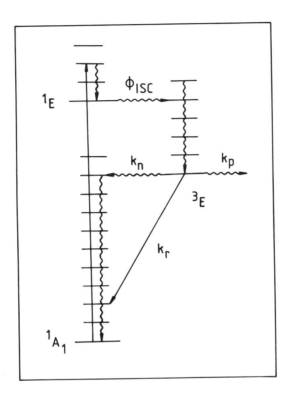

Figure 1. Jablonski diagram for the ligand-field photolysis of Rh(III) ammine complexes.

ϕ_p, can be simplified as shown in (3). In order to obtain k_p and its pressure

$$\phi_p = \phi_{ISC}\, k_p / (k_p + k_n + k_r)$$

$$\approx k_p / (k_p + k_n) = k_p \tau \qquad (3)$$

dependence, it is necessary to measure both ϕ_p and τ as a function of pressure. In terms of volumes of activation this can be expressed as in (4). This means that the volume of activation for the primary photochemical

$$\Delta V^{\#}_p = \Delta V^{\#}_{\phi} - \Delta V^{\#}_{\tau} \qquad (4)$$

reaction (k_p) is the difference in the volumes of activation determined from the pressure dependencies of ϕ_p and τ, the lifetime of the excited state.

A typical example of the pressure dependence of the photoaquation quantum yield for the $Rh(NH_3)_5Cl^{2+}$ complex is given in Figure 2. The two aquation reactions exhibit completely different pressure dependencies: an

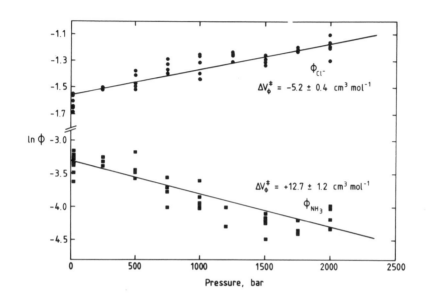

Figure 2. Pressure dependence of the reaction

$$Rh(NH_3)_5Cl^{2+} + H_2O \xrightarrow[LF]{h\nu} \begin{array}{l} Rh(NH_3)_5H_2O^{3+} + Cl^- \\ \\ trans\text{-}Rh(NH_3)_4(H_2O)Cl^{2+} + NH_3 \end{array}$$

increase in quantum yield with increasing pressure for the aquation of the chloride ligand and a decrease in quantum yield with increasing pressure for the aquation of ammonia. Similarly, the pressure dependence of the excited state lifetime was measured using pulsed laser techniques[1,3], and these are summarized along with the corresponding $\Delta V^{\#}_{\phi}$ values for a range of systems in Table I. From the available data it is also possible to determine the volume of activation for the non-radiative decay process by using eq. (5).

$$\Delta V^{\#}_n = \Delta V^{\#}_{\tau^{-1}} - \frac{\phi_1 \Delta V^{\#}_{\phi_1} + \phi_2 \Delta V^{\#}_{\phi_2}}{1 - \phi_1 - \phi_2} \tag{5}$$

Table I. Volumes of activation from photochemical and photophysical measurements on the photosolvolysis reactions of Rh(III) ammine complexes at 25°C

Solvent	Complex	Photosolvolysis product	$\Delta V^{\#}_{\phi}$	$\Delta V^{\#}_{\tau-1}$	$\Delta V^{\#}_{p}$	$\Delta V^{\#}_{n}$	Ref.
H_2O	$Rh(NH_3)_5Cl^{2+}$	$Rh(NH_3)_5H_2O^{3+}$	-5.2 ± 0.4	$(-3.4)^a$	-8.6 ± 1.6	$(-2.6)^a$	3
		$\underline{trans}-Rh(NH_3)_4(H_2O)Cl^{2+}$	$+12.7 \pm 1.2$		$+9.3 \pm 1.9$		
D_2O	$Rh(ND_3)_5Cl^{2+}$	$Rh(ND_3)_5D_2O^{3+}$	-4.2 ± 0.5	-3.5 ± 1.1	-7.7 ± 1.6	-2.6 ± 1.0	3
		$\underline{trans}-Rh(ND_3)_4(D_2O)Cl^{2+}$	$+9.5 \pm 1.6$		$+6.0 \pm 2.2$		
H_2O	$Rh(NH_3)_5Br^{2+}$	$Rh(NH_3)_5H_2O^{3+}$	-10.3 ± 1.2	$(+3.5)^a$	-6.8 ± 1.6	$(+2.5)^a$	3
		$\underline{trans}-Rh(NH_3)_4(H_2O)Br^{2+}$	$+4.6 \pm 0.6$		$+8.1 \pm 1.2$		
D_2O	$Rh(ND_3)_5Br^{2+}$	$Rh(ND_3)_5D_2O^{3+}$	-9.4 ± 1.5	$+4.1 \pm 0.6$	-5.3 ± 1.8	$+2.5 \pm 1.2$	3
		$\underline{trans}-Rh(ND_3)_4(D_2O)Br^{2+}$	$+3.4 \pm 0.5$		$+7.5 \pm 1.1$		
H_2O	$\underline{trans}-Rh(NH_3)_4Cl_2^+$	$\underline{trans}-Rh(NH_3)_4(H_2O)Cl^{2+}$	$+2.5 \pm 0.5$	b	$+2.8 \pm 0.6$	$\sim 0^a$	4
H_2O	$\underline{trans}-Rh(NH_3)_4Br_2^+$	$\underline{trans}-Rh(NH_3)_4(H_2O)Br^{2+}$	$+3.4 \pm 0.7$	b	$+2.9 \pm 0.7$	$\sim 0^a$	4
H_2O	$Rh(NH_3)_6^{3+}$	$Rh(NH_3)_5H_2O^{3+}$	$+3.7 \pm 0.5$	b	$+3.9 \pm 0.5$	$\sim 0^a$	5
H_2O	$Rh(NH_3)_5I^{2+}$	$\underline{trans}-Rh(NH_3)_4(H_2O)I^{2+}$	$+0.3 \pm 0.1$	b	$+1.4 \pm 0.9$	$\sim 0^a$	5
H_2O	$Rh(NH_3)_5SO_4^+$	$Rh(NH_3)_5H_2O^{3+}$	-2.7 ± 0.4	b	-3.9 ± 0.6	$\sim 0^a$	5
FMA	$Rh(NH_3)_5Cl^{2+}$	$Rh(NH_3)_5FMA^{3+}$	-4.6 ± 0.7	-0.3 ± 0.4	-4.9 ± 1.1	$+0.2 \pm 0.5$	6
		$\underline{trans}-Rh(NH_3)_4(FMA)Cl^{2+}$	$+4.2 \pm 0.9$		$+3.9 \pm 1.3$		
DMF	$Rh(NH_3)_5Cl^{2+}$	$\underline{trans}-Rh(NH_3)_4(DMF)Cl^{2+}$	$+6.3 \pm 0.9$	$+1.3 \pm 0.2$	$+7.6 \pm 1.1$	$+0.7 \pm 0.3$	6
Me_2SO	$Rh(NH_3)_5Cl^{2+}$	$Rh(NH_3)_5Me_2SO^{3+}$	-7.8 ± 1.8	-1 ± 1	-8.9 ± 2.7	-1 ± 1	6
		$\underline{trans}-Rh(NH_3)_4(Me_2SO)Cl^{2+}$	$+4.4 \pm 0.9$		$+3.3 \pm 1.8$		

[a] Assumed value - see reference   [b] Not measured - see Discussion

The non—radiative deactivation process exhibits a rather small pressure dependence for the investigated reactions. Throughout the series of pentaammine complexes in Table I the solvolysis of NH_3 is accompanied by a positive $\Delta V^{\#}_p$ value, whereas the solvolysis of X^{n-} exhibits negative values for $\Delta V^{\#}_p$ throughout the series of X^{n-} and solvents. These data can be interpreted in a qualitative way by considering the positive $\Delta V^{\#}_p$ values for solvolysis of NH_3 as evidence for a dissociative mechanism, which can also account for the negative $\Delta V^{\#}_p$ values for the solvolysis of X^{n-} since bond breakage is then accompanied by charge creation and a significant increase in electrostriction, i.e. an overruling negative effect of $\Delta V^{\#}_{solv}$ compared to a positive contribution from $\Delta V^{\#}_{intr}$. However, a more correct way to interpret these $\Delta V^{\#}_p$ data is to construct a volume profile for the photochemical reaction, as mentioned in the Introduction. In order to do this we need the partial molar volume of the Rh(III) ammine complex in the lowest excited triplet state from which the reaction occurs.

The partial molar volumes of the reactant and product species in the ground state were all measured with the aid of a densitometer and the overall reaction volume could be calculated as explained before[3] (see contributions

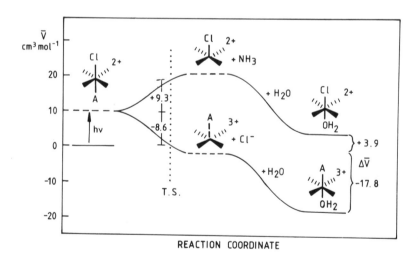

Figure 3. Volume profile for the reaction[1]

$$Rh(NH_3)_5Cl^{2+} + H_2O \xrightarrow[LF]{h\nu} \begin{array}{l} trans\text{-}Rh(NH_3)_4(H_2O)Cl^{2+} + NH_3 \\[2ex] Rh(NH_3)_5H_2O^{3+} + Cl^- \end{array}$$

by le Noble and van Eldik). For instance $\Delta\bar{V}$ for the aquation of $Rh(NH_3)_5Cl^{2+}$ is 3.9 and -17.8 cm^3 mol^{-1} for the release of NH_3 and Cl^-, respectively. These numbers (a difference of nearly 22 cm^3 mol^{-1}) show the same trend as the $\Delta V^{\#}_p$ values for this complex with a difference of 17 cm^3 mol^{-1}. The corresponding volume profile is presented in Figure 3 based on the assumption that the ligand field excited triplet state has a partial molar volume approximately 10 cm^3 mol^{-1} larger than the ground state molecule, based on an increase in bond length of 0.1 Å as observed in low temperature spectroscopic measurements[7]. According to this profile both photoaquation reactions involve a dissociative (D) reaction mode and it is quite possible that the excess volume of the excited state is focused along the axis of the bond cleavage process to produce a "ground state" like five-coordinate intermediate. This would then account for the general observations that the $\Delta V^{\#}_p$ values for these D reactions are significantly less positive than those generally found for related thermal reactions[8].

The results in Table I for non-aqueous solvents exhibit a similar trend as ascribed above for the water case, and can be interpreted in a similar way. Another observation is that those complexes showing larger $\Delta V^{\#}_p$ values are those undergoing these excited state reactions with the slower rates,

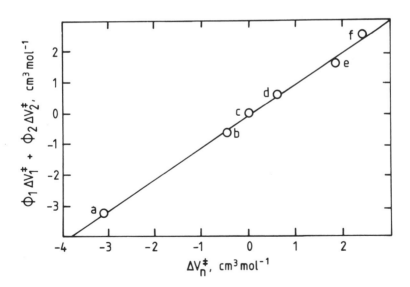

Figure 4. Plot of $\phi_1\Delta V^{\#}_1 + \phi_2\Delta V^{\#}_2$ versus $\Delta V^{\#}_n$ for the photosolvolysis of $Rh(NH_3)_5Cl^{2+}$ and $Rh(NH_3)_5Br^{2+}$ in various solvents[1]
Cl: a - D_2O; b - formamide; c - Me_2SO; d - DMF
Br: e - H_2O; f - D_2O

i.e. lower quantum yields. This can probably be attributed to the later transition states of the slower reactions, which should allow various rearrangements to make greater contributions to the observed activation parameters[1]. Finally, the results in Table I show that $\Delta V^{\#}_n$ has a small absolute value in all cases, indicating that earlier assumptions that $\Delta V^{\#}_n$ ~ 0 were not that unrealistic. It is interesting to note that there is a fairly good correlation between $\Delta V^{\#}_n$ and the sum of the products of $\Delta V^{\#}_p$ and ϕ_p, expressed as $\phi_1 \Delta V^{\#}_1 + \phi_2 \Delta V^{\#}_2$, in Figure 4. This may indicate the parallel character of k_p and k_n in terms of a strong coupling effect.

To summarize our findings for ligand field photolysis reactions of Cr(III) and Rh(III) ammine complexes, we would like to point to two difficulties with the interpretation of the $\Delta V^{\#}_p$ data. First of all is the actual meaning of the negative values reported for the solvolysis of X^{n-}, which could also be an indication of an associative substitution mode. For this reason we are presently investigating the ligand substitution reactions of low valency metal ion complexes of the type $M(CO)_6$, $M(CO)_5L$ and $M(CO)_4LL$, where M = Cr, Mo and W. In these systems no major changes in solvation are expected and the measured $\Delta V^{\#}_p$ data should represent the intrinsic contribution. Secondly, the uncertainty concerning the partial molar volume of the lowest energy excited state species and the location of the excess volume in the transition state makes the realistic construction of a volume profile rather uncertain. However, we have good reason to believe that this problem is not so severe in the case of charge-transfer photochemistry as seen from the discussion in the next section. Despite these uncertainties, we thoroughly believe that the cited results do demonstrate the ability of this technique to provide information on the dynamics of excited state species.

A few data have been reported for the photosolvolysis of Co(III) complexes. The photoaquation of $Co(CN)_6^{3-}$ produces $Co(CN)_5H_2O^{2-}$ and CN^-, and the quantum yield decreases slightly with increasing pressure[9] such that $\Delta V^{\#}_p = +2.0$ cm^3 mol^{-1} based on the assumption that all the other deactivation paths are pressure insensitive. A volume profile treatment revealed that a D mechanism should have a $\Delta V^{\#}_p$ as large as +9.4 cm^3 mol^{-1}, and an I_d mechanism was favoured[9]. However, in the light of the uncertainty in the partial molar volume of the excited state the experimental result may well be in agreement with a D mechanism. A very similar pressure dependence was reported[10] for the photoaquation of $Co(NH_3)_5Br^{2+}$. In this reaction the dissociative release of Br^- is also accompanied by charge creation, similar to that reported for the Rh(III) analogue. Drickamer and co-worker[11] studied the photoaquation of $Fe(CN)_6^{4-}$, which proceeds according to the overall reaction in (6). They report values of +6.2 and +5.1 cm^3 mol^{-1} for

$$Fe(CN)_6^{3-} + H_2O \underset{\Delta}{\overset{h\nu}{\rightleftharpoons}} Fe(CN)_5H_2O^{2-} + CN^- \qquad (6)$$

$\Delta V^{\#}{}_{\phi}$ in water and 20 % $EtOH-H_2O$, which results in +7.7 and +8.7 cm^3 mol^{-1} for $\Delta V^{\#}{}_p$ in the two solvents, respectively. The thermal back reaction showed a $\Delta V^{\#}$ of +13.5 $\pm$ 1.5 cm^3 mol^{-1}. Both these values are in good agreement with dissociative activation for the photochemical and thermal processes.

3. EFFECTS OF PRESSURE ON NON-RADIATIVE DEACTIVATION

We have already referred to the volumes of activation for the non-radiative deactivation of Rh(III) ammine complexes as summarized in Table I. Although it is not our intention to go into this topic in too much detail, since this has been done elsewhere[1,12], there is some very important information that must be mentioned here since it has direct consequences for the sections to follow. The pressure dependence of the emission lifetime of $Ru(bpy)_3^{2+}$ (bpy = 2,2´-bipyridine) in aqueous solution has been studied by two groups[13,14]. This complex shows little or no unimolecular photochemistry and the emission quantum yields are smaller than 0.1, so that the dominant deactivation process is non-radiative, i.e. $k_d = k_n$. In the earlier work[13] it was reported that $\Delta V^{\#}{}_n = -0.9 \pm 0.2$ cm^3 mol^{-1} at 18°C, from which it was generally concluded that non-radiative deactivation is essentially independent of pressure. However, in the recent work[14], the small negative $\Delta V^{\#}{}_n$ value was confirmed, but the authors found a dramatic increase in $\Delta V^{\#}{}_n$ with increasing temperature. For instance, $\Delta V^{\#}{}_n = +8.7$ and +7.5 cm^3 mol^{-1} for the deactivation in acetonitrile at 45°C and in water at 70°C, respectively. A similar result was found for the deactivation of Ru-$(phen)_3^{2+}$ for which $\Delta V^{\#}{}_n$ increases from +2.0 to +10.6 cm^3 mol^{-1} on going from 2 to 70°C. This behaviour was explained in terms of two competing non-radiative deactivation processes, one involving slow deactivation directly from the MLCT state, the second via a thermally activated higher energy ligand field state from which non-radiative deactivation is extremely rapid. This mechanism is presented schematically in Figure 5. From the available data it was estimated that $\Delta V^{\#}{}_n$ (CT $\rightarrow$ GS) = -2.2, -1.0 and -1.7 cm^3 mol^{-1} in H_2O, D_2O and MeCN, compared to $\Delta V^{\#}{}_n$ (CT $\rightarrow$ LF $\rightarrow$ GS) = +9.7, +9.1 and +10.1 cm^3 mol^{-1} in the same solvents, respectively. The ca. 10 $cm^3 mol^{-1}$ more positive value for deactivation via the LF state demonstrates very nicely that the partial molar volume of this state is significantly larger than that of the CT state. Furthermore, this difference is very close to the increased volume referred to in the previous section.

DiBenedetto et al.[15] studied the pressure tuning of the emission spectra and lifetimes of $Ir(Mephen)_2Cl_2^+$ and $Ir(bpy)_2Cl_2^+$ in DMF, where Mephen = 5,6-dimethyl-1,10-phenanthroline. From the pressure dependence of the emission spectra it could be calculated that the MLCT and LF states exhibit a volume difference of 4.2 $\pm$ 0.8 and 4.1 $\pm$ 0.8 cm^3 mol^{-1} for the Mephen and bpy complexes, respectively, with the LF state being higher in volume. This once again demonstrates the general concept of higher partial molar volumes for the LF excited states, probably due to the population of σ^*_{ML} orbitals and leading to bond extensions.

It can therefore be concluded that CT excited states are expected to be of a significantly lower volume than the LF excited states, and that they may in fact be very close in volume to that of the ground state species. Naturally, this observation has important meaning for the interpretation of the effect of pressure on charge transfer photochemical processes.

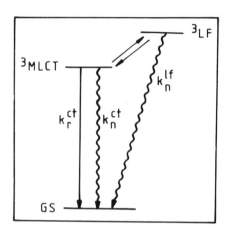

Figure 5. Model describing the proposed mechanism for decay of the MLCT state of Ru(bpy)$_3^{2+}$ and related complexes[1].

4. PHOTOCHEMICAL CHARGE TRANSFER REACTIONS OF TRANSITION METAL COMPLEXES

Only a few charge transfer photochemical reactions have been investigated under pressure. Kirk and Porter[13] found a $\Delta V^{\#}_{\phi}$ value of +4.8 cm^3 mol^{-1} for the CT photolysis of Co(NH$_3$)$_5$Br^{2+} and suggested the formation of a caged radical pair Co^{II}(Br$^{\cdot}$) from the LMCT state. Dissociation of this radical pair to the reaction products could account for the increase in volume as observed in the positive $\Delta V^{\#}_{\phi}$. However, a definitive interpretation of this data is hampered by the absence of information regarding the processes that lead to the formation and decay of the caged radical pair.

In a recent study in our laboratories[16], we investigated the CT photochemistry of trans-Pt(CN)$_4$(N$_3$)$_2^{2-}$ as a function of pressure. The overall reaction involves reductive elimination of azide to produce Pt-(CN)$_4^{2-}$ and nitrogen[17,18]. Irradiation at 313 nm produces LMCT excitation, and the quantum yield in water and ethanol as solvent decreases significantly with increasing pressure. The resulting $\Delta V^{\#}_p$ values are +8.1 $\pm$ 0.4 and +14.3 $\pm$ 0.9 cm^3 mol^{-1} for water and ethanol, respectively, calculated from the slope of plots of $\ln(\phi/1-\phi)$ versus pressure on the assumption that the non-radiative deactivation rate constants are independent of pressure (see scheme (7)). In this treatment it was assumed

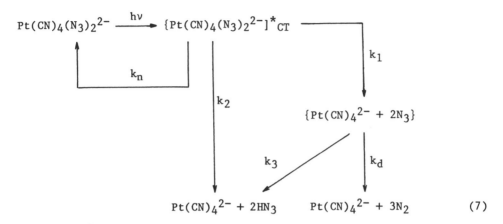

$$(7)$$

that the key photochemical reaction step occurs from the CT state directly populated during irradiation. The significantly positive $\Delta V^{\#}_p$ values clearly underline the dissociative nature of the photochemical reaction in both solvents. The difference in the values is significant and requires a detailed interpretation. We prefer a CT excited state model as outlined in (7) due to the wavelength independence of the observed quantum yield, and the non-unitary yield for radical pair formation observed in the present system. The reactive state is a thermally equilibrated ligand-to-metal CT excited state that can undergo non-radiative deactivation (k_n) and simultaneous Pt-N$_3$ bond cleavage (k_1) to produce a caged radical species, which subsequently decomposes to $Pt(CN)_4^{2-}$ and N_2 in water and ethanol ($k_d = 7 \times 10^9$ M^{-1} s^{-1} in water) or interacts with ethanol to produce HN$_3$ and ethanol radicals via the abstraction of hydrogen (k_3). The additional deactivation route for the CT excited state in ethanol can account for the higher quantum yield and $\Delta V^{\#}_p$ values observed in ethanol[16]. It follows that bond breakage plays an inportant role in both the deactivation routes k_1 and k_2. No major solvational changes are expected since there is no overall change in charge although the metal center is reduced from Pt(IV) to Pt(II). Alternative interpretations[16] also point to the dissociative decomposition of the caged radical pair.

The photooxidation of $Fe(CN)_5NO^{2-}$ has been studied by numerous investigators (see literature cited in ref. 19), and photooxidation of the metal center (i.e. MLCT) accompanied by solvation of the NO ligand was reported to be the major reaction mode in aqueous and non-aqueous liquid media (S = solvent in (8)). Little efforts have been made to study the intimate molecular nature of reaction (8). Arguments have been reported in

$$Fe(CN)_5NO^{2-} \xrightarrow[S]{h\nu} Fe^{III}(CN)_5S^{2-} + NO \qquad (8)$$

favour of both associative and dissociative reaction modes[20]. We found the quantum yield for the production of $Fe^{III}(CN)_5S^{2-}$ to depend on the irradiation wavelength (i.e. energy) and the nature of the solvent (H_2O, MeOH, Me_2SO, MeCN, DMF and C_5H_5N). The latter dependence could be described in terms of the Gutmann donor numbers and viscosity. The quantum yield decreases with increasing donor number and viscosity of the medium. A series of measurements in water–glycerol mixtures indicated that the quantum yield parallels the fluidity of the medium, and a plot of ϕ^{-1} versus η is linear up to 50 % glycerol, in the range where no observable ligand exchange involving glycerol occurred. A summary of the values of $\Delta V^{\#}_{\phi}$ calculated from plots of ln ϕ versus pressure (see typical examples in Figure 6) is given in Table II. These values seem to be fairly wavelength independent

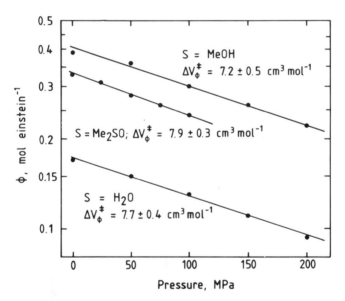

Figure 6. Plots of ln ϕ versus pressure for the photooxidation of $Fe(CN)_5NO^{2-}$ according to the overall reaction[19]

$$Fe(CN)_5NO^2 + S \xrightarrow[CT]{h\nu} Fe^{III}(CN)_5S^{2-} + NO$$

and only the quantum yield is significantly affected by the solvent. The latter trend correlates with the fluidity of the medium as mentioned above,

TABLE II. Values of $\Delta V^{\#}_{\phi}$ and $\Delta V^{\#}_{p}$ for the reaction

$$Fe(CN)_5NO^{2-} + S \xrightarrow{h\nu} Fe^{III}(CN)_5S^{2-} + NO$$

as a function of solvent and excitation energy

Solvent	λ_{irr} nm	ϕ [a] mol einstein^{-1}	$\Delta V^{\#}_{\phi}$ cm^3 mol^{-1}	$\Delta V^{\#}_{p}$ cm^3 mol^{-1}
H$_2$O	436	0.17	+7.7 ± 0.4	+8.8 ± 0.4
	313	0.37	+5.5 ± 0.7	+7.8 ± 1.0
MeOH	436	0.39	+7.2 ± 0.5	+10.3 ± 0.6
	313	0.63	+6.5 ± 0.8	+13.0 ± 1.9
Me$_2$SO	436	0.33	+7.9 ± 0.3	+11.1 ± 0.4
	405	0.39	+7.5 ± 1.1	+11.4 ± 1.6
	313	0.42	+8.4 ± 1.3	+14.1 ± 1.0

[a]Values at ambient pressure, i.e. 0.1 MPa

and points to the participation of a cage recombination mechanism as outlined in (9). In this scheme k_2 represents the rate constant for the radical pair formation step.

$$(S = \text{solvent}) \qquad Fe^{III}(CN)_5S^{2-} + NO \qquad (9)$$

The observed photooxidation yield for the mechanism outlined in (9) is given by the expression in (10), where ϕ_0 presents the primary quantum yield for radical pair formation/bond cleavage. ϕ_0 and k_3 are expected to be

$$\phi = \left(\frac{k_2}{k_1 + k_2}\right)\left(\frac{k_4}{k_3 + k_4}\right) = \phi_0 \left(\frac{k_4}{k_3 + k_4}\right) \qquad (10)$$

independent of viscosity, whereas k_4 is expected to decrease with increasing viscosity. It is therefore not surprising that ϕ^{-1} depends
linearly on the viscosity of the medium as reported above, with an inter-
cept ϕ_0^{-1}. The value of ϕ_0 is such that $\phi_{H_2O}/\phi_0 = 0.82$, indicating that
no significant cage effects are present in pure water. However, this ratio
deviates significantly from unity for the other solvents and cage effects do
play a significant role in those cases. The observed pressure effects and
apparent volumes of activation, $\Delta V^{\#}_{\phi}$, mainly present the contribution from
k_2, i.e. the formation of the caged radical pair, and support the
dissociative nature of this reaction. If we assume that k_1 for non-radiative
deactivation exhibits a minor pressure dependence[1,3,6], then $\Delta V^{\#}_p$ for the
k_2 step can be estimated from a plot of $\ln(\phi/1-\phi)$ versus pressure in the usual
way. The corresponding values in Table II show a very similar trend as $\Delta V^{\#}_{\phi}$,
and the larger values found in MeOH and Me_2SO are partly ascribed to the
pressure dependence of the viscosity of the solvent. The expected increase
in viscosity with increasing pressure, especially for Me_2SO, will result in
a decrease in ϕ and a more positive $\Delta V^{\#}$ value. The results definitely rule
out the possibility of an associative reaction mode[20] and reveal the
intimate nature of the dynamics of the excited state species.

In the final example, CT excitation results in a linkage isomerization
reaction, namely from an S- to an O-bonded sulfinato complex of
cobalt(III)[21,22]. This is a rather unique reaction in that the O-bonded
species exhibits a thermal back reaction to the S-bonded species, such that
a pressure dependence study of both the forward and back reactions should
enable the construction of a volume profile for the overall process
indicated in (11). It was shown[21,22] that the photoreactive state is a

$$(en)_2Co\begin{array}{c} O\ \ \ O \\ \diagdown S \diagup \\ \diagup \ \ \ \diagdown CH_2 \\ \ \ \ \ \ \ \ \ | \\ \diagdown \ \ \ \ \ CH_2 \\ NH_2 \diagup \end{array} \quad \underset{\Delta}{\overset{h\nu, CT}{\rightleftharpoons}} \quad (en)_2Co \begin{array}{c} O \\ \ \ \ \ \| \\ O - S \diagup \\ \diagup \ \ \ \ \ \ \diagdown \\ \ \ \ \ \ \ \ \ \ \ CH_2 \\ \diagdown \ \ \ \ \diagup \\ NH_2 - CH_2 \end{array} \qquad (11)$$

LMCT state corresponding to the $S(\sigma) \rightarrow Co(\sigma^*)$ transition at 288 nm. The
quantum yield for the forward reaction and the rate constant for the thermal
back reaction at 60°C exhibit opposite pressure dependencies as illustrated
by the data in Figure 7. These results indicate that the transition states
for the two processes must be completely different, which is not surprising
considering the different energetics involved. The following reaction
scheme in which CoSOON and CoOSON represent the S- and O-bonded isomers in
(12), respectively, can account for the pressure dependence of ϕ. The CT
excited state can either return to the ground state via non-radiative
deactivation (k_1), or produce a caged radical pair (k_2) which can
subsequently undergo cage recombination to produce the O-bonded (k_3) or S-

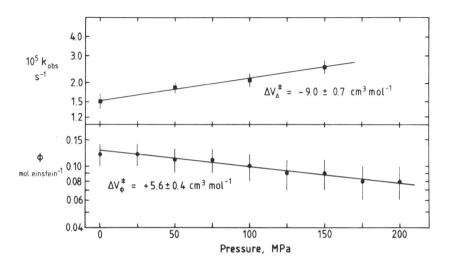

Figure 7. Pressure dependencies of the quantum yield for the forward and of the rate constant for the thermal back reaction in the overall process

$$Co(SOON)^{2+} \underset{\Delta}{\overset{h\nu, CT}{\rightleftharpoons}} Co(OSON)^{2+}$$

bonded (k_4) isomer. Radiative deactivation is omitted from the scheme since no emission could be detected. As a first approximation we can assume that cage recombination mainly leads to the O-bonded isomer and k_4 is small and

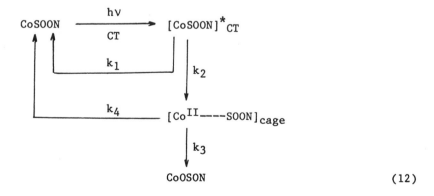

(12)

does not contribute towards ϕ. This can be rationalized assuming a weak bond between the oxygen lone pair and Co^{2+}. In this case the quantum yield is given by $\phi = k_2/(k_1 + k_2)$ or $\phi/(1 - \phi) = k_2/k_1$. From the plots of ln ϕ versus pressure (Figure 7) and $\ln(\phi/1 - \phi)$ versus pressure it follows that $\Delta V^{\#}_{\phi} = +5.8 \pm 0.5$ and $\Delta V^{\#}_{p} = \Delta V^{\#}(k_2) = +6.5 \pm 0.6$ cm^3 mol^{-1}, respectively. These positive values not only underline the dissociative nature of the photochemical process, but also prove the CT character of the reactive state. The equivalent reaction from a LF excited state would require a ring-opening step accompanied by significant charge creation and a large negative solvational contribution towards the observed $\Delta V^{\#}$ (compare discussion for thermal back reaction). If one does not accept k_1 to be indepedent of pressure, it follows that $\Delta V^{\#}_{p} = \Delta V^{\#}(k_2) - \Delta V^{\#}(k_1)$. Our earlier work[6] has demonstrated that $\Delta V^{\#}(k_2)$ and $\Delta V^{\#}(k_1)$ are expected to be of the same sign due to the strong-coupling effect. It follows that $\Delta V^{\#}(k_2) > \Delta V^{\#}(k_1)$ and the process must be dissociatively activated.

The value of ca. 6 cm^3 mol^{-1} for bond cleavage in the excited state is quite realistic for a process in which no significant charge creation occurs. The fairly low value of $\Delta V^{\#}$ demonstrates that no Co(II) (low spin) $\longrightarrow$ Co(II) (high spin) transition is likely to occur since high spin Co(II) complexes have significantly larger bond lengths which should lead to significantly larger $\Delta V^{\#}$ values. It is important to note that CT states are expected to be much less distorted from the ground state configuration than LF states as discussed in the previous section, such that the CT state can be expected to have a volume slightly larger than, or close to that of the ground state. If the cage recombination step is included in the scheme, the expression for the quantum yield modifies to $\phi = (k_2/k_1 + k_2)(k_3/k_3 + k_4)$. The rate constants for the cage recombination processes (k_3 and k_4) are expected to exhibit very similar pressure dependencies, such that the pressue dependence of ϕ will once again depend on that of k_1 and k_2 as treated before.

The thermal back reaction exhibits a fairly large and negative volume of activation, which can be ascribed to either an associative attack by a water molecule, or a ring-opening reaction of the sulfinate ligand accompanied by significant charge creation. The associative attack of a water molecule on a Co(III) center is rather unlikely in the light of much evidence in favour of an I_d or D mechanism for closely related complexes[24,25]. However, ring-opening of the sulfinate chelate will lead to a transition state of the type $[(en)_2Co^{3+}-NOSO^-]^{\#}$, which is expected to show an increase in electrostriction due to the creation of charges. An increase in the volume due to an intrinsic contribution from bond breakage (ring-opening) could be overruled by the negative solvational contribution to result in an overall $\Delta V^{\#}$ value of -9 cm^3 mol^{-1}. The postulated transition state also accounts for the formation of the five-membered ring since the lone pair of electrons is located on the sulfur atom. A very similar result was reported for the trans to cis isomerization of trans-$Cr(C_2O_4)_2(H_2O)_2^-$, for which $\Delta V^{\#} = -16 \pm 1$ cm^3 mol^{-1} and all the available evidence is consistent with a rate-determining opening of the oxalate ring[26].

We conclude from these measurements that both processes (forward and reverse) occur via ring-opening of the sulfinate ligand. No charge creation occurs in the case of the photochemical reaction, whereas significant charge creation in the thermal process results in a significantly more negative $\Delta V^{\#}$. In order to combine these data in an overall volume profile, the partial molar volumes of the reactant and product species were measured and turned out to be $\bar{V}(CoSOON) = 143.8 \pm 1.8$ cm^3 mol^{-1} and $\bar{V}(CoOSON) = 145.7 \pm 1.6$ cm^3 mol^{-1}. The slightly larger value for the latter species is in agreement with the six- instead of five-membered ring. In the volume profile in Figure 8 it was assumed that the CT state has a partial molar volume close to that of the ground state species as discussed above. The profile clearly demonstrates the difference of at least 13 cm^3 mol^{-1} between the two transition states and the effect of electrostriction during the thermal back reaction. In addition, this volume difference partly originates from the difference between Co^{II} and Co^{III}. To our knowledge this is the first example of a volume profile for a combined photochemical forward and thermal back reaction.

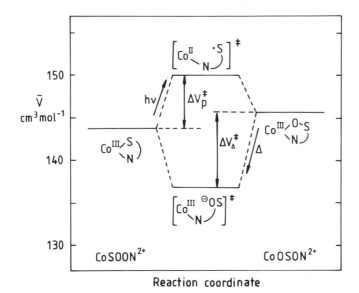

Figure 8. Volume profile for the system

$$Co(SOON)^{2+} \underset{\Delta}{\overset{h\nu,CT}{\rightleftharpoons}} Co(OSON)^{2+}$$

5. PHOTOCHEMICAL ISOMERIZATION REACTIONS OF TRANSITION METAL COMPLEXES

In the previous section we presented an example of a system in which CT excitation results in linkage isomerization of a sulfinato complex of Co(III). There are many ligand field photochemical reactions that result in geometrical isomerization, and especially the cis/trans photoisomerization of rhodium(III) ammine complexes has received significant attention during recent years. These systems are fairly well understood in terms of ligand dissociation from the hexacoordinate excited state prior to stereorearrangement, and several closely related theoretical models have been proposed to account for these effects. The product stereochemistries are principally interpreted in terms of the comparative energies of the square-pyramidal, five-coordinate apical (A^*) and basal (B^*) intermediates shown in Figure 9. These intermediates were proposed to be triplet excited states capable of isomerizaton on a time scale competitive with deactivation, followed by trapping of a solvent molecule. An important aspect of these models is that the stronger σ-donor ligand should show a strong site preference for the basal position in the five-coordinate intermediate. In this way the stereochemistry of the isomerization reactions could be predicted[27-29].

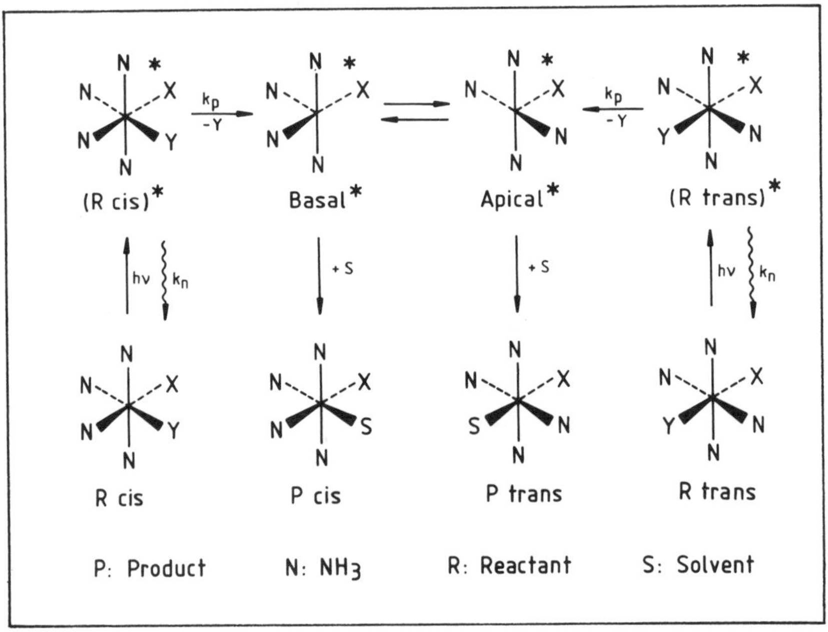

Figure 9. General scheme for the photochemical cis/trans isomerization of rhodium(III) ammine complexes

The pressure dependence of a series of such photoisomerization reactions was investigated along with the partial molar volumes of the reactant and product species[30]. The volumes of activation for the photochemical reaction, $\Delta V^{\#}_{p}$, were estimated from plots of $\ln(\phi/1-\phi)$ versus pressure on the basis of the ligand field excitation pattern reported for such substitution reactions of rhodium(III) complexes[3]. A summary of the values of $\Delta V^{\#}_{p}$ and the overall reaction volume $\Delta \bar{V}$ is given in Table III. In most of the cases the absolute value of $\Delta V^{\#}_{p}$ is very small and this complicates its interpretation. Similarly, $\Delta \bar{V}$ is also small and exhibits no specific correlation with $\Delta V^{\#}_{p}$. A detailed discussion of the data[30] indicated that the volume data do not provide a definitive description of the mechanism of the process but do underline the dissociative (I_d or D) nature of the process if a general mechanism is operable. Similar arguments to those presented before were adopted to access the partial molar volume of the LF triplet state[30]. It follows that the interpretation of small pressure effects in photochemical reactions is subjected to much uncertainty due to the complexity of the overall reaction scheme and various processes that can account for the observed effects.

TABLE III. Volumes of activation for the photoisomerization reactions of a series of tetraamminerhodium(III) complexes in aqueous solution

Complex	Principal product	$\Delta V^{\#}_{p}$ $cm^3\ mol^{-1}$	$\Delta \bar{V}$ $cm^3\ mol^{-1}$
cis-$Rh(NH_3)_4Cl_2^+$	$trans$-$Rh(NH_3)_4(H_2O)Cl^{2+}$	-3.5 ± 0.3	$+2.2 \pm 1.8$
cis-$Rh(NH_3)_4(H_2O)Cl^{2+}$	$trans$-$Rh(NH_3)_4(H_2O)Cl^{2+}$	0.0 ± 0.4	$+2.5 \pm 2.3$
cis-$Rh(NH_3)_4Br_2^+$	$trans$-$Rh(NH_3)_4(H_2O)Br^{2+}$	-2.3 ± 0.3	$+6.4 \pm 1.9$
cis-$Rh(NH_3)_4Br_2^+$	"$Rh(NH_3)_3(H_2O)Br_2^{+}$"	$+9.3 \pm 0.8$	$+0.7$
cis-$Rh(NH_3)_4(H_2O)Br^{2+}$	$trans$-$Rh(NH_3)_4(H_2O)Br^{2+}$	-1.0 ± 0.4	$+0.5 \pm 2.4$
$trans$-$Rh(NH_3)_4(OH)Cl^+$	cis-$Rh(NH_3)_4(OH)_2^+$	-8.8 ± 0.7	n.a.

n.a. - not available

6. CONCLUDING REMARKS

In the previous sections we have tried to draw together the reported studies using pressure effects to investigate the dynamics of excited state species in order to contribute to the mechanistic understanding of photochemical reactions. The selected examples clearly show that volumes of activation for excited state processes can provide valuable insight into the mechanistic interpretation of excited state reactions. It is not surprising that in some cases such effects have introduced ambiguities into the existing understanding of the investigated systems, and alternative possibilities must be considered. Naturally it should be kept in mind that $\Delta V^{\#}$ data alone cannot provide a definite description of an excited state mechanism[1]. Some of the difficulties lie in the unknown partial molar volumes of excited state species and the difficulty to separate intrinsic and solvational contributions towards $\Delta V^{\#}$.

In many cases high pressure techniques are employed to obtain further information on previously investigated reactions. The accuracy of the instrumentation employed to measure the pressure dependence of the quantum yield or rate constant for the photochemical process is such that it can also pinpoint information about the system not observed before. A typical example is our recent study[31] of the photoaquation of cis-$Rh(bpy)_2Cl_2^{+}$ in which differential spectrophotometry was used to determine the production of cis-$Rh(bpy)_2(H_2O)Cl^{2+}$. The observed differential spectrum of an irradiated sample versus an unirradiated sample exhibited small deviations from the theoretically expected spectrum for the conversion of the cis-dichloro to the cis-chloroaquo species. These deviations would never show up in the normal UV-VIS spectra, but this method is extremely sensitive to pick up any small deviations. Further studies[32] finally resulted in the identification of the species responsible for the observed deviations, which turned out to be trans-$Rh(bpy)_2(H_2O)Cl^{2+}$. This surprising result was supported by NMR measurements and a structure determination on a crystallized sample[33]. This is the first time that a trans-bisbipyridine complex of rhodium(III) was isolated and forced us to reconsider the excited state dynamics of such species to account for the formation of this product.

We are optimistic that such studies as described in this paper will continue to contribute to our understanding of the fundamental chemistry and the dynamic behaviour of excited state species of transition metal complexes in solution.

ACKNOWLEDGEMENTS

The author gratefully acknowledges financial support from the Deutsche Forschungsgemeinschaft, the Fonds der Chemischen Industrie, the Max-Buchner-Forschungsstiftung and the Scientific affairs Divison of NATO. He sincerely appreciates the fine collaboration with the group of Prof. Peter C. Ford at the University of California, Santa Barbara, and with those cited on the papers from this laboratory.

7. REFERENCES

1. P.C. Ford, in High Pressure Inorganic Chemistry: Kinetics and Mechanisms, R. van Eldik (Ed.), Elsevier, Chapter 6 (1986).
2. K. Angermann, R. van Eldik, H. Kelm and F. Wasgestian, Inorg. Chem., 20, 955 (1981).
3. W. Weber, R. van Eldik, H. Kelm, J. DiBenedetto, Y. Ducommun, H. Offen and P.C. Ford, Inorg. Chem., 22, 623 (1983).
4. L.H. Skibsted, W. Weber, R. van Eldik, H. Kelm and P.C. Ford, Inorg. Chem., 22, 541 (1983).
5. W. Weber and R. van Eldik, Inorg. Chim. Acta, 85, 147 (1984).
6. W. Weber, J. DiBenedetto, H. Offen, R. van Eldik and P.C. Ford, Inorg. Chem., 23, 2033 (1984).
7. R.B. Wilson and E.I. Solomon, J. Am. Chem. Soc., 102, 4085 (1980); K. Hakamata, A. Urushlyama and H. Kupka, J. Phys. Chem., 85, 1983 (1981).
8. R. van Eldik, in High Pressure Inorganic Chemistry: Kinetics and Mechanisms, R. van Eldik (Ed.), Elsevier, Chapter 3 (1986).
9. K. Angermann, R. van Eldik, H. Kelm and F. Wasgestian, Inorg. Chim Acta, 49, 247 (1981).
10. A.D. Kirk, C. Namasivayam, G.B. Porter, M.A. Rampi-Scandola and A. Simmons, J. Phys. Chem., 87, 3108 (1983).
11. M.I. Finston and H.G. Drickamer, J. Phys. Chem., 85, 50 (1981).
12. J. DiBenedetto and P.C. Ford, Coord. Chem. Rev., 64, 361 (1985).
13. A.D. Kirk and G.B. Porter, J. Phys. Chem., 84, 2998 (1980).
14. M.L. Fetterolf and H.W. Offen, J. Phys. Chem., 89, 3320 (1985); 90, 1828 (1986).
15. J. DiBenedetto, R.J. Watts and P.C. Ford, Inorg. Chem., 23, 3039 (1984).
16. W. Weber and R. van Eldik, Inorg. Chim. Acta, 111, 129 (1986).
17. A. Vogler, A. Kern and J. Huettermann, Angew. Chem., Int. Ed. Engl., 17, 524 (1978).
18. A. Vogler, A. Kern and B. Fusseder, Z. Naturforsch., Teil B, 33, 1352 (1978).
19. G. Stochel, R. van Eldik and Z. Stasicka, Inorg. Chem., in press.
20. S.K. Wolfe and J.H. Swinehart, Inorg. Chem., 14, 1049 (1975), and references cited therein.
21. H. Maecke, V. Houlding and A.W. Adamson, J. Am. Chem. Soc., 102, 6888 (1980).
22. V. Houldng, H. Maecke and A.W. Adamson, Inorg. Chem., 20, 4279 (1981).
23. W. Weber, H. Maecke and R. van Eldik, Inorg. Chem., in press.
24. R. van Eldik, D.A. Palmer and H. Kelm, Inorg. Chem., 18, 1520 (1979).
25. T.W. Swaddle, Inorg. Bioinorg. Mech., 2, 95 (1983).

26. P.L. Kendall, G.A. Lawrance and D.R. Stranks, Inorg. Chem., 17, 1166 (1978).
27. L.H. Skibsted and P.C. Ford, Inorg. Chem., 19, 1828 (1980).
28. L.G. Vanquickenborne and A. Ceulemans, Inorg. Chem., 17, 2730 (1978).
29. K.F. Purcell, S.F. Clark and J.D. Petersen, Inorg. Chem., 19, 2183 (1980).
30. L.H. Skibsted, W. Weber, R. van Eldik, H. Kelm and P.C. Ford, Inorg. Chem., 22, 541 (1983).
31. S. Wieland, J. DiBenedetto, R. van Eldik and P.C. Ford, Inorg. Chem., in press.
32. G. Krueger, Diplomarbeit, University of Frankfurt, 1986.
33. J.G. Leipoldt, G. Krueger and R. van Eldik, unpublished results.

SECTION III:

BIOCHEMICAL ASPECTS

HIGH–PRESSURE STUDIES OF BIOMEMBRANES BY VIBRATIONAL SPECTROSCOPY

P.T.T. Wong
Division of Chemistry
National Research Council of Canada
Ottawa, Ontario, Canada K1A 0R6

ABSTRACT. Pressure-induced modifications of structural and dynamic properties in a series of lamellar, micellar and reversed micellar model biomembranes in water have been investigated by infrared and Raman spectroscopy. Examples of pressure-induced inter- and intrachain ordering processes, configuration distortion and changes in interchain packing in these systems are presented. Results are also given to demonstrate that high-pressure infrared spectroscopy is an efficient method to detect the presence of chain interdigitation in the bilayers and the presence of hydrogen bonds in the interfacial zone of biomembrane lipids.

1. INTRODUCTION

Biomembranes are ubiquitous structures in all living organisms. The surface of every cell in plants, animals and bacteria is covered by a plasma membrane which separates the cell from its environment. In addition to the plasma membrane, virtually all subcellular organelles are made of or surrounded by biomembranes. Biomembranes are composed primarily of lipids and proteins. The membranes in different types of organisms and tissues are composed of different types of lipids and proteins.

Most of our knowledge concerning the properties of biomembranes has been obtained from experiments using model membrane systems. One of the most important model membrane systems is the liposome which is commonly prepared by dispersion of dried lipid into excess water to form closed multilamellar bilayers. The number of bilayers can be reduced to one or two by exposure to ultrasonic radiation.

Vibrational spectra of a model membrane system consist of bands and lines arising from the transitions between vibrational levels of various types of intramolecular and intermolecular vibrations in the ground electronic state. They are observed experimentally as infrared and Raman spectra. Many infrared and Raman spectral parameters, particularly the frequencies, widths, intensities, shapes, and splittings of the infrared and Raman bands are very sensitive to the structural

R. van Eldik and J. Jonas (eds.), High Pressure Chemistry and Biochemistry, 381–400.

and dynamic properties of membrane lipid molecules, as well as functional groups in the molecule (1,2). Therefore, vibrational spectroscopy is one of the most powerful physical techniques in the study of biomembranes. It provides not only macroscopic but also microscopic information on the structural and dynamic properties of biomembranes at the molecular level, as well as at the functional group level. Table I summarizes the correlations between the infrared and Raman spectral parameters and the structural and dynamic properties in model biomembranes.

TABLE I.

Spectral parameters*	Structural and dynamic properties
Spectral parameter discontinuities (IR and R)	Phase transition and critical pressures
H_{2930}/H_{2880} (R)	Gauche conformer population
H_{1080}/H_{1130} (R)	Gauche conformer population
Widths of anisotropic components (R)	Reorientational fluctuations
Widths of isotropic components (R)	Interchain interactions
H_{2880}/H_{2850} (R)	Interchain interactions
$\nu_S CH_2$ band shape (R)	Interchain packing
Correlation field splitting (IR and R)	Interchain packing
$\nu C=O$ frequency (IR and R)	Hydrogen bond and conformation in interfacial region
νCN and νPO_2^- frequency (IR and R)	Hydrogen bond and conformation of head groups
$I\gamma'CH_2/I\gamma CH_2$ (IR)	Orientation of neighboring chains
$\nu C=C$ frequency and $I\nu C=C$ (R)	Nature of C=C double bonds
$I\delta=CH/I\tau CH_2$ (IR)	Degree of unsaturation
$H\delta'CH_2/H\delta CH_2$ (IR)	Interdigitation packing

*Ha/Hb: peak height ratio between band a and b; Ia/Ib = integrated intensity ratio between band a and b; ν = stretching mode; δ and δ' = bending mode and its correlation field component; γ and γ' = rocking mode and its correlation field component; (R) = Raman spectrum; and (IR) = infrared spectrum.

Vibrational spectra of a series of fully hydrated liposomes of a large number of lipids with saturated and unsaturated chains, different chain lengths, different head groups, and different linkages between head groups and hydrocarbon chains have been studied systematically as a function of pressure in this laboratory. The use of the pressure parameter in these studies provides information about the perturbation of pressure on the physical properties of these systems, as well as the interactions between these systems and other biomolecules and water at ambient conditions. Some representative examples from these results are given here to illustrate the kinds of problems in model biomembranes to which high-pressure infrared and Raman spectroscopy can provide solutions.

2. PRESSURE-INDUCED STRUCTURAL MODIFICATIONS IN LAMELLAR MEMBRANES

2.1. Phase Behavior and Critical Pressures

Pressure-induced phase transitions in aqueous multibilayer dispersions of 1,2-diacyl phosphatidylcholine (diacyl PC) have been studied extensively. Above the main transition temperature, all the aqueous diacyl PC's are in the liquid crystalline state (LC). When external pressure is applied to the LC phase, it undergoes a series of five phase transitions as indicated by the discontinuities in the pressure dependences of vibrational spectral parameters (1-6). These transitions were ascribed to the LC/GI, GI/GII, GII/GIII, GIII/GIV and GIV/GV transitions. Consequently, a total of five high-pressure gel phases have been found (GI-GV). The critical pressures for these transitions are dependent on the acyl chain length in diacyl PC's. The critical pressure of a corresponding phase transition is lower for a longer acyl chain. The critical pressures of these transitions for three of the most common model membrane systems i.e. aqueous multibilayer dispersions of 1,2-dimyristoylphosphatidylcholine (DMPC), 1,2-dipalmitoyl-phosphatidylcholine (DPPC) and 1,2-distearoylphosphatidylcholine (DSPC) are compared in Table II.

TABLE II.

	Critical pressure at 28°C (kbar)				
	LC/GI	GI/GII	GII/GIII	GIII/GIV	GIV/GV
DMPC	0.15	1	2.6	8	–
DPPC	0.17[a]	0.6[a]	1.7	4.8	15
DSPC	–	–	1.1	3.1	–

[a]Critical pressure at 47°C

The difference among these three lipids is the length of the acyl chains. The number of carbon atoms in the acyl chains is 14,16 and 18 for DMPC, DPPC and DSPC, respectively.

2.2. Structural and Dynamic Changes at Critical Pressures

The changes in the structural and dynamic properties associated with pressure-induced phase transitions in aqueous diacyl PC's have been monitored by the changes in the infrared and Raman spectral parameters. These results together with the X-ray diffraction results of the LC,GI and GII phases (7,8 and refs. therein) are summarized in Fig. 1.

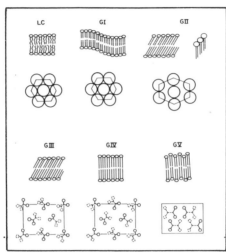

Figure 1. Representations of the interchain structure in the LC phase and GI—GV gel phases. From ref. 9.

The strong gauche C—C stretching band and the weak trans C—C stretching band and also the large peak height ratio H_{2930}/H_{2880} in the Raman spectra of the LC phase indicate that the acyl chains in this phase are highly disordered, containing a large number of gauche conformers. The interchain orientation is also disordered in this phase and the rate of reorientational fluctuations is extremely high as indicated by the large band width of the anisotropic component of the Raman bands (3). In the LC phase, the average orientation of the chains is perpendicular to the lipid—water interface and the acyl chains are packed in an one—dimensional ordered hexagonal lattice.

At the LC/GI transition, the number of gauche conformers is reduced (10). The rate of the interchain reorientational mobility decreases continuously in the GI gel phase with increasing pressure. The interchain interaction in this phase is, however, only slightly affected by pressure due to the instability of the lattice arising from distortion. The GI phase exhibits a two—dimensional structure consisting of lipid lamellae distorted by a periodic ripple. The packing of the acyl chains in this phase is similar to that in the LC phase with an hexagonal lattice.

In the GII phase, the acyl chains are mainly extended. As the pressure further increases in the GII phase, the interchain reorientational rate decreases gradually and the interchain interaction increases sharply. The GII phase consists of one—dimensionally ordered lipid bilayer in which the acyl chains are tilted with respect to the bilayer normal and packed in a distorted hexagonal lattice.

The acyl chains are in a fully extended state and the interchain orientations are highly ordered in the GIII, GIV and GV phases. The correlation field splitting in their infrared and Raman spectra, the number of correlation field components, the infrared and Raman selection rules and the changes in the shape of Raman bands suggest that the interchain packings in the GIII,GIV and GV phases are that of solid monoclinic DMPC dihydrate (1,3) and perpendicular and parallel ortho-rhombic fatty acids (1,2,5,6), respectively.

2.3. Mechanism of the Phase Modifications in Lamellar Membranes

In the LC phase, the average cross-sectional area of the two acyl chains is enlarged to the dimension comparable with that of the choline head group due to the large number of gauche bonds in the chains (1,3). This fulfills the requirement for the orientation of acyl chains perpendicular to the bilayer plane. Due to the rapid reorientational fluctuations along the chain axis (3), the lipid molecules in this phase appear as circular rods and thus favour an hexagonal packing.

The average dimension of the cross-section of two acyl chains in the GI phase decreases to that slightly smaller than that of the choline head group as a result of a decrease in the number of gauche conformers in the chains (10). The requirement of a mutual accommoda-tion of the head group lattice and the chain matrix leads to a distor-tion of the head group packing by a periodic ripple in this phase. The rate of reorientational fluctuations is still relatively high (3,6,7) and the long chain molecules remain in a circular rod shape in this phase. Therefore, the interchain packing in this phase resembles that in the LC phase with an hexagonal lattice.

The conformational structure in the acyl chains of the GII phase is highly ordered (1,3,10), and thus the dimension of the cross-section of two acyl chains in each molecule is further reduced. The chains in this phase tilt with respect to the bilayer normal as a direct conse-quence of the area mismatch between two acyl chains and the head group in each molecule. Although the interchain reorientation is still disordered in this phase, the reorientational rate of the chains is significantly reduced with respect to that in the GI and LC phases. Statistically the lipid molecules in this phase are no longer in a perfect circular rod shape, and the interchain packing is distorted from a perfect hexagonal lattice.

Not only the intrachain conformation but also the interchain orientation is highly ordered in the GIII phase (1,6). Consequently, the tilted configuration of the chains in the GII phase remains in the GIII phase. Moreover, the chains are packed in such a way that the zig-zag planes of the neighboring methylene chains are nearly perpen-dicular to each other, in order to accommodate the more spacious, layer-aligned head group lattice. The tilting and the zig-zag chain perpendicular packing in the GIII phase are equivalent to that in a perpendicular monoclinic subcell (11).

As the pressure is further increased, the area of each head group in the GIV phase is likely to be squeezed to that comparable with the effective cross-sectional area of the methylene chains packed in mutually perpendicular zig-zag planes. Consequently, the tilting of the chains with respect to the bilayer surface is no longer required for area matching between the head group and the chains, and the lipid molecules in the GIV phase are packed in a lattice resembling the perpendicular orthorhombic lattice (11).

In the GV phase, the orientation between the acyl chains in each
lipid molecule is forced towards the parallel by pressure and thus the
cross-sectional area of two acyl chains in each molecule is further
reduced.   To match this, an alternate displacement of the head group
with respect to the bilayer plane is required.
     The results from these high-pressure vibrational spectroscopic
studies clearly show that the pressure-induced structural phase transitions in aqueous diacyl PC's involve modifications in the acyl chain
packing.   These modifications are driven by changes in the mutual
accommodation between the area of the choline head group and the cross-
sectional area of the acyl chains as a result of the pressure-induced
intrachain conformational and interchain reorientational ordering
processes.

3.   INTERCHAIN CONFIGURATION DISTORTION

The integrated intensities of the two infrared active correlation field
components of the CH_2 rocking mode, I_γ, have been shown to be related
to the angle θ between the methylene plane and the a axis of the unit
cell as (12) $I_{\gamma'}/I_\gamma = \tan^2\theta$, where γ' is the high-frequency component.
     The integrated infrared intensity ratios $I_{\gamma'}/I_\gamma$ of aqueous DPPC
have been measured over the pressure range 2 to 25 kbar (2) and are
shown in Fig. 2.   The intensity ratio decreases with increasing pressure and shows a change of slope at 4.8 and 15 kbar, where the GII/GIV
and GIV/GV transitions take place, respectively, and remains almost
constant at pressures above 15 kbar.   There are two molecules and thus
four methylene chains in each unit cell in the GIII phase of DPPC,
oriented alternately nearly perpendicular and parallel to each other,
similar to the situation in the monoclinic DMPC dihydrate solid (13).
The correlation between two parallel chains gives zero infrared intensity to the correlation field band and thus the intensity ratio $I_{\gamma'}/I_\gamma$
is also zero.   For the correlation between two perpendicular chains,
the value of $I_{\gamma'}/I_\gamma$ is maximum and is close to one.   Consequently, for
a system with an alternately perpendicular and parallel orientation of
neighboring chains, the $I_{\gamma'}/I_\gamma$ value is also zero.

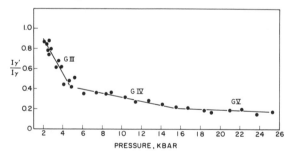

Figure 2. Pressure dependence of the integrated intensity ratio
$I_{\gamma'}/I_\gamma$.   From ref. 2.

At 2 kbar the ratio $I_{\gamma'}/I_{\gamma}$ of the GIII phase is 0.87. This high intensity ratio indicates that the parallel chain orientation is hardly present in this phase. Raman spectroscopic results (1,5,6) have demonstrated that the reorientational fluctuations of the DPPC molecules along the chain axis are significant in the GIII phase at 2 kbar, and that the rate of these fluctuations decreases with increasing pressure; the fluctuations are almost completely damped before the GIII/GIV transition at 4.8 kbar. Therefore, the observed $I_{\gamma'}/I_{\gamma}$ value of 0.87 at 2 kbar is mainly due to the correlation between two acyl chains within each DPPC molecule, which orient almost perpendicular to each other, while the correlation between the parallel chains of neighboring DPPC molecules is insignificant due to the reorientational fluctuations of the molecules. The decrease in the $I_{\gamma'}/I_{\gamma}$ ratio in the GIV and GV phases may be carried by two pressure effects, i.e. an increase in the parallel correlation between chains of neighboring DPPC molecules, and/or a decrease in the plane orientation angle between two chains within each DPPC molecule. At low pressure the contribution of the first effect is significant, whereas at higher pressures the second effect may dominate.

4. PRESSURE EFFECTS ON THE MOLAR VOLUME CHANGE AT THE CRITICAL TEMPERATURE

The pressure effects on the critical temperature of the main transition, T_m, between the GI gel phase and the liquid crystalline LC phase of aqueous bilayer dispersions of lecithins have been studied with a wide variety of physical methods (14). All the results indicate that dT_m/dp is independent of pressure over the pressure range 1-304 bar. This pressure effect can be expressed by the Clausius-Clapeyron relation

$$\frac{1}{T_m}\frac{dT_m}{dp} = \frac{\Delta V_m}{\Delta H_m}$$

where ΔV_m, ΔH_m are the molar volume change and the enthalpy change, respectively, at T_m. Since experimental results show that T_m increases with increasing pressure while dT_m/dp is constant (14), the ratio $\Delta V_m/\Delta H_m$ must decrease with increasing pressure, implying either a decrease in ΔV_m, or an increase in ΔH_m or both. However, several differential scanning calorimetry and dilatometry measurements have indicated that ΔH_m and ΔV_m are invariant with pressure. This is probably because the pressure effects on ΔH_m and ΔV_m are small and the pressure applied in these measurements was not high enough to observe a significant change in ΔH_m and ΔV_m.

The main source of the volume change at the main transition in lipid bilayers is a loosening in the packing of the acyl chains arising essentially from the intramolecular conformational and intermolecular orientational disordering processes. Many Raman spectral parameters are very sensitive to these disordering processes. Therefore, we have investigated the pressure dependences of the thermotropic main

transition of aqueous DPPC and DMPC bilayers in the pressure range up
to 1210 bar by Raman spectroscopy.

The critical temperature, T_m, of DPPC at several pressures obtained from the discontinuities in the temperature dependences of Raman
spectral parameters at the corresponding pressures are plotted vs.
pressure in Fig. 3. The pressure dependence of the critical temperature T_p of the pretransition from the GII to the GI gel phase in DPPC
is also shown in Fig. 3. A linear pressure dependence of both T_m and
T_p is evident from Fig. 3. The slopes of the transition temperatures
vs. pressure give a dT_m/dp value of 20.8°C kbar^{-1} and a dT_p/dp value of
16.2°C kbar^{-1} for DPPC and a dT_m/dp value of 20.1°C kbar^{-1} for DMPC.

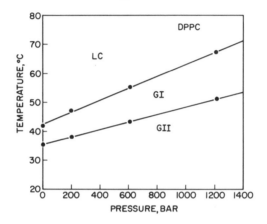

Figure 3. Pressure dependence of the critical temperatures T_m and T_p
in DPPC. From ref. 10.

Figure 4 shows the temperature dependences of several representative spectral parameters of DPPC, i.e. the CH_2 symmetric stretching
frequency, v_sCH_2, the peak height ratio between the asymmetric and
symmetric CH_2 stretching bands, H_{2880}/H_{2845}, and the peak height ratio
between the disorder allowed 2925 cm^{-1} band and the asymmetric CH_2
stretching band, H_{2925}/H_{2880}, measured at 1 and 1210 bar pressures. A
discontinuity at T_m and a change of slope at T_p are clearly observed in
these spectral parameters. While the v_sCH_2 and H_{2925}/H_{2880} values
increase, the H_{2880}/H_{2845} value decreases dramatically at T_m.

The spectral parameters v_sCH_2, H_{2925}/H_{2880} and H_{2880}/H_{2845} are
very sensitive to the loosening of the chain packing, population of
gauche conformers and interchain interactions, respectively (1). At
higher pressure, the changes in all these parameters at T_m are samller
than those at lower pressure. Therefore, the loosening of the chain
packing and the decrease in interchain interactions at T_m from the GI
gel phase to the LC phase is smaller at higher pressure. Moreover, the
population of gauche conformers in the gel phase near T_m is considerably larger at higher pressure and thus the volume of the gel phase at
T_m is larger at higher pressure. Consequently, it is expected that the
volume change at the transition is smaller at higher pressure.

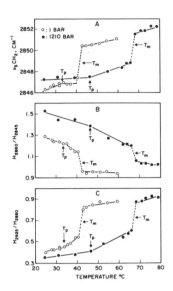

Figure 4. Temperature dependences of $\nu_s CH_2$ frequency, H_{2880}/H_{2845} and H_{2925}/H_{2880} of DPPC at 1 atm and 1210 bar pressures.

Furthermore, the enthalpy change associated with these structural and dynamic changes at T_m should be smaller, not larger, at higher pressure.

The enthalpy change at T_m of aqueous DPPC and DMPC has been measured extensively, and found to be constant between 1 and 138 bar pressure within experimental error (10 and refs. therein). The experimental ΔH_m values generally accepted for aqueous DPPC and DMPC are 8.74 and 5.44 kcal mol^{-1}, respectively. With these values and the dT_m/dp

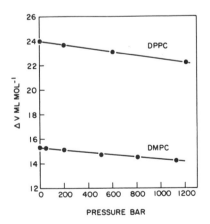

Figure 5. Pressure dependence of the critical volume change at T_m in aqueous bilayers of DPPC and DMPC. From ref. 10.

and T_m values the values of the volume change, ΔV_m, at the main endothermic phase transition in DPPC and DMPC as a fuction of pressure were
determined and are shown in Fig. 5. Although the decrease in ΔH_m with
increasing pressure is expected to be small, it would certainly affect
the evaluation of the pressure dependence of ΔV_m. Therefore the
pressure dependence of ΔV_m obtained here is slightly underestimated.

5. MECHANISM FOR THE FORMATION OF THE LAMELLAR SUBGEL PHASE

Aqueous DPPC bilayers exist in the distorted hexagonal GII gel phase at
ambient conditions. Upon gradual cooling the GII phase converts to the
GIII gel phase at -30°C (15). However, when the GII phase is incubated
at temperatures between -8 and 6°C at atmospheric pressure for a long
period of time, the GII phase converts to the SG subgel phase (16).
The SG phase transforms back to GII phase at 15°C (16). The transition
from the GII phase to the GIII phase can also be induced by an external
pressure of 1.7 kbar at room temperature (1,6), whereas external
pressure prevents the transition from the GII phase to the SG phase
(17,18). Once the SG phase is formed, external pressure can stabilize
the SG phase up to 46.3 kbar at temperatures above the subtransition
temperature of 15°C.

We have used high-pressure Raman spectroscopy to investigate the
mechanism for the formation of the SG phase and the nature of the
complex phase relationship among the GII, GIII and SG phases of aqueous
DPPC bilayers (19). Figure 6 shows the Raman spectra in the region of
the CH_2 bending modes of incubated and non-incubated aqueous DPPC samples at several pressures. The correlation field band near 1410 cm^{-1}
is present in the spectra of non-incubated DPPC (Fig. 6B) but is absent

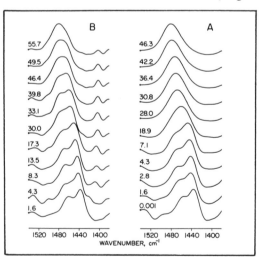

Figure 6. Raman spectra of fully hydrated DPPC bilayers in the CH_2
bending region at several pressures (A. spectra for incubatd and B.
spectra for nonincubated DPPC samples). From ref. 19.

from the spectra of incubated DPPC (Fig. 6A). In the Raman spectra of solid n-alkanes, the correlation field band is only observed in ordered n-alkanes with orientationally non-equivalent chains in the unit cell, such as those with monoclinic or orthorhombic structure. On the other hand, the correlation field band is absent from the Raman spectra of n-alkanes with orientationally equivalent chains in the unit cell, such as those with triclinic structure (1,20). Figure 7 shows the pressure profile of the Raman spectra of the triclinic solid n-hexadecane (A) and the orthorhombic n-pentadecane (B). A comparison of the Raman spectra in Fig. 6 and Fig. 7 clearly shows that the acyl chains of each molecule in the SG phase are parallel to each other similar to the parallel chain packing in the triclinic n-hexadecane, whereas the chains in the GIII phase are nearly perpendicular to each other.

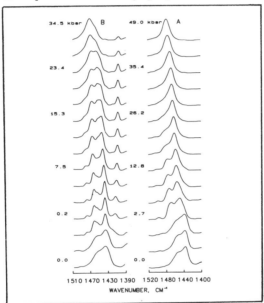

Figure 7. Raman spectra of n-hexadecane (A) and n-pentadecane (B) in the CH_2 bending region at several pressures.

Figure 8 illustrates the structure changes associated with the GII/SG and GII/GIII phase transitions. The GII/GIII transiton involves the damping of the inter- and intrachain rotational reorientations so that the orientation of all the acyl chains becomes highly ordered in the GIII phase. The GII/SG transition not only involves the damping of the reorientational fluctuations of the chains but also the change in the orientation of the two acyl chains within each DPPC molecule from nearly perpendicular to parallel to each other. This reorientation of acyl chains within each molecule is a slow process and only takes place over a limited temperature range (i.e. -8 to +6°C). At temperatures above this range, the mobility of the chains, including the torsion and twisting motions of each acyl chain, becomes too large to form an

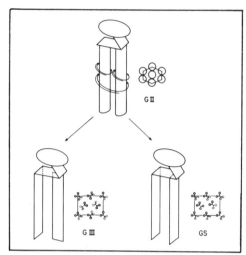

Figure 8. Schematic representations of the structural changes associated with the GII/SG and GII/GIII transitions.

orientationally ordered phase such as the SG phase. On the other hand, at temperatures below this range, the orientational mobility of the acyl chains, as well as that of the entire DPPC molecule, is highly restricted. In fact, at -30°C the reorientational fluctuations among the DPPC molecules are almost completely damped and the GII phase converts to the highly ordered GIII phase in which the acyl chains of each DPPC molecule remain nearly perpendicular to each other due to the restricted chain reorientation at this low temperature. It is well documented (1) that pressure slows down the mobility of the chains. At 1.7 kbar and room temperature the rotational reorientations of DPPC molecules are almost completely damped and the GII phase transforms to the highly ordered GIII phase. Therefore, due to the pressure restriction of the reorientation of the acyl chains in each DPPC molecule from nearly perpendicular to parallel, pressure does prohibit the transformation of the GII phase to the SG phase. On the other hand, once the SG phase is formed, pressure also restricts the reorientation of the acyl chains from parallel back to perpendicular to each other, and keeps the reorientational fluctuations highly damped. Consequently, the parallel orientation of the acyl chains in the SG phase is "locked in" and stabilized by external pressure even at temperatures above the subtransition temperature (15°C). It is apparent that the mechanism of the GII/SG and GII/GIII transitions is governed by the intra- and intermolecular reorientational mobility of the acyl chains.

6. PRESSURE INDUCED STRUCTURAL MODIFICATIONS IN MICELLAR AND REVERSED
    MICELLAR MEMBRANES

Some natural lipids such as lysolecithins adopt a micellar structure in water at ambient conditions (21). In these micellar aggregations, the molecules are arranged in a sphere, a prolate spheroid or a cylindroid with the polar head-group on the outside and in contact with bulk water. There are also some lipids such as phosphatidylethanolamines that form reversed micelles in water (22). In these aggregates the surfactant of lipid molecules also form spherical or cylindroid micelles, however, with the polar head-groups buried in the interior.

At low temperature, both the disordered micellar and reversed micellar phases convert into the lamellar coagel phase (21,22). This temperature-induced transition is between two entirely different structures and exhibits quite different characteristics from those in the temperature-induced transition from the disordered LC phase to the GI gel phase in aqueous bilayer dispersions.

When external pressure is applied to the LC phase, it transforms into the GI gel phase as described in section 1. Presumably, pressure also induces transformations from the disordered micellar and reversed micellar phases to the lamellar coagel phases with different characteristics from those in the pressure-induced transformation between the LC phase and the GI gel phase in aqueous bilayer dispersions. To investigate this we have measured the Raman and infrared spectra as a funciton of pressure of aqueous 1-palmitoyllysophosphatidylcholine (lyso PPC) and dioleoylphosphatidylethanolamine (DOPE), which are in the micellar and reversed micellar state, respectively, at ambient conditions (21,22).

6.1. Micellar Membranes

The clear micellar solution of lyso PPC transforms into an opaque coagel phase at 1.9 kbar. The transformation is clearly evident in the drastic changes in the shape of Raman bands at 1.9 kbar from that typical of the conformationally disordered micellar phase to that typical of the coagel phase (23,24). The transformation at this pressure is further evident from the discontinuities in all the spectral parameters such as the peak height ratio between the antisymmetric and the symmetric CH_2 stretching modes H_{2880}/H_{2850} as shown in Fig. 9.

When pressure is released the coagel phase does not reconvert to the micellar phase until the pressure is reduced to 0.8 kbar. This high hysteresis reflects the drastic change between two entirely different structures i.e. the cylindrical aggregation in the micellar phase and the lamellar structure in the coagel phase. This high hysteresis has not been observed in the LC/GI transition in diacyl PC's.

Spectral changes and discontinuities in the pressure dependences of most of the spectral parameters were also observed at 4 and 15 kbar, which indicate that structural phase transitions take place at these pressures. Thus, a total of 3 coagel phases (CI-CIII) have been found. The spectral parameters of these coagel phases show that the chain packing and interchain order of the highly rigid CI coagel phase are similar to those of the GIII gel phase of DPPC. Furthermore, the CII and CIII coagel phase of lyso PPC share the chain packing and the strong interchain interactions with the GIV and GV gel phases of DPPC. The phases corresponding to the less ordered and less rigid GI and GII gel phases of the diacyl PC's do not exist in lyso PPC. This has been

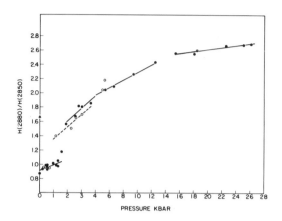

Figure 9. Pressure-dependence of the ratio between the peak-height of
the antisymmetric and symmetric methylene stretching bands of aqueous
lyso PPC. The solid circles were obtained by increasing the pressure,
and the open circles were obtained by decreasing the pressure. The
ratio for solid lyso PPC at atmospheric pressure is shown as a solid
triangle. From ref. 21.

attributed to the result of an interdigitated structure in the lyso PPC
coagel phases (21).

6.2. Reversed Micellar Membranes

Figure 10 shows the pressure profile of the frequency and width of the
$C=O$ stretching band of DOPE. A discontinuity, marked by a change of
slope, is observed in the pressure dependences of these and other spectral parameters at 9 kbar. These spectral changes and discontinuities
correspond to those at the thermotropic transition between the reversed
micellar and the lamellar coagel phase of a series of phosphatidylethanolamines (22,25). Consequently, the pressure required for the transformation of the reversed micellar state to the lamellar state in DOPE
is 9 kbar, which is 7.1 kbar higher than that for the transformation of
the micellar state to the lamellar state in lyso PPC, although their
transformation temperatures are about the same (~5°C) (21,22).

7. HYDROGEN-BOND INTERACTIONS IN THE INTERFACIAL ZONE OF MEMBRANE LIPIDS

There are proton-accepting groups such as the carbonyl group, $C=O$, in
the interfacial regions of most membrane lipids. Consequently, one
would naturally assume that molecules with proton-donating groups such
as water and cholesterol are anchored in biomembranes by hydrogen-
bonding to the interfacial regions of membrane lipids. In fact, some
NMR and infrared spectroscopic studies have suggested the existence of
such hydrogen-bonding in lipid associations. However, some other

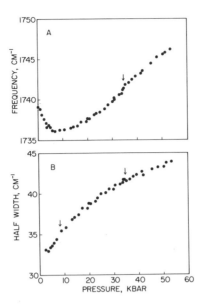

Figure 10. Pressure dependence of the frequency (A) and halfwidth (B)
of the C=O stretching band of aqueous DOPE dispersions. From ref. 22.

reports have discounted such direct hydrogen-bonding in recombinant
lipid systems (26 and refs. therein). Therefore, this significant
biological phenomenon is still not well understood.

We have developed a high-pressure infrared spectroscopic technique
to detect the presence of hydrogen-bonding in the interfacial zone of
membrane lipids. The water soluble lipid, triacetyl glycerol, was
chosen for this study (26). The only polar group in this lipid capable
of forming hydrogen-bonds with water is the ester carbonyl group, C=O,
in the lipid interface. The solubility of this lipid in water is
solely a consequence of hydrogen-bond formation between the polar C=O
group and water.

Figure 11 shows the pressure dependences of the C=O stretching
frequency of neat (liquid) triacetyl glycerol and its 1:15 (v/v)
solution in H_2O and D_2O. The C=O stretching frequency decreases from
1748 cm^{-1} to 1732 and 1728 cm^{-1} when it forms hydrogen-bonds with H_2O
and D_2O, respectively. As shown by the detailed frequency vs. pressure
profile of the C=O band of triacetyl glycerol in D_2O (Fig. 11), the
frequency of the hydrogen-bonded C=O group decreases further with
increasing pressure due to the strengthening of the hydrogen-bonding by
pressure. At 20 kbar, the frequency of the C=O stretching band shows a
sharp increase to 1744 cm^{-1}, a frequency value characteristic for a
free C=O stretching band. The further slight increase in this frequency between 20 and 30 kbar parallels that observed for the free C=O
band. This abrupt increase in the frequency of the hydrogen-bonded C=O
group at 20 kbar which literally snaps back to the value of a free C=O
band indicates that water preferentially binds to itself and thus with-

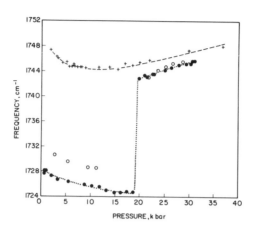

Figure 11. Pressure dependence of the C=O stretching frequencies of
neat triacetyl glycerol (crosses) and of solutions in D_2O (filled
circles) and H_2O (open circles). From ref. 26

draws the bound water molecules from the interfacial zone of triacetyl
glycerol at this pressure (release pressure). The pressure effect on
the hydrogen bond strength between water molecules has been estimated
by the pressure shift of the OH or OD stretching frequency of water
molecules (27). Consequently, the relative strength of the hydrogen-
bond interaction in the interfacial region of membrane lipids can be
estimated by the relative magnitude of their release pressures.
        We have used this high-pressure infrared spectroscopic technique
to investigate the possibility of water penetration into the interfacial zone in a series of model biomembranes. No hydrogen-bonding
between water molecules and the interfacial C=O groups has been found
in model biomembranes with either lamellar, micellar, or reversed
micellar structure (2,22, unpublished work from this laboratory).
        For the hydrogen-bond interaction of biological molecules contain-
ing OH proton-donating groups with the interfacial C=O groups of biomembranes, we have used high-pressure, which enhances hydrogen-bonding,
to investigate the infrared spectral characteristics of the free and
associated OH and C=O moieties. The model lipid 1,2-dipalmitoyl-sn-
glycerol (DPG) was chosen for this study because the single OH group in
DPG is known to be hydrogen-bonded intermolecularly to one of the
interfacial carbonyls (28, and refs. therein).
        The infrared spectra of DPG shown in Fig. 12 for the C=O stretch-
ing (A) and the OH stretching modes (B) were measured between ambient
pressure and 42.5 kbar. The two infrared bands at 1708 and 1733 cm^{-1}
in Fig. 12A (frequencies at 0.001 kbar) are due to the hydrogen-bonded
C=O and to the free C=O group, respectively. The strong OH stretching
band at 3498 cm^{-1} in Fig. 12B is due to the associated OH moiety. The
frequency of this band at atmospheric pressure is about 100 cm^{-1} lower
than that of the free OH group and is extremely sensitive to pressure.
It shifts monotonically to lower frequencies with increasing pressure
(Fig. 13). The decrease in the OH stretching frequency with pressure

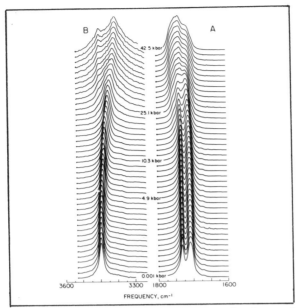

Figure 12. Infrared spectra of 1,2-dipalmitoyl glycerol (DPG) in the region of the C=O stretching modes (A) and in the region of the hydroxyl stretching modes (B) at several pressures between 1 bar and 42.5 kbar. From ref. 28.

is the consequence of the pressure strengthening of the hydrogen-bond interaction.

At 25 kbar, all the bands split into two and discontinuities are observed in the pressure dependences of frequencies of these bands (Figs. 12 and 13), which strongly suggests that a structural phase transition takes place at this pressure. Two OH stretching bands are found in this high-pressure phase. The frequency of the new OH

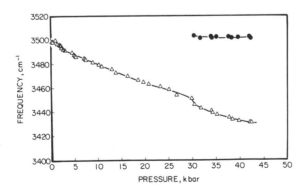

Figure 13. Pressure dependences of the OH stretching frequencies of DPG. From ref. 28.

stretching band at 25 kbar is still about 100 cm^{-1} below that of a free OH group at atmospheric pressure. Consequently in this high-pressure phase of DPG all the OH groups are hydrogen-bonded and two types of hydrogen-bonds with slightly different strengths co-exist.

The results from this study demonstrate that a combination of the high-pressure technique with infrared spectroscopy provides a means to investigate hydrogen-bond formation in water molecules and biomolecules containing OH moieties with the C=O groups in the interfacial zone of biomembranes.

8.   INTERDIGITATION IN LAMELLAR MEMBRANES

Infrared spectra of aqueous dispersions of DPPC and its ether-linked analog, 1,2-dihexadecyl-sn-glycero-3-phosphocholine (DHPC), were measured at 28°C as a function of pressure (29). Although these two lipids differ only in the linkage to the hydrocarbon chains, significant differences were observed in the barotropic behaviour of their infrared spectra. X-ray diffraction studies (30) indicate that the main difference in the interchain structure between these two lipid systems is that the interchain packing in the lamellar gel phase becomes fully interdigitated in the ether-linked DHPC. Consequently differences in the pressure dependences of various spectral parameters between these two lipid systems may be useful as a test of interdigitation. The most notable differences are the magnitudes of the pressure-induced correlation field splittings of the scissoring and rocking modes, and the relative intensities of the corresponding component bands.

Figure 14 compares the pressure-induced correlation field splitting of the methylene scissoring band δCH_2 (A) and the methylene rocking band γCH_2 (B), in DHPC and DPPC. Not only can the correlation

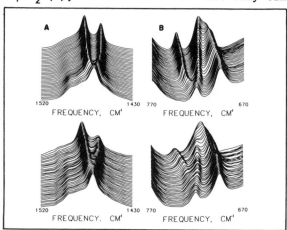

Figure 14.   Stacked contour plots of the infrared spectra of aqueous DHPC (top half) and DPPC (bottom half) in the CH_2 scissoring region (A) and in the CH_2 rocking region (B). From ref. 29.

field component band be resolved at a lower pressure in interdigitated DHPC (1.2 kbar, as compared to 2.2 kbar in DPPC), but the initial magnitude of the correlation field splitting in DHPC, particularly below 9 kbar, is significantly greater than that observed in DPPC.

At all pressures where the correlation field component band $\delta'CH_2$ can be resolved, the relative peak height intensity ratio $R=I\delta'/I\delta$ is greater in DPPC than in DHPC as shown in Fig. 15.

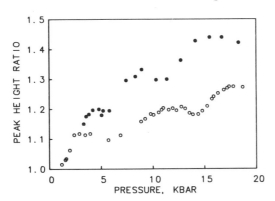

Figure 15. Pressure dependence of the peak height intensity ratio $R=I\delta'/I\delta$ in DHPC (open circles) and DPPC (filled circles). From ref. 29.

The specific characteristics of these spectral parameters are common to biomembranes with interdigitated lamellar structures and were also observed in other known interdigitators, such as the gel phase of 1,3-DPPC and the coagel phase of lyso PPC (unpublished work from this laboratory). Therefore, the parameter $R=I\delta'/I\delta$ may be used as a diagnostic tool for the presence of chain interdigitation. In fully interdigitated bilayers, the R parameter will exhibit a pressure dependence very similar to that of DHPC.

9. REFERENCES

1. P.T.T. Wong, Ann. Rev. Biophys. Bioeng. **13**, 1 (1984).
2. P.T.T. Wong and H.H. Mantsch, J. Chem. Phys. **83**, 3268 (1985).
3. P.T.T. Wong, W.F. Murphy and H.H. Mantsch, J. Chem. Phys. **76**, 5230 (1982).
4. P.T.T. Wong and D.J. Moffatt, Applied Spectrosc. **37**, 85 (1983).
5. P.T.T. Wong and H.H. Mantsch, J. Chem. Phys. **81**, 6367 (1984).
6. P.T.T. Wong and H.H. Mantsch, J. Phys. Chem. **89**, 883 (1985).
7. M.J. Janiak, D.M. Small and G.G. Shipley, J. Biol. Chem. **25**, 6068 (1979).
8. J. Stamatoff, B. Feuer, H.J. Guggenheim, G. Tellez and T. Yamane, Biophys. J. **38**, 217 (1982).
9. P.T.T. Wong, Physica **139** and **140** B, 847 (1986).
10. P.T.T. Wong and H.H. Mantsch, Biochemistry **24**, 4091 (1985).

11. S. Abrahamsson, B. Dohlin, H. Löfgren and I. Pascher, Progress in Chemistry of Fats and Other Lipids **16**, 125 (1978).

12. R.G. Snyder, J. Mol. Spectrosc. **7**, 116 (1961).

13. R.H. Pearson, Nature **281**, 499 (1979).

14. A.G. Macdonald, Philosophical Trans. Royal Soc. London, B**304**, 47 (1984).

15. P.T.T. Wong and H.H. Mantsch, Biochim. Biophys. Acta **732**, 92 (1983).

16. S.C. Chen, J.M. Sturtevant and B.J. Gaffney, Proc. Natl. Acad. Sci., U.S.A. **77**, 5060 (1980).

17. P.T.T. Wong, E.C. Mushayakarara and H.H. Mantsch, J. Raman Spectrosc. **16**, 427 (1985).

18. W.-C. Wu, P.L.-G. Chong and C.-H. Huang, Biophys. J. **47**, 237 (1985).

19. E.C. Mushayakarara, P.T.T. Wong and H.H. Mantsch, Biophys. J. **49**, 1199 (1986).

20. F.J. Boerio and J.L. Koenig, J. Chem. Phys. **52**, 3425 (1970).

21. P.T.T. Wong and H.H. Mantsch, J. Raman Spectrosc. **17**, 335 (1986).

22. P.T.Wong, S.F. Weng and H.H. Mantsch, J. Chem. Phys, in press.

23. P.T.T. Wong and H.H. Mantsch, J. Chem. Phys. **78**, 7362 (1983).

24. P.T.T. Wong and H.H. Mantsch, J. Phys. Chem. **87**, 2436 (1983).

25. H.H. Mantsch, A. Martin and D.G. Cameron, Biochemistry **20**, 3138 (1981).

26. E.C. Mushayakarara, P.T.T. Wong and H.H. Mantsch, Biochem. Biophys. Res. Commun. **134**, 140 (1986).

27. D.D. Klug and E. Whalley, J. Chem. Phys. **81**, 1220 (1984).

28. E.C. Mushayakarara, P.T.T. Wong and H.H. Mantsch, Biochim. Biophys. Acta **857**, 259 (1986).

29. D.J. Siminovitch, P.T.T. Wong and H.H. Mantsch, Biophys. J. (1987) in press.

30. M.J. Ruocco, D.J. Siminovitch and R.G. Griffin, Biochemistry **24**, 2406 (1985).

DISSOCIATION OF OLIGOMERIC PROTEINS BY HYDROSTATIC PRESSURE

Gregorio Weber
School of Chemical Sciences,
University of Illinois,
1209 W California St.
Urbana,IL 61801, U.S.A

ABSTRACT. The methods of study of proteins under high pressure, and the effects of pressure upon proteins made up of a single peptide chain are briefly reviewed. The effects upon oligomeric proteins are examined in detail: The existence of time-dependent changes in the conformation of the dissociated subunits, termed a "conformational drift", is revealed by the temporary decreases in subunit affinity, enzymic activity and changed spectroscopic properties of the aggregates formed after decompression. Its significance in relation to the equilibria established under pressure is discussed and the insuficiency of conventional descriptions of the chemical equilibrium for these cases is noted.

1. SPECTROSCOPIC METHODOLOGY

The effects of high pressure upon proteins and other molecules of biological interest have been detected, up to the present, almost exclusively by means of optical spectroscopic methods (Heremans, 1982;Weber & Drickamer, 1983). We shall briefly discuss their application to our topic. Absorption and emission of light follow the Franck-Condon principle, according to which these processes occur in times which are too short for any change in nuclear coordinates to occur. It follows from it that the changes in intensity and spectrum of absorption with pressure are determined by the thermodynamic equilibrium distribution of the absorbing species. The fluorescence properties are those of the fluorophore molecules in a different electronic state, the fluorescent state, with a lifetime that seldom exceeds a few nanoseconds. The difference in electronic distribution in this state with respect to the ground state leads to changes, sometimes of considerable magnitude, in the interaction of the excited molecules with their surroundings and the limited lifetime of the excited state determines the extent to which the new equilibrium is approached. The advantage of fluorescence over absorption

401

R. van Eldik and J. Jonas (eds.), High Pressure Chemistry and Biochemistry, 401–420.
© *1987 by D. Reidel Publishing Company.*

spectroscopy resides in the greater sensitivity of the emission to the environment of the chromophore and in the lower concentration range to which the technique can be applied. Absorption determinations of acceptable accuracy require chromophore concentrations of the order of 10^{-5}M while the relevant fluorescent properties may be measured, in fluorophores with quantum yields of 0.1 and higher, at concentrations of 10^{-9}M. Additionally the fluorescence contains information on time-dependent events that take place between absorption of light and its reemission. The most notable of these are molecular rotations which can be revealed by measurements of both stationary and time-dependent fluorescence polarization.

The effects of pressure upon the fluorescence properties are most commonly followed by measurements of spectral distribution, yield and polarization of the emission. The absorption and fluorescence of organic molecules in solution consist usually of unresolved, approximately gaussian bands with half-widths of 500 to 2000 cm^{-1}. It is usual to characterize the fluorescence spectrum by its average wavenumber $\langle v_\omega \rangle$ defined by the equation

$$\langle v_\omega \rangle = \sum F_i v_i / \sum F_i \qquad\qquad (1)$$

where F_i is the photon intensity emitted at wavenumber v_i. Under the most favorable circumstances $\langle v_\omega \rangle$ can be measured with a precision of ± 10 cm^{-1}, more typically with a precision of ± 30–40 cm^{-1}. The usefulness of $\langle v_\omega \rangle$ comes from the fact that displacements of the fluorescence spectrum result from changes in the interaction of the fluorophore with its surroundings and are therefore subjected to energy conservation: The energy gained in the molecular interactions equals the loss of energy of the excited state so that the magnitude of the spectral shift gives an indication of the extent, and sometimes of the character, of the changed interactions of the fluorophore (e.g. Macgregor & Weber 1981).

The changes in polarization of fluorescence reflect, in general, those of the rotational rate of the fluorophore. The phenomenology is described, for the case of free rotations with spherical symmetry, by the Perrin (1926) equation:

$$1/p-1/3= (1/p_o-1/3)(1+3\tau/\rho) \qquad\qquad (2)$$

where p is the observed polarization, p_o the polarizartion in the absence of motion, ρ the Debye rotational relaxation time and τ the fluorescence lifetime. When the fluorophore corresponds to a specific group forming part of a macromolecule, and this is the only case of interest for us, the changes in fluorescence polarization can originate in motions of the macromolecule as a whole (overall rotation) or in a rotation of the fluorophore about one or more of the covalent bonds that unite it to the larger particle (local rotation). Changes in overall rotation in a protein are to be expected when the protein splits into smaller compact units, as can be the case with oligomeric molecules, while changes in local rotations follow from local or general unfolding of the peptide chain. When the fluorophore has a lifetime that is short in comparison with the rotational relaxation time of the macromolecule no appreciable overall motion can take place during the fluorescence lifetime and any depolarization of the fluorescence must then be attributed to the increase in local motion. On the other hand when the fluorescence lifetime is long, on the order of tens of nanoseconds as is the case with dansyl and other attached fluorophores, splitting of the large particle into smaller ones, as well as local or general unfolding of the peptide chain, will produce changes in the polarization.

2. PRESSURE EFFECTS: PROTEINS WITH A SINGLE PEPTIDE CHAIN.

Since the early observation of reversible changes in serum albumin solutions subjected to pressure (Suzuki et al,1963) a dozen or so proteins made up of a single peptide chain have been studied employing either absorption or fluorescence spectroscopy. The changes in the instensity of the emission and those in the spectral distribution of the absorption and emission under pressure indicate that appreciable changes in conformation occur at relatively high pressures, 5 kbar and higher. In most cases a plateau in the spectral changes is reached below 10 kbar but in azotobacter flavodoxin Visser et al.(1977) have shown that only the inital changes can be seen, starting to appear at about 11 kbar. In almost all instances the spectral changes are rapidly reversed upon release of the pressure. The generality and reversible character of the spectral changes has lead to the belief that at high hydrostatic pressure a large change in conformation must take place, and we speak of a "pressure denaturation" of proteins by analogy with the better characterized changes that follow application of high temperatures or the addition of large amounts of specific chemicals like guanidine or urea to protein solutions. The fluorescence observations permit us not only to determine the range of pressures involved but they also furnish some indication of the nature of the changes undergone by the protein under hydrostatic pressure. It is well known that, in fluorophores in which the electric dipole moment of the fluorescent state is larger than that of the ground state, absorption and emission shift to the red as the polarity of the medium increases (Lippert,1957) . Tryptophan, and other indole derivatives, belong to this class of chromophores. In proteins the emission maximum of tryptophan is found to cover a range of wavelengths from 320 nm (azurin) to 355 nm (albumin). The emission in azurin corresponds to a tryptophan placed in the interior of the protein surrounded by hydrophobic residues and that of albumin to a tryptophan residue in contact with water. Crystallographic studies have shown that in these and other cases the polarity of the tryptophan environment correlates well with the energy of the fluorescence emission. It is uniformly observed that the spectral changes in intensity under pressure are accompanied by a shift of the emission to longer wavelengths indicating that at the higher pressures the native environment of the tryptophan is replaced by one of considerably greater polarity. The simplest interpretation of this general finding is that at high pressure water penetrates to the interior of the protein and water molecules are then to be found at distances of the tryptophan residues sufficiently small to permit strong interaction with the field of the dipole fluorophore. Hydrodynamic measurements are indispensable to form a clear picture of the structural changes. In particular we would like to know both the translational and the rotational diffusion coefficients of the pressure-denatured forms. There is at present a single observation of the latter quantity (Chryssomallis et al. 1981) obtained from measurements of polarization of the fluorescence of dansyl conjugates of lysozyme. The measurements indicate an abrupt increase in the rotational relaxation time of about 60% in the pressure range in which other spectral changes are observed (5.5-7 kbar), an increase fully consistent with swelling of the lysozyme molecule by water penetration. Although it is possible to form at present a preliminary picture of the character of the pressure denaturation changes in single-chain proteins we lack direct observations of many properties that are indispensable to formulate a reliable theory of pressure denaturation. Measurements of optical activity and high resolution NMR and Raman spectroscopies are necessary and certainly possible and should become available in the near future (Heremans,1986).

For the purpose of analyzing the effects of hydrostatic pressure upon oligomeric proteins the most important conclusion to be derived from observations of single-chain proteins is that conspicuous changes in the conformation of compact peptide chains are observed only at pressures higher than about 5 kbar.

3. EFFECTS UPON OLIGOMERIC PROTEINS

Josephs & Harrington (1967) observed the apparent decrease in the sedimentation constant of myosin solutions as the rotor speed of the ultracentrifuge was increased and deduced that the excess hydrostatic pressure associated with the spinning of the sample caused the dissociation of myosin aggregates. Since that date similar sedimentation behaviour has been demonstrated in a number of oligomeric proteins. Heremans and coworkers (1974) demostrated by light scattering measurements that both casein and glutamate dehydrogenase were dissociated by applications of high pressure to their solutions. The ultracentrifuge, as at present designed, results in pressures inferior to 200 bars which in many cases are insufficient to induce dissociation. Light scattering can detect fragmentation of large particles but it is clearly insensitive for the study of dissociation of dimers and tetramers with molecular weights below 200 kDa.

3.1 Importance of the fluorescence methods.

Consider a dimer-monomer equilibrium, certainly the simplest chemical equilibrium between an oligomer and its subunits. Calling a the degree of dissociation of the dimer, K(p) the dissociation constant at pressure p and C the total concentration of protein, as dimer, we have

$$K(p)= 4a^2C/(1-a) \tag{3}$$

Moreover

$$K(p)=K_{mtm}*exp(pdV^o/RT) \tag{4}$$

where K_{mtm} is the dissociation constant at atmospheric pressure and dV^o the standard change in volume upon association of the monomers to form the dimer. If we take $dV^o= 150$ ml mol^{-1} and p=2.3 kbar eq.(4) gives $K(p)/K_{mtm}= 1.1$ 10^6. For a typical value of $K_{mtm}=10^{-7}$M a micromolar solution has, by eq.(3), a degree of dissociation of 0.016 at atmospheric pressure and 0.996 at 2.3 kbar. Thus, on application of a pressure at which we expect only minimal direct effects upon either the dimer or monomers, we can pass from almost complete association to virtually complete dissociation. Detection of the changes in solution properties at micromolar concentrations of protein is virtually limited to changes of the fluorescence emission. We have employed measurements of the stationary fluorescence polarization of both the intrinsic protein fluorescence and the fluorescence from covalent conjugates of the protein with dansyl fluorophores. These two kinds of observations do

not yield quite the same information: as the lifetime of the
tryptophan fluorescence of proteins is found in the range of
2-6 ns and the rotational relaxation time of protein
subunits commonly range from 40 to 100 ns the polarization
of the fluorescence of the tryptophan is hardly influenced
by the overall rotation of the proteins and is determined
almost completely by local motions of this residue or of a
small part of the protein to which it is attached. The
existence of local motions in covalently bound fluorophores
has been clear since the earliest observations on
fluorescent conjugates of proteins and small fluorophores
(Weber,1952; Wahl & Weber,1967) and similar conclusions have
been derived more recently from observations of the
intrinsic fluorescence polarization of proteins: The
polarization values recorded in glycerol-water mixtures
sufficiently viscous to abolish completely the overall
motions of the protein are still inferior to the
polarization of the motionless fluorophores, tyrosine and
tryptophan (Lakovicz & Weber,1980; Kasprszak & Weber,1982;
Rholam et al.,1985). The dissociation of the protein into
subunits commonly results in conspicuous decrease of the
fluorescence polarization of the tryptophan (Xu & Weber
,1982; Paladini & Weber, 1981; King & Weber,1986a) thus
indicating that local motions of this amino acid residue
become more active in the separated subunits. If the
aggregate and dissociated subunits are each characterized by
a fixed value of the fluorescence polarization we should be
able to estimate the degree of dissociation as a function of
pressure and thus calculate the standard change in volume in
the dissociation reaction, whether the change is due to the
decrease in volume of the particles or to the increase in
the local rotations or to both these causes. By the same
token any other spectroscopic property that shows clear
limiting values at atmospheric pressure and at high pressure
and changes monotonically in this pressure range can be used
to obtain similar information. However, we must be sure that
the observed changes arise from the actual dissociation and
not from another condition independent of the dissociation
i.e. a first order reaction that changes intensity, spectral
shift or polarization of the fluorescence. The distinction
is easily made: a dissociation reaction must depend upon the
protein concentration in a predictable way (Paladini &
Weber,1981; Weber(1986); Silva et al (1986)) while a first
order reaction is concentration independent. We thus
consider indispensable to determine the concentration
dependence of the changes under pressure before pronouncing
these as representative of the dissociation. In this way we
have been able to show that both the intrinsic tryptophan
polarization and the polarization of dansyl conjugates of
enolase have equivalent concentration dependence (Paladini &
Weber,1981). Similarly Silva et al.(1986) have demonstrated

that the dissociation of B_m dimers of tryprophan synthase
may be followed equally well by measurements of the spectral
shift upon dissociation or by changes in the fluorescence of
the natural prosthetic fluorophore (pyridoxal phosphate) or
of the fluorescence polarization of the dansylated and
pyridoxal-reduced holoprotein.

3.2 Generality and causes of the pressure-dissociation.

The dissociation of protein aggregates by application of
hydrostatic pressure has been found to be a very general
phenomenon. The indefinite aggregates of myosin, actin or
tubulin as well as the oligomeric proteins formed of two or
more subunits undergo dissociation under pressure. It is
then expected that the causes of this behaviour are to be
found in very general features of the protein structures.
Three main causes may be listed: Paladini & Weber (1981)
proposed as one cause of the dissociation the existence of
voids, or "dead spaces" between the subunits which would
disappear when the neighbouring subunit is replaced by
solvent. Zamyatnin (1972) has estimated from dilatometric
data that folding of an extended peptide chain into a
globular protein takes place with an increase in volume of
ca. 2%. X-ray crystallography has identified in globular
proteins cavities of varied size, most of which do not
contain solvent; they are discussed in a recent paper of
Rashin et al (1986). Additionally Lewis & Rees (1985) note
that the intersubunit surfaces in superoxide dismutase and
in concanavalin are significantly more uneven than the
surfaces of these proteins that contact the solvent. A
second influence, and probably a very important one, arises
from the existence of salt linkages at the intersubunit
boundaries. This has been verified by X-ray analysis but can
also be derived from the observation that extremes of pH, as
well as large increases in ionic strength of the solvent,
result in dissociation of many oligomeric aggregates. Upon
dissociation the charges at the newly formed surfaces
undergo hydration with the corresponding decrease in volume
of the system by electrostriction. A third cause of decrease
in volume upon dissociation should follow from the
replacement of interactions between non-polar groups at the
two sides of the boundary by interaction of each with the
water dipoles. However, as Boje & Hvidt (1971) have pointed
out, the decrease in volume of non polar molecules upon
disolution in water, though particularly noticeable in very
dilute solutions can markedly decrease, and even invert,
with increase in concentration, and the density of non polar
groups in proteins is such that they may be best represented
by the latter case. Also the "interior" of the protein
contains a large number of peptide dipoles and cannot be
considered an apolar medium (Warshel et al,1984; Macgregor &

Weber, 1986). Thus, although all the causes that we can
easily enumerate will favor dissociation of the aggregates
under pressure we are for the moment unable to decide as to
their relative importance, both in general and in any
particular case.

3.3 Conformational drift of the separated subunits.

The most interesting finding to emerge from all the studies
made up to now of the dissociation of oligomeric proteins
under pressure is that separation of the protein subunits is
followed by conspicuous conformation changes. This
conclusion can be reached from at least three different
kinds of observations:
1. The curve giving the dependence of the degree of
dissociation with increasing applied pressure differs from
the curve obtained on decreasing the pressure from that at
which complete dissociation is reached. The latter curve is

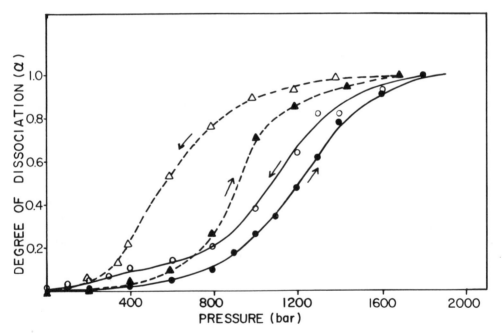

Figure 1. Hysteresis effects observed in the compression and
decompression of solutions of the B₂ subunit of trytptophan
synthase at two temperatures, 5° C (triangles)and 25° C
(circles). From Silva et al.,1986.

systematically displaced to lower pressures with respect to
the former indicating that the reassociated species have

decreased subunit affinity. This characteristic hysteretic behaviour is seen clearly both in the dissociation of the B_m subunit of tryptophan synthase (Fig 1) and in that of porcine or bovine lactate dehydrogenase (King & Weber, 1986a).

2. On rapidly diminishing the pressure of complete dissociation to reach atmospheric pressure reassociation of subunits takes place "immediately" (Figure 4) but the recovered protein shows reduced enzyme activity. This increases with time and reaches, in most cases, the value before compression. Such an effect is particularly evident in the lactate dehydrogenases where recovery of activity may take many hours (King & Weber, 1986a); it is also noticeable in the B_m subunit of tryptophan synthase where recovery has a half time of about 10 minutes (Silva et al, 1986).

3. Other properties that depend upon protein conformation like the fluorescence spectrum of tryptophan or the fluorescence yield of pyridoxal phosphate in B_m tryptophan synthase show slow recovery on decompression but over times periods that differ conspicuously from both the reassociation of the subunits and the regain of enzyme activity (Fig. 2).

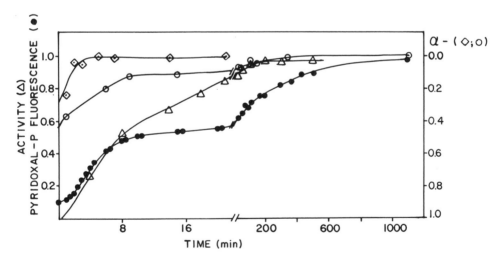

Figure 2. Recovery of various properties after return to atmospheric pressure from 2.3 kbar in solutions of B_m dimer of tryptophan synthase. Degree of dissociation (right side scale) was measured by polarization of dansyl fluorescence (rombs) and center of spectral mass (circles). From Silva et al.,1986.

From these observations we conclude that upon dissociation of the aggregate the subunits undergo a progressive change in conformation losing partially the affinity for each other and that upon reassociation a conformational adjustment that restores the original properties of the aggregate takes place. There is little reason to doubt that the equilibria established at the different pressures involves average forms of the aggregate and the subunits that vary with the degree of dissociation. From the fact that the recovery of different properties takes place following different time courses one is forced to conclude that the free subunits exist in a variety of conformations rather than in a unique one, thus undergoing after separation a process of "conformational drift".

Few data exist at present of the rates of association and dissociation of the particles. Silva et al (1986) have provided an estimate of these for B_2 tryptophan synthase by a study of the changes in fluorescence polarization of dansyl conjugates of the reduced holoprotein. In this species the Schiff base linkage of pyridoxal phosphate to the protein is reduced to suppress the fluorescence which would otherwise interfere with that of the dansyl fluorophore. Observations of the fluorescence shift of tryptophan under pressure, as well as the changes in dansyl polarization indicate that in this reduced and dansylated protein dV^o is the same as in the intact holoprotein (160 ml mol^{-1}) while K_{atm} is only slightly decreased ,6.10^{-11}M as opposed to 4.10^{-10}M in the holoprtein. When the pressure is suddenly raised from atmospheric to either 1.6 kbar, which results in 60% dissociation, or 2.2 kbar which yields virtually complete dissociation (Fig 3), half of the total decrease in fluorescence polarization takes place in about 12 minutes, which corresponds to a rate of dissociation of $10^{-3}s^{-1}$. When after establishment of the equilibrium at 1.6 kbar the pressure is rapidly shifted to 2.2 kbar complete dissociation is seen already at the first measurement, that is after 1 or 2 minutes (Fig 4). It is therefore evident that the increase in the dissociation constant with pressure must be greatly dependent upon an increase in the rate of dissociation, thus following a rule that applies to chemical equilibria of many biological molecules (e.g Weber, 1975).In contradistinction decompression is followed by reassociation within the dead time of the measurements (1-2 min) indicating a minimal rate of association of 10^4 $mol^{-1}s^{-1}$.

The observed rates of association and dissociation are consistent with the determined dissociation constant at atmospheric pressure, ca.10^{-7}M. The slow rate of dissociation, that results in a time for stabilization of about 40-50 min after application of pressure has also been seen in bovine and porcine lactate dehydrogenases (King & Weber,1986a), which also exhibit "immediate" reassociation

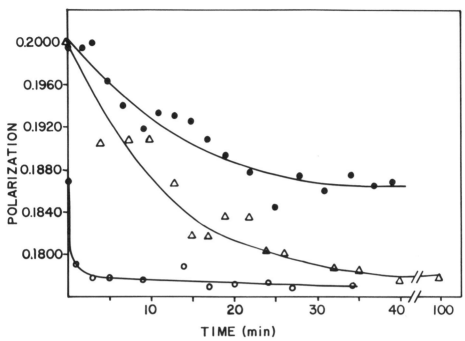

Figure 3.   Time course of the fluorescence polarization of a
conjugate   of   2-dimethylamino   naphthalene   5-sulfochloride
with pyridoxal-reduced B_2 tryptophan   synthase after   rapid
increase in pressure: Atmospheric-->1.6 kbar (dark circles),
atmospheric-->2.2 kbar   (triangles) and   1.6 kbar-->2.2 kbar
(white circles). From Silva et al.,1986.

on decompression.   A   time of equilibration of 1 or 2 hours
was found   by Xu and Weber   (1981) for   the dissociation of
enolase by dilution at atmospheric pressure.
Figure 5 shows a   plot of   the logarithm of calculated half
lives of   dimer, $\langle t_D \rangle$ and monomer, $\langle t_M \rangle$ of the B_2 subunit of
tryptophan synthase.   They were   derived on   the   assumption
that the change in K(p) with pressure is entirely due to the
increase in the rate of dissociation k_{-} so that

$$\langle t_D \rangle = 1/k_{-} = 1/\{k_{-atm} * \exp(pdV^o/RT)\} \qquad (5)$$

with $k_{-atm} = 1.5\ 10^{-3} s^{-1}$, and the additional relation

$$\langle t_M \rangle / \langle t_D \rangle = a/(1-a) \qquad (6)$$

The figure shows that for dissociation smaller than 0.5 both
$\langle t_M \rangle$ and $\langle t_D \rangle$ decrease   with pressure   in   almost   parallel
fashion but   for a>0,5 $\langle t_M \rangle$ remains virtually constant while
$\log\langle t_D \rangle$ continues to decrease linearly with pressure. The

figure provides an explanation of the dependence of the conformational drift upon the degree of dissociation: At

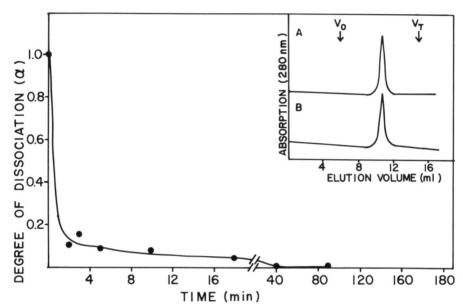

Figure 4. Time course of the reassociation of the conjugate of figure 3 after rapid reduction of pressure:2.2 kbar--> atmospheric. Inset: Molecular sieve elution patterns of the preparation before pressurization (upper trace) and 2 minutes after decompression (lower trace). From Silva et al.,1986.

less than half dissociation the almost equivalence of $\langle t_D \rangle$ and $\langle t_M \rangle$ results in recovery from the drift upon reassociation of the monomers into the dimer, but for higher degrees of dissociation the continuously decreasing ratio $\langle t_D \rangle / \langle t_M \rangle$ results in the incomplete recovery of the original properties of the dimer after reassociation.

3.4 Character of the dissociation equilibria.

All these observations lead us into the question of the mechanism of the dissociation both at atmospheric and higher pressures. The dissociation into subunits requires diffusion of the particles away from each other and the state favorable to dissociation must therefore have a lifetime sufficiently long to permit such diffusion: The minimum distance, dx at which we can consider that the particles resulting from the dissociation become independent of each other will be when separated by a layer of water molecules, that is about 5A units. From Einstein's diffusion equation

$$<dx^2>=2D\ dt \tag{7}$$

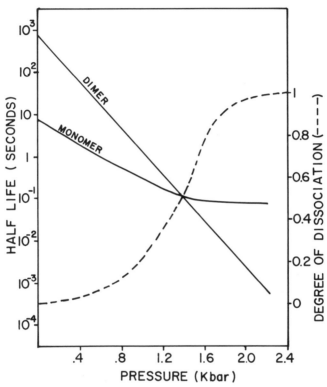

Figure 5. Plots of the logarithm of the half-lives of the dimer and monomer of the B_2 subunit of tryptophan synthase, calculated by means of eqs. (5) and (6), and of the degree of dissociation (---), against the applied pressure.

A diffusion coefficient D of 10^{-6} cm^2s^{-1}, which corresponds to a particle of some 35 kDa, gives dt> 1 ns. Thus the "dissociating state" that permits the particles to diffuse away from each other must have a lifetime longer than 1 ns. Clearly this state cannot be one of transient accumulation of energy in some degree of freedom, as it is now clear that such a state, typically a vibrational state of some sort, would not last beyond some picoseconds. What is required is a transient fluctuation of structure endowed with an appropriate lifetime. A reasonable candidate (Weber, 1986) is a fluctuation of charge that produces a greatly diminished interaction energy at the boundary or even an

actual electrostatic repulsion. Salt linkages are known to
exist at the intersubunit boundaries e.g in hemoglobin
(Bolton & Perutz,1970) or in lactate dehydrogenase (Holbrook
et al,1975) and fluctuations of charge, affecting
particularly the carboxyl groups at the intersubunit
boundary must take place. It is also known that the half-
life of a protonated carboxyl group is a few microseconds,
quite sufficient to permit separation of the particles by
diffusion.
Fig.6 depicts a possible scheme for the various steps of
the equilibrium between a dimer carrying a quadrupole
disposition of charges at the boundary between the subunits
and the constitutive monomers. Since the dissociation is

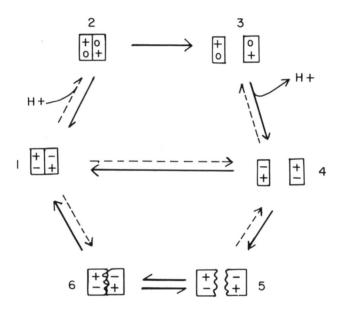

Figure 6. Schematic representation of the steps of a dimer-
monomer equilibrium that dissociates following a charge
fluctuation that abolishes the negative charges of the
quadrupole at the boundary, and in which the separated
subunits are subject to conformational drift.

brought about by a relatively rare fluctuation of boundary
charges that results in repulsion there is no reason to
expect that reassociation will only occur when the same form
is adopted; this form actually opposes monomer association,
which we further know to occur at a rate not much lower
than the rate of monomer encounters. The violation of
detailed balance involved in the scheme is discussed
elsewhere in detail (Weber,1986). At low degrees of

dissociation the equivalent lifetime of dimer and monomers (Fig.5) may not permit the conformational drift in the monomers to influence to any serious extent the average structure of the dimer and association of monomers takes place as indicated in the upper half of the diagram. At large degrees of dissociation the decreased lifetime of the dimers permits a significant degree of conformational drift to persist after reassociation and yields the modified dimer shown in the lower half of the diagram. Upper and lower halves of the diagram convey the concept that we are dealing with protein forms to which different chemical potentials must be attributed according to the degree of dissociation at which the equilibrium becomes established. This conclusion is at variance with the proposition, employed since the original formulation by Gibbs, that the chemical potentials are constants, without referemce to the particular extent of reaction at which the chemical equilibrium is established. It is to be noticed that a similar conclusion as regards the variability of the chemical potential was reached by Xu and Weber (1982) in a study of the dissociation by dilution at atmospheric pressure of another dimer protein, yeast enolase. It is further strengthened by other observations apart from those already mentioned:

1. In comparing the pressure dependence of the polarization of covalent adducts of lactate dehydrogenase with 2-diethylamino naphthalene 5-sulfochlride with the pressure dependence of the intrinsic protein fluorescence King & Weber (1986a) have noticed that dV° equals ca. 170 ml/mol in the former case and close to twice this value in the latter. The polarization values of the dansyl conjugates reflect primarily the average volume of the observed particles and may be expected to correspond closely to the degree of dissociation. The observed difference with the tryptophan fluorescence (Fig 7) evidently indicate that the changes in conformation responsible for the increase in the local motions of the tryptophan do not exactly correspond to the appearence of the dissociated form; instead they take place over a restricted range of degrees of dissociation and lead to an incorrectly high value of dV°. The difference in dV° arises from employment of two methods of observation that yield different averages because they preferentialy weight either the local rotations of the fluorophore or the overall rotations of the particles.

2. In a study of the effect of pressure upon the dissociation of the B_{2} dimer of the tryptophan synthase from E.Coli, Silva et al.(1986) have shown that both holo and apoprotein have similar values of dV°, 160 ml/mol, but that the apoprotein has not only a diminished free energy of association but exhibits a very asymmetric dependencce of the degree of dissociation upon pressure. While 900 bars are

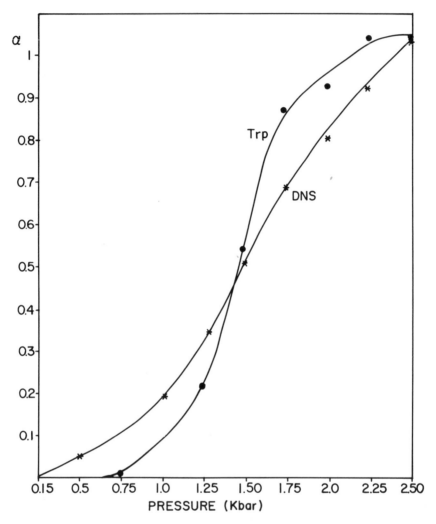

Figure 7. Plots of degree of dissociation of lactate dehydrogenase from pig heart as derived from measurements of fluorescence polarization under pressure. Trp:Tryptophan fluorescence. DNS: fluorescence of the conjugate with 2-diethylamino naphthalene 5-sulfochloride. Redrawn from data of L.King: Doctoral dissertation, University of Illinois, 1985.

required to increase the degree of dissociation from 0.1 to 0.5 less than a quarter of this value is sufficient to pass from dissociation 0.5 to 0.9 (Fig 8). In contradistinction the holoprotein shows a minimal assymmetry (Fig.1). As indicated in the figure, the hysteresis observed upon complete dissociation of the apoprotein is very marked, while it is virtually absent if the maximum dissociation reached is 0.5. This difference in behaviour corresponds to that predicted by the upper and lower halves of the diagram of figure 6.

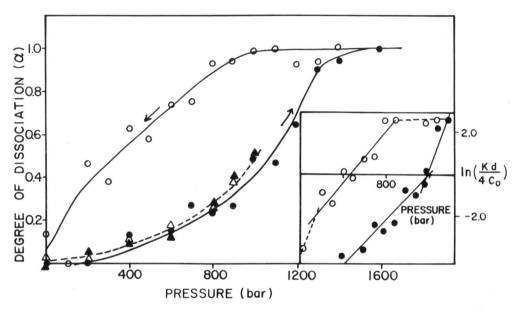

Figure 8. Compression and decompression plots of the apoprotein of B₂ tryptophan synthase as followed by the center of mass of the emission. Circles: Reversal of pressure after full dissociation (1.5 kbar). Triangles: Reversal of pressure after 50% dissociation (0.9 kbar). Inset: Logarithmic plot of the dissociation constant against pressure. From Silva et al.,1986.

3.5 The effects of temperature and pressure.

In both the lactate dehydrogenases and in B₂ tryptophan synthase decreasing the temperature results in a decrease in subunit affinity and the hysteresis phenomena become more marked at the lower temperature (Fig 1). The ratio of the dissociation constants at 5°C and 25°C, at atmospheric pressure, reaches 60 for lactate dehydrogenase and about 10 for B₂ tryptophan synthase. Evidently in this temperature range decrease in temperature and application of pressure produce similar effects. At atmospheric pressure micromolar solutions of lactate dehydrogenase lose most of the enzymic activity after a couple of weeks at cold-room temperature (4° C) but regain it completely after a few hours at room temperature (25° C). The degree of dissociation remains very small, virtually undetectable, during the cold inactivation period and the subsequent room temperature activation. Both processes are concentration dependent: the rate of cold inactivation increases with decreasing concentration and the

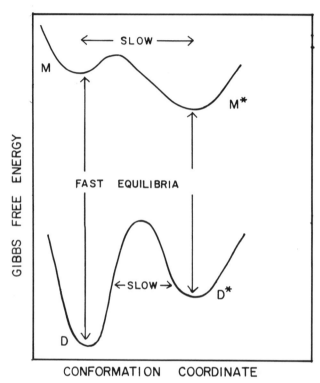

Figure 9. Free energy diagram of a dimer molecule that undergoes conformational drift of the separated subunits.D,M dimer and monomer of the native species; D*,M* conformationally drifted species.From King & Weber,1986b.

rate of room-temperature reactivation increases with concentration (King & Weber, 1986b). Cryoinactivation has been observed with many oligomeric proteins while no single-chain enzyme has been reported to be cold sensitive. King & Weber (1986b) have proposed that cryoinactivation of lactate dehydrogenase is brought about by a process of dissociation, conformational drift of the separated subunits and reassociation into inactive aggregates similar to those observed after decompression. At low temperature the inactive aggregates predominate owing to the expected high energy of activation required to convert the drifted aggregates into the original native species, while at room temperature reactivation is rapid enough to maintain the enzymic activity and the higher subunit affinity characteristic of the original aggregate. A free energy scheme that describes this formulation is shown in figure 9.

REFERENCES

Boje L. & Hvidt A.(1971) Biopolymers **11**,2357-2364.

Bolton W. & Perutz M.F.(1970) Nature(London) **228**,551-552.

Chryssomallis G.,Torgerson P.M.,Drickamer H.G. & Weber G.(1981) Biochemistry **20**,3955-3959.

Heremans K.A.H.(1974) Proc.4th.Intl. Conference on High Pressure Kyoto. pp 627-630.

Heremans K.A.H.(1982) Ann.Rev.Biophys.Bioengeng.**11**,1-21.

Heremans K.A.H.(1986) Nato ASI on "High Pressure Chemistry and Biochemistry", Corfu 1986.

Holbrook J.J.,Liljas A.,Steindel S.J. & Rossman M.G.(1975) The Enzymes (P.D.Boyer Ed.) pp 191-292. Academic,New York.

Josephs R. & Harrington W.F.(1967) Proc.Natl.Acad.Sci.USA **58**,1587-1594.

Kasprzak A. & Weber G.(1982) Biochemistry **21**,5924-5927.

King Lan & Weber G.(1986a) Biochemistry **25**,3632-3637.

King Lan & Weber G.(1986b) Biochemistry **25**,3637-3640.

Lakowicz J.R. & Weber G.(1980) Biophys.J. **10**,591-601.

Lewis M. & Rees D.C.(1985) Science,**230**,1163-1165.

Lippert E.(1957) Z.Elektrochem. **61**,952-975.

Macgregor R.B. & Weber G.(1981) Ann.N.Y.Acad.Sci. **366**,140-154.

Macgregor R.B. & Weber G.(1986) Nature(London) **319**,70-73.

Paladini A.A. & Weber G.(1981)Biochemistry,**20**,2587-2593.

Perrin F.(1926) Ann.Physique.Rad.**7**,390-398.

Rashin A.A.,Iofin M. & Honig B.(1986) Biochemistry **25**,3619-3625.

Silva J.L.,Miles E.W & Weber G.(1986) Biochemistry **25**,5780-5786.

Suzuki K.,Miyosawa Y. & Suzuki C.(1963) Archs.Biochem. Biophys. **101**,225-228.

Wahl Ph. & Weber G. (1967) J.Mol.Biol.**30**,371-382.

Warshel A.,Russell S.T. & Churg A.K. (1984) Proc.Natl.Acad. Sci.USA **81**,4785-4789.

Weber G.(1952) Biochem.J. **51**,145-167.

Weber G.(1975) Adv.Prot.Chem.**29**,1-68.

Weber G. & Drickamer H.G.(1983) Quart.Rev.Biophys.**16**,89-112.

Weber G. (1986) Biochemistry **25**,3626-3631.

Xu G.-J & Weber G.(1982) Proc.Natl.Acad.Sci.USA **79**,5268-5271.

Zamyatnin A.A.(1972) Prog.Biophys.Molec.Biol.**24**,107-123.

PRESSURE EFFECTS ON THE REACTIONS OF HEME PROTEINS

Karel Heremans
Department of Chemistry
Katholieke Universiteit Leuven
B-3030 Leuven
Belgium

ABSTRACT. A short review is given of the effects of pressure on proteins that are relevant for studies on monomeric heme proteins. The coupling of spin equilibria and ligand binding processes is discussed for metmyoglobin. The difference between horse and sperm whale metmyoglobin points to the steric control of the protein on the ligand binding process. For imidazole a difference in compressibility is observed between the high-spin and the low-spin compound. In a final section the conformational change of horse cytochrome c at alkaline pH is discussed.

1. INTRODUCTION

There are a number of reasons why one becomes interested in the study of high pressure effects on biological systems. First, a considerable amount of the biomass on our planet is found in a marine environment where the pressure is higher than 1 bar. The question of adaptation to this medium is complicated by the fact that also the temperature is lower than ambient. Secondly, the high pressure nervous syndrome (HPNS) which has important practical consequences, is being studied intensively. Two references treat these topics in detail (1,2).

The third approach is the physical-chemical. Here pressure is used as a physical parameter that determines the free energy content of a system like temperature, electric field and magnetic field. This is the approach taken in this review. We concentrate on heme proteins which belong to an important class of biomacromolecules in view of their fysiological role: myoglobin and cytochrome c. From all the biological topics this is perhaps the most related to organic and inorganic chemistry at high pressures since covalent bonds are formed and broken during

R. van Eldik and J. Jonas (eds.), High Pressure Chemistry and Biochemistry, 421–445.
© 1987 by D. Reidel Publishing Company.

the reaction of ligand binding. Changes in the electronic
configuration of the iron atom also play an important part
in the reactions of spin equilibria and redox reactions.
Some of these processes can be studied on inorganic model
systems. This enables one to estimate the contribution of
the protein to the reactions of the iron center. A
discussion of hemoglobin and other hemeproteins can be
found in a recently published monograph (3).

Pressure studies on other biochemical reactions and
systems have been reviewed elsewhere (4-7). At this school
some of these topics are covered by Gregorio Weber and
Patrick Wong. Aspects of the three approaches were covered
at a recent meeting, the proceedings of which are in
preparation (8).

2. PRESSURE EFFECTS ON PROTEINS

2.1 General considerations

Regnard was the first to observe that pressures below 1
kbar do not irreversibly affect enzyme processes in
bacteria (9). Bridgman, on the other hand, found that a
pressure of 8 kbar irreversibly coagulates proteins (10).
In the last decade or two it has been found that the
reversibility of the effects observed at high pressures may
depend on the nature of the protein as well as on the
solvent conditions. It was also observed that pressure
affects the interaction of small substrate-like molecules
with enzymes and that the quaternary structure of many
proteins is very sensitive to pressure.

While it is outside the scope of this review to
discuss all these processes in detail, we intend to
indicate the conditions where the experiments on heme
proteins have been performed.

2.2 Protein compressibility

The volume of a protein in solution consists of three
contributions (11): a) the constitutive volume of the
atoms, b) the void volume that is due to the imperfect
packing of the atoms in the interior of the protein, and c)
the volume decrease that results from the hydration of
peptide bonds and the amino acid residues. Since the
partial specific volume of a protein can be calculated from
the constitutive atomic volume, contributions from b) and
c) seem to cancel each other.

Experimentally it is possible to obtain
compressibilities of proteins from classical compression
experiments (12), ultrasonic velocity data (13-16) or the
pressure dependence of reaction or activation volumes.

Ultrasonics gives the adiabatic compressibility (β_S), the other methods give the isothermal compressibility (β_T). The two quantities are related with the following expression (13)

$$\beta_S = \beta_T \ (Cv/Cp)$$

where Cv and Cp are the heat capacity at constant volume and constant pressure respectively. The relation can be transformed to an expression which contains the thermal expansivity (α) and the density (d)

$$\beta_S = \beta_T - \alpha^2 \ T \ / \ d \ Cp$$

Unfortunately these quantities are only available for a limited number of proteins so that only adiabatic compressibilities are available. The results for globular proteins (5 - 15/Mbar) are lower than those observed for liquids and solid polymers but larger than those of metals and covalent solids. While a number of experimental techniques reveal a rather flexible protein interior, this has been interpreted as an indication for the solid-like interior of a protein consisting of domains (16). These domains would correspond to regions of secondary structure. Regions of α-helix and β-structure would have a relatively low compressibility compared to regions between these structures. This observation has important consequences for the interpretation of pressure effects on heme proteins. As will be discussed in the following paragraphs, pressure effects on heme proteins are mainly experienced at the interface between helix structures and the heme macrocycle.

Hemmes et al. (14,17) have considered the relaxation contributions to protein compressibilities. Since proton transfer processes give large apparent relaxation compressibilities they might contribute a considerable amount to measured protein compressibilities.

2.3 Protein denaturation

Pressure induced protein denaturation has been studied in detail for a number of single chain proteins with a number of techniques. It was found that temperature and solvent conditions play a role, so that the phase diagram (p, T, pH) may become quite complex. From the thermodynamic point of view this behaviour arises from the fact that the native and the denatured protein have a different compressibility, thermal expansivity and heat capacity. From the integration of the temperature and pressure dependence of the free energy

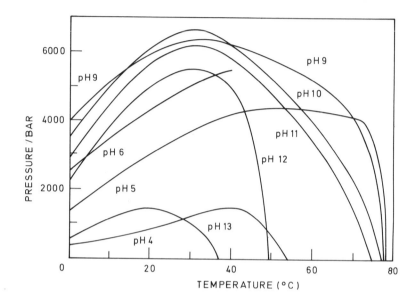

Figure 1. Experimental phase diagram for the reversible denaturation of metmyoglobin (30).

$$d(\Delta G) = - \Delta S \ dT \ + \ \Delta V \ dp$$

it follows that

$$\Delta G = \Delta Go + \Delta Vo \ (p-po) - \Delta So \ (T-To)$$
$$+ \ (\Delta\beta/2) \ (p-po)^2 \ + \ (\Delta Cp/2To) \ (T-To)^2$$
$$+ \ \Delta\alpha \ (p-po)(T-To)$$

In this expression po = 1 bar and To = 273 K. In Fig. 1 the experimental phase diagram for metmyoglobin is shown as a function of T, p and pH (30). At high temperature and pressure, the protein is in the denatured state. A detailed interpretation of the changes of the thermodynamic parameters will be given in section 3.3. The volume changes for the denaturation, however, are always small and of the order of less than 1% of the protein volume which ranges between 10.000 and 20.000 ml/mol depending on the protein. Recently a high pressure Raman study was performed on lysozyme (18). At 5.5 kbar irreversible denaturation and precipitation o ccurs, which corresponds to changes observed in fluorescence polarization studies (6). The Raman experiments also support the dynamic light scattering data, which indicate increasing intermolecular interactions with increasing pressure (19). While it remains to be

investigated whether this is a general phenomenon, these
data indicate that care has to be taken in the
interpretation of results obtained with one single
technique.

In contrast to the presumed global changes which
occur in protein denaturation, pressure may also induce
localized conformational changes. A typical example is the
disruption of the salt bridge in chymotrypsin (20). The
salt bridge stabilizes the active structure of the enzyme.
Its disruption blocks the active site so that binding of
the substrate becomes impossible. Another example of a
localised conformation change will be discussed in section
5.1: the alkaline isomerization of cytochrome c. Spin
equilibria in heme proteins are also examples and will be
discussed in section 3.3 in detail.

The interaction of small molecules (usually dyes)
with proteins, a process which mimics substrate enzyme
interactions, are also pressure sensitive (21,22). The
process is also referred to as ligand binding, but it need
not be the same as ligand substitution in inorganic
chemistry. The interaction of a small molecule is
noncovalent in nature and refers to the specific
recognition with well defined sites near or at the protein
surface. The reaction volume is usually small (less than 10
ml/mol) if no conformational change of the protein is
involved in the binding process.

The effect of pressure on the quaternary structure of
proteins is one of the best documented systems (4,5,6).
With a few exceptions, pressure disrupts protein-protein
interactions. The reaction volumes may be quite large. In
some systems the volume changes are quite strongly pressure
dependent. This implies a change in compressibility between
the subunits and the quaternary structure (23,24).

From the review of the pressure data in this section,
we may conclude that the main effects, in the pressure
range up to about 2 kbar, are to be expected on small
localized conformational changes in proteins and
protein-protein interactions. All phenomena to be discussed
in this chapter are restricted to single chain proteins. A
final word of caution should be concerning the effect of
pressure on buffers used in biological work. Unless one is
sure that the pH does not affect the processes under study,
amine buffers should be employed (25).

3. IRON SPIN EQUILIBRIA

3.1 Biological relevance

The possibility that changes in the spin state of the iron
in hemoglobin might be at the heart of the cooperative

binding of oxygen, has stimulated research both on model
and on protein systems (26,27). In deoxyhemoglobin, the
heme Fe(II)centers are five coordinated and high-spin. In
oxyhemoglobin, the Fe(II)centers are six coordinated and
low-spin. There are linked changes in the quaternary
structure between deoxy- and oxyhemoglobin. In the
deoxyform the structure is designated tense (T), in the
oxyform the structure is relaxed (R).

deoxy-Fe(II)Hb	oxy-Fe(II)Hb
tense (T)	relaxed (R)
high-spin	low-spin
S=2	S=0

Stereochemical studies of the deoxyform have revealed
that the iron is 0.06 nm outside the plane of the porphyrin
macrocycle. The iron is displaced towards the proximal
histidine side. The nitrogen of this histidine, which is
coordinated to the iron atom, is 0.27 nm from the porhyrin
plane. The geometry is close to that observed in model
compounds. For the oxyform, no structural data on
hemoglobin are available. The oxygen adduct of a picket
fence model compound shows that the iron atom is in the
porphyrin plane and that the nitrogen from histidine moves
0.06 nm towards the plane. This is a consequence of a
shortening of the Fe-N bond by about 0.01 nm. This bond
shortening should be accompanied by a decrease in molar
volume. High pressure is therefore expected to promote the
low-spin state provided that the system is not too far from
equilibrium.

For Fe(III) derivatives of these compounds, high-spin
(S=5/2) and low spin (S=1/2) states may be in true thermal
equilibrium under certain conditions. In biological systems
it is not always possible to change the conditions such
that the system is at a favorable equilibrium position.
Some derivatives of the oxidized state of hemoglobin, then
called methemoglobin, show such a favorable behaviour. The
same is true for metmyoglobin, the oxidized derivative of
myoglobin. For pressure studies metmyoglobin has the
advantage of being a single chain protein.

3.2 An inorganic model system.

McGarvey et al. (28) have measured reaction and activation
volumes for the spin interconversion of two Fe(II)
complexes in several nonaqueous solvents. The investigated
complexes are Fe(II)tris(Hpyim) and Fe(II)tris(Hpybim),
where Hpyim = 2-(2-pyridyl)imidazole and Hpybim =
2-(2-pyridyl)benzimidazole. The relaxation of the spin
equilibrium

TABLE I

Reaction and activation volumes (ml/mol) for the low-spin to high-spin interconversion ($^1A = {^5}T$) in Fe^{2+} complexes in several solvents (28).				
Complex	Solvent	ΔV	$\Delta V_{15}^{\#}$	$\Delta V_{51}^{\#}$
$Fe(Hpyim)_3{}^{2+}$	MeCN	14.3	8.9	-5.4
	Me_2CO	10.3	4.9	-5.4
	MeOH-20%MeCN	5.3	0.0	-5.3
$Fe(Hpybim)_3{}^{2+}$	MeCN	12.4	5.9	-6.4
	Me_2CO	9.6	4.7	-4.9
	MeOH-20%MeCN	4.3	0.2	-4.1

low-spin (singlet) $\rightleftharpoons$ high-spin (quintet)

was followed spectrophotometrically in the region 360-500 nm, after photochemical perturbation at 530 nm by a Q-switched Nd^{3+}/glass laser. The relaxation times span the range 120-160 ns at 1 bar and decrease with increasing pressure. Reaction volumes were determined in separate equilibrium measurements. The results are shown in Table I. For both complexes the reaction and activation volumes for the formation of the quintet state show a marked solvent dependence. The activation volume for the formation of the singlet state is almost solvent independent and constant for both complexes. This indicates that the transition and quintet states are solvated to a similar extent. The major contribution to the activation volume in the step from the quintet to singlet state is therefore the shortening of the Fe-N ligand bond. Finally, the small solvent dependence trend in the activation volume for the quintet to singlet transition for the Fe(II)tris(Hpybim) complex is not surprising in view of the greater steric demands and consequences on the Fe-N bond lengthening of the benzimidazol ligand. For the reverse process, from singlet to quintet, the activation volume should reflect the opposing contributions of intrinsic volume expansion and solvent contraction. Both the reaction and activation volume for the low spin to high spin transition correlate with the nucleophilic properties of the solvent. This suggests that the singlet state is less strongly solvated than the transition and the quintet states. This may be due to a lowering of the effective charges on the ligand due to π electron back donation from the metal orbitals, an effect that is most pronounced for the singlet isomer.

These results demonstrate that the solvent plays an

important role in spin equilibria. In this respect the
study of model compounds may help us in our understanding
of the role of the protein on spin equilibria in heme
proteins. Mechanistic considerations based on activation
entropies have to be made with caution (28).

3.3 Spin equilibria in myoglobin and its derivatives

Myoglobin is a heme protein composed of a single
polypeptide chain. When we want to estimate the
contribution of the protein matrix to the pressure induced
spin transitions in proteins, it is obvious to start with a
simple protein without quaternary structure.

 Pressure induced spectral changes in myoglobin and
some of its derivatives were first reported by Kauzmann and
coworkers (29). For oxymyoglobin, pressure induces changes
in the visible spectrum which are characteristic for a heme
protein with a ferric iron atom in the low-spin
configuration. This means that pressure denatures the
protein with concomitant oxydation of the iron. The
spectrum of deoxymyoglobin, a five coordinated ferro
compound, at 6 kbar, represents a reversibly denatured
deoxymyoglobin. The spectra indicate a behaviour of the
protein which is quite different from what is observed for
the other derivatives.

 In their first report on metmyoglobin fluoride, a
high-spin derivative, Zipp et al. (30a) reported a pressure
induced change in the spectrum which indicates that the
fluoride ligand is displaced by a nitrogen atom from an
imidazole side chain of the protein. The reaction

metmyoglobin fluoride(HS) $\rightleftharpoons$ high pressure conformation(LS)

is essentially complete at 6.5 kbar at room temperature and
neutral pH. The designation "high pressure conformation"
represents a protein where the fluoride has been replaced
by a nitrogen from the protein. Similar results were
obtained for metmyoglobin, where the sixth ligand is a

$$\begin{array}{c} H \\ \diagdown \\ \quad O-Fe-N \\ \diagup \\ H \end{array} \qquad \text{His F8}$$

water molecule, and for metmyoglobinazide.
Metmyoglobincyanide on the contrary, which is entirely
low-spin, does not show any changes up to 8 kbar. The
volume changes have been studied in great detail for
metmyoglobin. They are between - 50 and - 100 ml/mol
depending on the pH, temperature and pressure (30). These
reaction volumes indicate the substantial contri- bution
from the change in protein conformation as discussed in

section 2.3.
     We now turn our attention to spin transitions for
which the sixth ligand does not alter. In general, we
consider the reaction

metmyoglobin-L (HS) $\rightleftharpoons$ metmyoglobin-L (LS)

where L=H_2O, OH^-, N_3^-, Imidazole, CN^-. The spectral changes
are much smaller and the pressure used to study the changes
in the equilibrium position are less than 2 kbar, well
below the pressure induced changes described in the
previous paragraph. Messana et al. (31) calculated a
reaction volume of -12.5 ml/mol heme for the high- to
low-spin conversion of sperm whale metmyoglobinazide from
optical spectroscopy in the visible region. Morishima et
al. (32) studied the same process with NMR observing the
hyperfine shifts of the heme methyl signals. They obtained
- 15 ml/mol for the same reaction. They observed similar
changes in the NMR spectrum under pressure for other
metmyglobin derivatives. Morishima and Hara (33) also
observed pressure induced structural changes in the heme
environment of the low-spin cyanide complex of
metmyoglobin. The metmyoglobincyanide complex from sperm
whale is more stable than the same complex from horse
metmyoglobin. In the vicinity of the heme there are two
amino acid substitutions. We come back to this interesting
result in section 4.4 where the difference between both
proteins becomes even more striking. The NMR results also
show that the protein is more affected from the distal
side, the ligand binding side of the porphyrin ring, than
from the proximal side. More generally, the observed
reaction volume reflect the contribution from the protein
to the process of spin transition.
     Recently we determined the reaction volumes for the
spin equilibria of several complexes of metmyoglobin. The
results are given in Table II. Two points are of interest
with respect to these results. The first is that the
reaction volumes correlate with the general trend in the
entropy changes. Correlations between enthalpy and entropy
changes have been attributed to stereochemical changes of
the protein during spin changes (34). Volume changes would
therefore also reflect changes in the protein. If we
compare the magnitude of the volume changes obtained for
metmyoglobin and the model compounds, no conclusions can
obviously been drawn, since they are of the same order of
magnitude. It is more appropriate to compare the protein
results with the data obtained by McGarvey et al. (28) on
one single inorganic model complex as a function of the
solvent composition. As shown before, in these systems a
similar correlation is found between

TABLE II

Thermodynamic data for spin equilibria of sperm whale metmyoglobin (metMb)[a]			
	K	ΔH	ΔS	ΔV[b]
metMb(H$_2$O)	8.1	0.06	4.4	3
metMb(OH$^-$)	1.85	1.23	5.8	5
metMb(N$_3$$^-$)	0.2	3.8	9.6	9
metMb(Im)	0.16	10.8	32	12
[a]Sano et al.(35), K=hs/ls, ΔH/Kcal, ΔS/e.u. [b](63), ΔV/ml/mol			

volume and entropy changes. One is thus tempted to consider an analogy between solvent effects on inorganic models and the effect of the environment of the protein in the metmyoglobin systems. It is also of interest to note that several lines of experimental evidence suggest that core expansion of the porphyrin ring occurs in spin changes in heme proteins as well as in model compounds (26). Iron movement out of the porphyrin plane may not be a general mechanism nor a unique configuration for the high-spin state (27).

From the previous paragraph it is evident that changes in spin state may be quite complex reactions even in small model systems. For proteins, the contribution of the nature of the sixth ligand, our second point of interest, becomes very clear in the imidazole complex. The reaction volume for the spin transition is pressure dependent. If we fit the data to a quadratic equation

$$\ln K = \ln K_o + a P + b P^2$$

where $a = - \Delta V_o/RT$, and $b = \Delta\beta/2RT$, we calculate that the difference in compression between the high- and the low-spin state is $\Delta\beta = 10$ ml/kbar. We do not observe a pressure dependence of the reaction volume for the other derivatives. Two questions arise: Why do we observe the effect for imidazole and not for the other ligands, and how do we interpret these results. As to the first question, we note that imidazole, in contrast to other ligands such as azide, is bound perpendicular to the porphyrin plane. Thus the binding geometry of the sixth ligand is crucial in spin equilibria.

As to the interpretation of the change in compressibility, we note that the compressibilty of proteins has been related to volume fluctuations in the protein structure (36).

$$(\overline{\Delta V})^2 = kTV\beta_T$$

A similar expression relates the energy fluctuations with the heat capacity

$$(\overline{\Delta U})^2 = kT^2 mCv$$

where m and V are the mass and the volume of the system.
From the relation between the isothermal compressibility and the mean square volume fluctuations, we note that the high-spin configuration has a larger volume fluctuation than the low-spin form. The difference in compressibility is thus an indication of the dynamic nature of the protein structure. Changes in compressibility have been observed before for other protein reactions: the dissociation in subunits of enolase (23), the association of mellitin (24) and protein denaturation processes. Other high pressure studies that have been interpreted in terms of the dynamic nature of proteins are: high pressure studies on the rotational motion of side chains in proteins (37), and compressibility studies of proteins (15,16).

4. LIGAND BINDING

In view of the physiological functions of hemoglobin and myoglobin, much experimental work has been done to understand the binding of oxygen and several other ligands as model systems for the binding of natural substrates. When the three dimensional structure of myoglobin became available, it was evident that there is no obvious channel for potential ligands from the protein surrounding to the heme iron on the distal side. Structural changes in the protein are therefore necessary. One can easily imagine that some amino acid side chains must be forced away so that the ligand can enter the heme pocket. Recently, a molecular doorstop has been described which provides a possible pathway (38). In this section we review pressure studies on the binding of several ligands to heme proteins and a model system.

4.1 The binding of CO to heme

One obvious way to study the influence of the protein matrix on the binding of small molecules to the heme iron, is to remove the protein and to see what factors control the binding. Caldin and Hasinoff (39) studied the kinetics of the binding of CO to heme as a function of pressure, temperature and solvent by means of a laser flash-photolysis technique. The photodissociation and recombination process can be represented as follows

$$\text{heme-CO} \xrightarrow{\ h\nu\ } \text{heme} + \text{CO} \longrightarrow \text{heme-CO}$$

Under the experimental conditions concerned, the recombination is a pseudo-first-order process. Since the binding constant is very high, the rate constant for the dissociation can be neglected.

In ethylene glycol + water (80%v/v) the activation volume for the recombination is 2 ml/mol at 20°C. The contribution from the reaction volume resulting from the high-spin in the unliganded state to low-spin in the liganded state, is thought to be small. The authors envisage a dissociative process whereby a solvent molecule is partially lost as in the Eigen-Wilkins mechanism. In glycerol the activation volume is found to be 14 ml/mol at the same temperature. Here the data point to a diffusion controlled process.

4.2 The binding of O_2 and CO to myoglobin

In contrast to oxygen, carbon monoxide is a poison. This is a consequence of the fact that the binding constant of CO for myoglobin is about 200 times larger than for oxygen. For an isolated heme this difference is more than a factor hundred higher. This reduced affinity of CO for the protein with respect to the free heme, is thought to be the result of steric factors inside the heme pocket. While CO is bound perpendicular to the heme plane, there is bending due the presence of the distal histidine in myoglobin. The oxygen molecule is bound with an angle of about 112°. There is no Van der Waals contact, but a hydrogen bond is formed with the distal histidine (40). The activation volume for the binding of both ligands has been measured by Hasinoff (41) (see Table III). Note the difference in sign of the activation volume for CO and oxygen. For the binding of oxygen the reaction volume is - 2.9 ml/mol. For CO the binding constant is too large to be determined. Similar results have been obtained at temperatures as low as 160 K (42,43).

While the difference in activation volume may be the consequence of the difference in binding geometries of

TABLE III

Reaction and Activation volumes (ml/mol) for the binding of ligands to heme, sperm whale myoglobin (41).		
	ΔV	$\Delta V^{\#}$
Heme + CO	–	+2.0
Myoglobin + O_2	-2.9	+7.8
Myoglobin + CO	–	-8.9

these ligands, the manner in which both ligands find their way through the protein matrix towards the heme iron must also contribute. This is evident from the results obtained by Sorensen (43). He compared the binding of CO and oxygen to sperm whale and horse myoglobin. There are two amino acid substitutions in the vicinity of the heme, close to a proposed pathway for ligand entrance via the protein matrix (see section 4.4). Clearly a simple interpretation of the activation volumes, the differences between oxygen and CO, and the differences between two similar proteins, cannot interpreted quantitatively (42). In the following section we discuss binding studies with other ligands at room temperature. They show that the reaction volumes are also ligand dependent, and for the same ligand protein dependent.

4.3 The binding of azide and imidazole to metmyoglobin

Ogunmola, Kauzmann and Zipp (44) have measured volume changes for the reactions of various ligands to sperm whale metmyoglobin from the pressure dependence of the binding constants. Their results are summarized in Table IV. As the presentation of the reactions implies, ligand binding should be correctly defined as ligand substitution for metmyoglobin. Note that in myoglobin there is no sixth ligand.

metmyoglobin-H_2O + Ligand $\rightleftharpoons$ metmyoglobin-Ligand + H_2O

From the Table it can be seen that there is no correlation between the sign of the reaction volume and the changes in spin state which occur as a result of the binding. The nature of the ligand is clearly involved. In an attempt to further characterize the influence of the nature of the ligand and also of the protein, we have studied the effect of pressure on the binding and the kinetics of ligation of a charged (azide) and a neutral (imidazole) ligand on sperm whale and horse metmyoglobin.

TABLE IV

Reaction volumes (ml/mol) for the binding of ligands to
metmyoglobin (Mb)(44). Percentage of low-spin is for the
product formed. MbH_2O is 11% low-spin (45).

Reaction	pH	ΔV	%ls
$MbH_2O + N_3^- \rightleftharpoons MbN_3^- + H_2O$	7.0	-9.6	83
$MbH_2O + F^- \rightleftharpoons MbF^- + H_2O$	6.0	-3.3	3
$MbH_2O + formate \rightleftharpoons Mbformate + H_2O$	6.0	7.5	14
$MbH_2O + imidazole \rightleftharpoons Mbimidazole + H_2O$	6.0	0.0	86
$MbH_2O + OH^- \rightleftharpoons MbOH^- + H_2O$	8.9	11.0	35
$MbH_2O \rightleftharpoons MbOH^- + H^+$	8.9	-11.7	35

In Table V we summarize our experimental results
obtained with high pressure stopped-flow and equilibrium
measurements for the binding and kinetics of azide and
imidazole to sperm whale metmyoglobin. As pointed out
before, in the reactions considered the iron atoms in the
unliganded and liganded states do not have the same spin
configuration. When we interpret the reaction volume for
the ligand binding we must take the pressure dependence of
the spin equilibria of the protein compounds into account.
Consider the following reaction scheme:

$$metMb(W)hs + L \rightleftharpoons metMb(L)hs + W$$

$$metMb(W)ls + L \rightleftharpoons metMb(L)ls + W$$

The indexes (W) and (L) represent bound water and ligand
(azide or imidazole), respectively. We use metMb as an
abbreviation for metmyoglobin. We define the following
equilibrium constants:

$$K(W) = (metMb(W)ls) / (metMb(L)hs)$$

$$K(L) = (metMb(L)ls) / (metMb(L)hs)$$

These expressions are for the spin equilibria of the
unliganded and liganded forms of the protein. For the
binding we consider the possibility that the ligand can
bind to the high or the low spin state of the protein.

$$K(hL) = (metMb(L)hs) / (metMb(W)hs) (L)$$

$$K(lL) = (metMb(L)ls) / (metMb(W)ls) (L)$$

The following relationship exists between the equilibrium
constants:

TABLE V

Reaction and Activation volumes (ml/mol) for the binding of ligands to sperm whale metmyoglobin (swMb) and horse metmyoglobin (hMb) (63).				
	azide		imidazole	
	ΔV	$\Delta V^{\#}$	ΔV	$\Delta V^{\#}$
swMb	-9	+4	0	+8
hMb	-5	+11	+10	+14

$K(W) \cdot K(1L) = K(L) \cdot K(hL)$

The experimentally observed equilibrium constant K(obs) is a composite quantity of these constants:

$K(obs) = K(hL) \left[(1+K(L)/(1+K(W)))\right]$

In order to find the reaction volumes for ligand binding, we use the reaction volumes for the spin equilibria from section 3.3.

We now turn our attention to the reaction volumes for the ligand binding. The small volume changes observed by Hasinoff (see Table III) for the binding of oxygen to myoglobin can be attributed to spin changes but, in the absence of data on the reaction volume for the spin equilibria of deoxymyoglobin and oxymyoglobin, this conclusion is speculative.

For the data on the binding of azide and imidazole to metmyoglobin, we can be more specific. From the relation between the equilibrium constants we obtain the following expression for the reaction volumes:

$\Delta V(1L) + \Delta V(W) = \Delta V(hL) + \Delta V(L)$

It follows that if differences are observed for the reaction volumes of spin transitions of different protein compounds, thermodynamic data for the binding of the same ligand to either the high-spin or the low-spin configuration of the protein will differ. For azide the observed volume change on binding is negative. The reaction volume for the binding to the high-spin configuration is - 2 ml/mol, while it is - 8 ml/mol for the low-spin configuration. From the negative reaction volume for the binding we conclude that no electrostatic factors are involved. The formation of a bond may be sufficient to account for the negative sign, but geometric factors

certainly also explain the difference between the high-spin
and low-spin configuration. It should be kept in mind that
azide is bound to the iron in a similar way as oxygen.

     his F8   his F8

       There is no pressure effect on the observed binding
constant for imidazole. Here we invoke essentially the same
mechanism as for the azide binding, keeping in mind that
this ligand is larger and bound perpendicular to the heme
plane. X-ray diffraction studies of the imidazole complex
of metmyoglobin show that the distal histidine changes
position with the ligand in the pocket (46).

4.4 Comparison between horse and sperm whale metmyoglobin

Table V also compares the experimental observed reaction
volume for the binding of azide and imidazole to sperm
whale metmyoglobin and horse metmyoglobin. The reaction
volume for the binding of azide are negative, for imidazole
they are zero and positive. In general the reaction and
activation volumes are more positive for the reaction with
horse metmyoglobin. As noted before, there are two
substitutions in the vicinity of the heme group: the
replacement of a threonine residue in sperm whale
metmyoglobin for a valine in horse metmyoglobin, and the
replacement of arginine for lysine. Recently a channel has
been proposed (38) which is formed during the ligand
binding in the vicinity of these side chains. NMR
experiments have indicated that the replacement of
threonine by valine is not critical for ligand binding
(47).

	45			64	65	66	67	68
horse myoglobin	-lys	-	-	-his	-gly	-thr	-val	-val
sperm whale myoglobin	-arg	-	-	-his	-gly	-val	-thr	-val
dog myoglobin	-lys	-	-	-his	-gly	-val	-thr	-val
	CD3			E7	E8	E9	E10	E11

This could easily be tested with dog metmyoglobin which has
only the arginine lysine substitution. The difference in
activation and reaction volumes between the two proteins
observed for the binding of the same ligands points to a

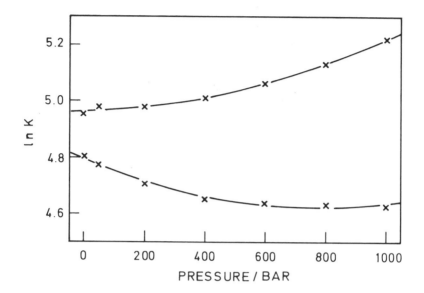

Figure 2. Pressure dependence of the equilibrium constant for the binding of imidazole to sperm whale- (upper) and horse metmyoglobin (lower curve) (3).

subtle role of the amino acid side chains packed around the heme. The striking difference between the activation volumes for the binding of CO and oxygen observed by Hasinoff (41) and Sorensen by low temperature kinetics under pressure (43), are consistent with this picture.

An interesting observation is that for the binding of imidazole to sperm whale and horse metmyoglobin, the reaction volume depend on the pressure (Fig. 2). This implies a difference in compressibility between the complex and the unliganded protein. The compressibility of the complex is larger than for metmyoglobin. However, a more detailed analysis shows that the change in compressibility is primarily due to the changes in spin state and not to the binding per se (see section 3.3). This supports the observed correlation between protein stability, as determined by chemical denaturation, and the spin state (48). Here we find that also the geometry of the ligand binding is important since no changes in compressibility are observed for azide binding.

4.5 Protein effects on ligand binding.

The binding of small molecules to the iron atom of the heme
inside the protein, put new problems on the interpretation
of the observed activation volumes. While a simple
bimolecular mechanism for the reaction of myoglobin (Mb)
with CO

Myoglobin + CO $\rightleftharpoons$ myoglobin-CO

is sufficient to account for the data at room temperature,
low temperature kinetics ask for a more elaborate scheme
such as

Mb + CO $\rightleftharpoons$ Mb::CO $\rightleftharpoons$ Mb:CO $\rightleftharpoons$ Mb-CO

Here Mb::CO represents the diffusion process of the ligand
through the protein matrix and Mb:CO indicates that the CO
molecule is present in the haem pocket but has not yet
formed the bond (Mb-CO) to the iron atom. When all these
processes are at equilibrium at room temperature, the
activation volume contains the reaction volumes for all the
intermediate processes (41).
      The experimental observation that the activation
parameters for ligand binding to myoglobin and metmyoglobin
depend on the geometry of the ligand, has also some
implication for recently proposed theoretical models for
protein reactions. Our data are more in favour of a free
volume model as proposed by Wagner (37) than the collision
model proposed by Karplus and McCammon (49). For the latter
model one would expect similar activation parameters for
different ligands since the theory only takes the changes
in internal viscosity of the protein into account.
      A fundamental difficulty remains the assessment of
the dielectric conditions inside the protein and the heme
pocket. While a protein interior is generally considered to
be quite hydrophobic, fluorescent probes and theoretical
considerations indicate that this may not be the case
(50,51). This strongly affects the interpretation of
reaction and activation volume parameters for reactions
that occur inside proteins.
      Another very striking example of the interplay
between the reaction of the iron and the structure of the
protein is given by the spin equilibrium of the imidazole
complex of metmyoglobin (see sections 3.3). The observation
is that the reaction volume for the spin transition is
pressure dependent, implying a difference in
compressibility between the high and low spin states of the
protein. For inorganic reactions this would be interpreted
in terms of solvation mechanisms. For protein reactions
this is interpreted in terms of fluctuating protein

structures. Since the isothermal compressibility is proportional to the volume fluctuations of a system, much is expected from this approach to study the dynamics of proteins (36). Unfortunately, as pointed out before, only the adiabatic compressibility is the easily accessible experimental quantity for proteins.

5. CONFORMATIONAL CHANGES IN CYTOCHROME C

5.1 The alkaline isomerization.

The biological role of cytochrome c is not related to the binding of ligands but to electron transfer. The oxidation-reduction potential of the protein drops from 260 mV at pH 7 to 120 mV at pH 10 with a pK of 9.3. This was interpreted as a pH induced conformational change. This alkaline isomerization process can be studied with fast reaction techniques. The pH dependent reduction of cytochrome c with ascorbate can be studied with the stopped-flow technique. At alkaline pH the reaction becomes biphasic consisting of a fast step, the reduction, and a slow step, a first-order isomerization. From this and other experiments it was concluded that the alkaline conformation cannot be reduced by ascorbate. The redox reaction

$$\text{Cytochrome } c^{3+} + Fe^{2+} \rightleftharpoons \text{Cytochrome } c^{2+} + Fe^{3+}$$

can be studied with the temperature-jump relaxation technique. Here Fe^{2+} and Fe^{3+} represent ferro- and ferrihexacyanide, respectively. Since the rate constants for this proces are pH independent it follows that the drop in redox potential is not related to the rate of electron transfer. Again an alkaline isomer of the protein is proposed, which is not reduced by ferrocyanide. Inclusion of the pH dependence of the redox potential leads to the following reaction scheme (52).

$$\text{Cytochrome } c^{3+} + H^+$$
$$\updownarrow$$
$$\text{H-Cytochrome } c^{3+} + Fe^{2+} \rightleftharpoons \text{H-Cytochrome } c^{2+} + Fe^{3+}$$

A further kinetic and thermodynamic characterization of the isomerization process reveals that the reaction is coupled to the breaking of the Fe-S bond. It was suggested that a nitrogen from the amino group of a lysine replaces the sulfur from methionine (53). This, however, has been questioned by several authors and an hydroxyl group has been proposed as an alternative (54). Whatever the molecular details, the alkaline isomerization of ferricytochrome c can thus be represented as follows:

H-Cytochrome c(Fe-S) $\rightleftharpoons$ Cytochrome c(Fe-S) + H$^+$

$\Updownarrow$

Cytochrome c(Fe-X)

    Thus the isomerization of cytochrome c gives us another possibility to study the substitution of the sixth ligand at the heme iron by a side chain of the protein. This is a strong field ligand since the ferric center remains low-spin.

```
  Me                                             |
    \     |                                      |
     \  S-Fe-N      his 18         H₂N-Fe-N      his 18
     /    |                        /    |
met 80    |                   lys 79    |

     pH=7                             pH=10
```

    Equilibrium measurements under pressure show that the pK of the transition increases with increasing pressure (20). The conformation with the Fe-S bond intact is thus stabilized at high pressure. The reaction volume for the ionization can only be obtained indirectly from a kinetic analysis, but the experimental error is large. The reaction volume is small, which is more consistent with the ionization of a lysine residue than a water molecule. The results of NMR experiments also favour the displacement of the methionine residue by a lysine residue (55). It is only above pH 12 that the lysine is replaced by hydroxide. This conclusion is also supported by the enthalpy and entropy changes (52).

    The activation volume for the isomerization points towards a dissociative mechanism. Time resolved Raman spectroscopy has shown that a transient species is formed that consists of high- and low-spin components (56). Our observed activation volume supports this.

5.2 Ligand binding to cytochrome c

In this section we consider the effect of pressure on the binding and the kinetics of azide and imidazole to ferri horse cytochrome c. In contrast to myoglobin and its derivatives, where the sixth coordination position of the iron is a solvent molecule, in cytochrome c the sixth ligand is part of the protein chain. Binding studies of ligands to this protein can therefore be considered as an alternative approach to study the influence of the protein on the ligand binding process. As indicated in the previous section, the Fe-S bond can be broken by a change to alkaline pH or an external ligand such as azide and imidazole. Table VI shows the results of our pressure studies (63). In contrast to the reaction of these ligands

TABLE VI

Reaction and activation volumes (ml/mol) for the binding of ligands to horse heart cytochrome c (63)		
	ΔV	$\Delta V^{\#}$
Cytochrome + imidazole	+ 7	+ 20
Cytochrome + azide	-13	+ 20

with myoglobin, the iron remains in the low-spin state. We relate the differences between the reaction volumes for azide and imidazole to the different geometrical arrangement of the ligand in the heme pocket as discussed for metmyoglobin. It is also of interest to note that, as observed for metmyoglobin, the reaction volume is more positive for imidazole than for azide. Note that the reaction volume may contain small contributions from changes in spin state. The kinetics are consistent with the following mechanism (57):

cytochrome(Fe-S) $\rightleftharpoons$ cytochrome(Fe::S)*

cytochrome(Fe::S)* + ligand $\rightleftharpoons$ cytochrome(Fe-L)

where the asterisk represent a conformation of the protein in which the Fe::S bond is already partially broken. The similar activation volumes also reflect the fact that for both ligands a bimolecular binding mechanism is to simple to explain the kinetics. The kinetics therefore primarily reflect the change in conformation due to the Fe-S bond breaking.

The reactions discussed in the previous paragraph are all for the oxidized state of the protein. There is no change in the redox state. The redox process itself takes place when a redox active reagent binds. The reaction volumes for this processes can only be obtained indirectly. The redox reactions of cytochrome c are not accompanied by gross conformational changes wich are detectable in terms of volume changes. This is more fully discussed elsewhere (58). Changes in compressibility have been observed by Eden et al. (15) and interpreted in terms of differences in volume fluctuations. The results have been questioned by Russian workers who find a difference which is an order of magnitude lower (59).

6. CONCLUSION

A valid question one might ask is what information, on the

molecular level, can be obtained from pressure experiments on biological molecules and their reactions. There is first the classical answer that reaction and activation volumes contain information which is not easy to get from reaction enthalpies and activation energies. But there is much more to learn especially when proteins are involved.

In view of the theoretical relationship between protein compressibility and volume fluctuations, an increasing interest in this physical quantity is expected. But one should keep in mind that the technique of sound velocity gives experimental data which contains several contributions and requires the separation of the relevant parameter (13-16,60,61). The same is true for any other method such as fluorescence transfer measurements which probe the distance between two residues in a protein (62). An extreme example of a change in compressibility is the difference between oxidized and reduced cytochrome c observed by Eden etal. (15).

Ligand binding processes continue to provide interesting results which indicate that our molecular models are probably to static to give realistic interpretations. Pressure studies have much to teach us on molecular substitution effects which become possible with genetic engineering experiments. In the meantime nature gives us already a few interesting possibilities as shown with the results obtained from horse and sperm whale myoglobin (63). Inorganic model systems are certainly also valid in this respect. Here solvation effects can be studied as shown for the spin transitions by McGarvey et al. (28). A final point is the theories we use to interpret the activation parameters. It is clear that steric effects which are difficult to incorporate in a theory, are very sensitive to pressure. The striking difference between the pressure effect on the binding of CO and O_2 to myoglobin is a classical example (41,43). In this way it becomes clear that acquisition of pressure tolerance probably involves only minor changes in the primary structure of some enzymes (64).

REFERENCES

1. Diving and Life at High Pressures, Phil. Trans. R. Soc. Lond. **B 304**, 1-197, 1984.
2. A.J.R. Pequeux and R. Gilles, Eds. High Pressure Effects on Selected Biological Systems, Springer-Verlag, Berlin, 1985.
3. K. Heremans in Inorganic High Pressure Chemistry, R. van Eldik, Ed. Elsevier, Amsterdam, 1986.
4. R. Jaenicke, Ann. Rev. Biophys. Bioeng., **10**, 1, 1981.

5. K. Heremans, Ann. Rev. Biophys. Bioeng., **11**, 1, 1982.
6. G. Weber and H.G. Drickamer, Q. Rev. Biophys., **16**, 89, 1983.
7. P.T.T. Wong, Ann. Rev. Biophys. Bioeng., **13**, 1, 1984.
8. H.W. Jannash, R.E. Marquis and A.M. Zimmerman, Eds. Current Perspectives in High Pressure Biology, Aacdemic Press, New York, in press.
9. P. Regnard, La vie dans les eaux, Masson, Paris, 1891.
10. P.W. Bridgman and J.B. Conant, Proc. Nat. Acad. Sci. USA, **15**, 680, 1929.
11. W. Kauzmann, Adv. Prot. Chem., **14**, 1, 1959.
12. S.A. Hawley, Methods in Enzymology, **49**, 14, 1978.
13. K. Gekko and H. Noguchi, J. Phys. Chem., **83**, 2706, 1979.
14. A.P. Sarvazyan and P. Hemmes, Biopolymers, **18**, 3015, 1979.
15. D. Eden, J.B. Matthew, J.J. Rosa and F.M. Richards, Proc. Nat. Acad. Sci., **79**, 815, 1982.
16. B. Gavish, E. Gratton and C.J. Hardy, Proc. Nat. Acad. Sci., **80**, 750, 1983.
17. A.P. Sarvazyan, D.P. Kharakoz and P. Hemmes, J. Phys. Chem., **83**, 1796, 1979.
18. K. Heremans, P.T.T. Wong, Chem. Phys. Letters, **118**, 101, 1985.
19. B. Nystrom and J. Roots, Makromol. Chem., **185**, 1441, 1984.
20. K. Heremans and J. Wauters, in B. Vodar and Ph. Marteau (Eds), High Pressure and Technology, Pergamon, Oxford, 845, 1980.
21. K. Heremans, J. Snauwaert, H. Vandersypen and Y. Van Nuland, Proceedings 4th Int. Conf. High Pressure, Kyoto, 623, 1974.
22. P.M. Torgerson, H.G. Drickamer and G. Weber, Biochemistry, **18**, 3079, 1979.
23. A.A. Paladini and G. Weber, Biochemistry, **20**, 2587, 1981.
24. R.B. Thompson and J.R. Lakowicz, Biochemistry, **23**, 3411, 1984.
25. R.C. Neuman, Jr., W. Kauzmann and A. Zipp, J. Phys. Chem., **77**, 2687, 1973.
26. W.R. Scheidt, I.A. Cohen and M.E. Kastner, Biochemistry, **18**, 3546, 1979.
27. Z.R. Korszun and K. Moffat, J. Mol. Biol., **145**, 815, 1981.
28. J.J. McGarvey, I. Lawthers, K. Heremans and H. Toftlund, J. Chem. Soc. Chem. Comm., 1575, 1984.
29. G.B. Ogunmola, A. Zipp, F. Chen and W. Kauzmann, Proc. Nat. Acad. Sci. USA, **74**, 1, 1977.
30. A. Zipp and W. Kauzmann, Biochemistry, **12**, 4217,

1973.

30a.A. Zipp, G. Ogunmola, R.C. Neuman, Jr. and W. Kauzmann, J. Am. Chem. Soc., **94**, 2541, 1972.

31. C. Messana, M. Cerdonio, P. Shenkin, R.W. Noble, G. Fermi, R.N. Perutz and M.F. Perutz, Biochemistry, **17**, 3652, 1978.

32. I. Morishima, S. Ogawa and H. Yamada, Biochemistry, **19**, 1569, 1980.

33. I. Morishima and M. Hara, Biochemistry, **22**, 4102, 1983.

34. J. Otsuka, Biochim. Biophys. Acta, **214**, 233, 1970.

35. T. Sano, J. Ishibashi, N. Ikeda, T. Yasunaga, S. Ogawa and I. Morishima, Biopolymers, **20**, 187, 1981.

36. A. Cooper, Proc. Nat. Acad. Sci. USA, **73**, 2740, 1976.

37. G. Wagner, FEBS Letters, **112**, 280, 1980.

38. D. Ringe, G.A. Petsko, D.E. Kerr and P.R. Ortiz de Montellano, Biochemistry, **23**, 2, 1984.

39. E.F. Caldin and B.B. Hasinoff, J. Chem. Soc. Farad. Trans. I, 515, 1975.

40. S.E.V. Phillips and B.P. Schoenborn, Nature, **292**, 81, 1981.

41. B.B. Hasinoff, Biochemistry, **13**, 3111, 1974.

42. H. Frauenfelder and P.G. Wolynes, Science, **229**, 337, 1985.

43. L.B. Sorensen, thesis, University of Illinois at Urbana-Champaign, 1980.

44. G.B. Ogunmola, W. Kauzmann and A. Zipp, Proc. Nat. Acad. Sci. USA, **73**, 4271, 1976.

45. D.W. Smith and R.J.P. Williams, Biochem. J., **110**, 297, 1968.

46. M. Bolognesi, E. Cannillo, P. Ascenzi, G.M. Giacometti, A. Merli and M. Brunori, J. Mol. Biol., **158**, 305, 1982.

47. J.T.J. Lecomte and G.N. La Mar, Biochemistry, **24**, 7388, 1985.

48. G. McLendon and K. Sandberg, J. Biol. Chem., **253**, 3913, 1978.

49. M. Karplus and J.A. McCammon, FEBS Letters, **131**, 34, 1981.

50. A. Warshel, S.T. Russell and A.K. Churg, Proc. Nat. Acad. Sci. USA, **81**, 4785, 1984.

51. R.B. Macgregor and G. Weber, Nature, **319**, 70, 1986.

52. L.A. Davis, A. Schejter and G.P. Hess, J. Biol. Chem., **249**, 2624, 1974.

53. D.O. Lambeth, K.L. Campbell, R. Zand and G. Palmer, J. Biol. Chem., **248**, 8130, 1973.

54. H.R. Bosshard, J. Mol. Biol., **153**, 1125, 1981.

55. I. Morishima, S. Ogawa, T. Yonezawa and T. Iizuka, Biochim. Biophys. Acta, **495**, 287, 1977.

56. T. Uno, Y. Nishimura and M. Tsuboi, Biochemistry,

23, 6802, 1984.
57. N. Sutin and J. Yandell, J. Biol. Chem., **247**, 6932, 1972.
58. K. Heremans, M. Bormans, J. Snauwaert and H. Vandersypen, Farad. Disc. Chem. Soc., **74**, 343, 1982.
59. D.P. Kharakoz and A.G. Mkhitaryan , Molecular Biology (in Russian), **20**, 396, 1986.
60. K.C.Cho, W.P.Leung, H.Y. Mok and C.L.Choy, Biochim. Biophys. Acta, **830**, 36, 1985.
61. W.P. Leung, K.C. Cho, Y.M. Lo and C.L. Choy, Biochim. Biophys. Acta, **870**, 148, 1986.
62. M.C. Marden, G. Hui Bon Hoa and F. Stetzkowski-Marden, Biophysical J., **49**, 619, 1986.
63. K. Heremans and M. Bormans, Physica **139 & 140B**, 870, 1986.
64. J. F. Siebenaller, Biochim. Biophys. Acta., **786**, 161, 1984.

SUBJECT INDEX

absorption edge 146
activation energies 96
activation parameters 311
activation volumes 282, 298
    - for aquation reactions 344
    - for base hydrolysis reactions 345
    - for complex formation 343
    - for ligand binding to cytochrome c 441
    - for ligand binding to heme 433
    - for ligand binding to metmyoglobin 435
    - for ligand binding to myoglobin 433
    - for photochemical reactions 357
    - for solvent exchange 323, 326, 343
addition reactions 351
adiabatic compressibility 52, 423
alkali halide melts 44, 47
alkali hydrides 102
alkaline isomerization 439
alkane 176
allylation 303
amorphous phase 3
anation reactions 340
Anderson process 137
anisotropic scattering intensity 193
apical intermediate 374
argon 27
    - shock compression 27
Asano equation 282
astrophysics 10
atomic phase binding energy 40
atomic properties 1
autoclave 21

Bakhuis-Roozeboom diagram 160
band-crossing effects 140
barophylic bacteria 6
basal intermediate 374
base promoted solvolysis 287
binary systems 108
  - water-methane 112

CONTRIBUTED PAPERS

A.M. de F. Palavra
Measurements of gaseous mixtures at high temperatures

J.R. Guedes
Dynamic properties of liquid nitric oxide at pressures up to 140 MPa

E.G. de Azevedo
Supercritical fluid extraction of metal and biological products

H. Uchtmann
Optical absorption of mercury near the critical point

T.V. Horton
The analytic fit to the thermodynamic equation of state for a
classical fluid with the alpha-exponential-six pair potential from
computer simulation studies to high pressure

R. Diguet
The dielectric and refractive properties of liquids and of binary
liquid mixtures: the Vedam linear relationships

B.A. Bilal
Electrochemical and spectroscopic study of complex formation in
hydrothermal systems up to 1 kbar and 250°C

D.A. Palmer
Metal complexation from solubility measurements: ferrous acetate
complexation thermodynamics to 250°C and 1250 bars

F.E. Prieto
A Universal equation of state at shock pressures

W. Dultz
The phase diagram of the alkali cyanides

A.L. Ruoff
Crystal structure determination at 2 M bars

R. Winter
Neutron scattering results on the structure factor of expanded liquid
cesium

E. Whalley
Chemical and physical kinetics at very high pressures

J. Burgess
Volumes of activation and solvent effects

H. Brackemann
Laser-induced high pressure polymerization of pure ethylene

J.W. Scheeren
High pressure promoted polar (2+2) cycloadditions of imines with
electron rich alkenes

L. Heremans
High pressure Raman spectroscopy of proteins

C. Balny
Application of cryobaro-enzymology

M.B. Boslough
Real-time studies of shock-induced chemical reactions

L. Helm
Typical example of a solvent exchange process at high pressure

Y. Ducommun
Pressure effects on complex formation reactions

M. Kotowski
Volumes of activation for substitution reactions of square-planar
complexes

I. Krack
Pressure dependence of the outer-sphere electron-transfer reaction
between Co(III) and Fe(II) complexes

A. Hvidt
Volume effects in aqueous solutions of amphiphilic molecules

H. Lentz
Melting pressure and compressibility of human blood

H. Ludwig
The influence of pressure on bacteria and viruses

J.A. Paciorek
Erythrocytes: changes in glycolytic enzymes in dives deeper than 300m

J. DiBenedetto
Pressure effects on non-radiative rates of inorganic excited state
reactions

S. Wieland
Photosubstitution reactions of Rh(III) complexes at elevated pressure

LIST OF PARTICIPANTS

Dr. Sitki Aygen
University of Frankfurt
Inst. for Physical Chemistry
Niederurseler Hang
6000 Frankfurt/Main
FRG

Dr. E.G. Azevedo
Centro Quimica Estrutura
Instituto Superior Tecnico
1000 Lisbon
Portugal

Dr. C. Balny
INSERM U 128
BP 5051
Montpellier Cedex
France

Susana F. Barreiros
Centro de Quimica Estrutural
Grupo III, IST
1096 Lisboa Codex
Portugal

Resat Benzer
Firat Un Fen-ed. Fak.
Kimya-BL
Elazig
Turkey

Prof. A.L. Bilal
Hahn-Meitner-Inst., Berlin
Gienickerstr. 100
1000 Berlin 39
FRG

Dr. Mark B. Boslough
Division 1131
Sandia National Laboratory
Albuquerque, NM 87185
USA

Holger Brackemann
Inst. f. Physikalische Chemie
Tammannstr. 6
3400 Goettingen
FRG

Dr. M.S. Bradley
045A Venable Hall
Department of Chemistry
University of North Carolina
Chapel Hill, NC 27514
USA

Dr. J. Burgess
Dept. of Chemistry
University of Leicester
Leicester LE1 7RH
England

Cedric Cossy
Inst. de chimie minerale
et analytique
Pl. du Chateau
CH 1005 Lausanne
Switzerland

M. Dellert
University of Frankfurt
Inst. for Physical Chemistry
Niederurseler Hang
6000 Frankfurt/Main
FRG

Dr. John DiBenedetto
2012F South Circle View Drive
Irvine, CA 92715
USA

Dr. R. Diquet
Universite Nancy
Faculte des Sciences Chimie
BP 239
54506 Vandoevre-Nancy
France

Prof. H.C. Drickamer
Dept. of Chemical Engineering
University of Illinois
1209 West California
Urbana, Ill. 61801
USA

Dr. Yves Ducommun
Inst. de chimie et analytique
Pl. du Chateau
CH 1005 Lausanne
Switzerland

Dr. Wolfgang Dultz
Universitaet des Saarlandes
FR 11.2 Experimentalphysik
Bau 22
6600 Saarbruecken
FRG

Prof. H.B. Dunford
Department of Chemistry
University of Alberta
Edmonton, Alberta
Canada T6G 2G2

Dr. Rudi van Eldik
University of Frankfurt
Inst. for Physical Chemistry
Niederurseler Hang
6000 Frankfurt/Main
FRG

Prof. E.U. Franck
Inst. f. Physikaische Chemie
Universitaet Karlsruhe
Postfach 6380
7500 Karlsruhe 1
FRG

A. Gerhard
University of Frankfurt
Inst. for Physical Chemistry
Niederurseler Hang
6000 Frankfurt/Main
FRG

Dr. G. Gillies
Dept. of Chem. Royal Holloway
and Bedford New College
Egham Hill
Egham, Surrey TW20 0EX
England

Dr. P. Grandinetti
Dept. of Chemistry
University of Illinois
Urbana, Ill. 61801
USA

H.J.R. Guedes
Dept. Quimica
FCT-UNL-Quinta da Torre
2825 Monte da Caparica
Portugal

Bjorn Hellquist
Oorganisk kemi 1
Kemicentrum, Box 124
22100 Lund
Sweden

Dr. Lothar Helm
Inst. de chimie minerale
et analytique
Pl. du Chateau
CH 1005 Lausanne
Switzerland

Prof. F. Hensel
Inst. f. Physikalische Chemie
Universitaet Marburg
Hans Meerwein Str.
3550 Marburg
FRG

Dr. K. Heremans
Katholieke Universiteit Leuven
Lab. v. Chem. & Biol. Dynamic
Celestijnenlaan 200 D
3030 Leuven
Belgium

Luc Heremans
Lab. voor Chemische Dynamica
Celestijnenlaan 200 D
B 3030 Leuven
Belgium

Timothy V. Horton
Dept. of Chemistry
University of Manchester
Manchester, M13 9PL
England

Dr. Aase Hvidt
Chemistry Department
H.C. Orsted Institute
Universitetsparken 5
DK 2100 Copenhagen 0
Denmark

Prof. J. Jonas
Dept. of Chemistry
University of Illinois
1209 West California
Urbana, Ill. 61801
USA

Dr. L.A. Kleintjens
DSM Research
Postbus 18
6160 MD Geleen
The Netherlands

Dr. Mirjana Kotowski
University of Frankfurt
Inst. for Physical Chemistry
Niederurseler Hang
6000 Frankfurt/Main
FRG

Ingeborg Krack
University of Frankfurt
Inst. for Physical Chemistry
Niederurseler Hang
6000 Frankfurt/Main
FRG

Jochen Kraft
University of Frankfurt
Inst. for Physical Chemistry
Niederurseler Hang
6000 Frankfurt/Main
FRG

Gerd Krueger
University of Frankfurt
Inst. for Physical Chemistry
Niederurseler Hang
6000 Frankfurt/Main
FRG

Dr. B. LeNeindre
CNRS LIMHP Univ. Paris Nord
Villetaneuse 93430
France

Prof. Dr. Harro Lentz
UCH, Fachbereich 8
Postfach 101240
D 5900 Siegen
FRG

Prof. N. Lewis
Department of Chemistry
University of Miami
Coral Gables, FL 33124
USA

Prof. Horst Ludwig
Inst. Angewandte
Physikalische Chemie
Im Neuenheimer Feld 253
6900 Heidelberg
FRG

Prof. A.E. Merbach
Inst. Chem. Minerale et Analyt.
Universite de Lausanne
Place du Chateau 3
CH 1005 Lausanne
Switzerland

Prof. W.J. le Noble
Department of Chemistry
SUNY at Stony Brook
Stony Brook, NY 11794
USA

Dr. Judy Ann Paciorek
Department of Chemistry
University of Bergen
Allegt 41
N 5000 Bergen
Norway

Dr. A.M. de Figueiredo Palavra
Centro de Quimica Estrutural
Instituto Superior Tecnico
Av. Rovisco Pias
Lisboa
Portugal

Dr. D.A. Palmer
Chemistry Division
Oak Ridge National Laboratory
Oak Ridge, TN 37830
USA

J. Pienaar
Inst. for Physical Chemistry
University of Frankfurt
Niederurseler Hang
6000 Frankfurt/Main
FRG

Pierre-Andre Pittet
Inst. de chimie minerale
et analytique
Pl. du Chateau
CH 1005 Lausanne
Switzerland

Dr. Fernando E. Prieto
Instituto de Fisica UNAM
Apartado Postal 20364
Delegacion Alvaro Obregon
01000 Mexico, DF
Mexico

Luis Paulo da S.N.M. Rebelo
Centro de Quimica Estrutural
Complex I, IST
1096 Lisboa Codex
Portugal

Dr. M. Ross
Lawrence Livermore Nat. Lab.
P.O. Box 808
Livermore, CA 94550
USA

Prof. Arthur L. Ruoff
Dept. Materials Science
& Engineering, Bard Hall
Cornell University
Ithaca, NY 14853,
USA

Pierre-Yves Sauvageat
Inst. de chimie minerale
et analytique
Pl. du Chateau
CH 1005 Lausanne
Switzerland

Dr. J.W. Scheeren
Laboratory of Organic Chemistry
Catholic University of Nijmegen
Toernooiveld
6525 ED Nijmegen
The Netherlands

Karen Schneider
University of Frankfurt
Institute for Physical Chemistry
Niederurseler Hang
6000 Frankfurt/Main
FRG

Dr. L.H. Sutcliffe
Chemistry Department
Royal Holloway & Bedford
New College
Egham, Surrey TW20 OEX
England

Dr. S. Suvachittanont
Chemistry Department
Prince of Songkla University
PO Box 3 Korhong
HAAD-YAI 90112
Thailand

Nurhan Turgut
Warsaw Technical University
Chemical Engineering Inst.
VL Warynskiego 1
00-645 Warsaw
Poland

Dr. H. Uchtmann
Physikalische Chemie
Hans Meerwein Str.
3550 Marburg
FRG

Prof. J. Voiron
Inst. National Polytechnique
46, avenue Felix-Viallet
38031 Grenoble Cedex
France

Prof. G. Weber
School of Chemical Sciences
University of Illinois
1209 West California
Urbana, Ill. 61801
USA

Renate Wegert
University of Frankfurt
Inst. for Physical Chemistry
Niederurseler Hang
6000 Frankfurt/Main
FRG

Dr. E. Whalley
Division of Chemistry
National Research Council
Ottawa
Canada K1A OR6

Stefan Wieland
University of Frankfurt
Inst. for Physical Chemistry
Niederurseler Hang
6000 Frankfurt/Main
FRG

A.J. Williams
Department of Chemistry
Royal Holloway and Bedford
New College, Egham Hill,
Egham, Surrey TW20 OEX
England

Dr. Roland Winter
Philipps Universitaet, FB 14
Auf den Lahnbergen
3550 Marburg,
FRG

H.G. Wissink
Akzo Zout Chemie Nederland B.V.
Boortorenweg 20
7554 Rs Hengelo (o)
Nederland

Dr. P. Wong
Nat. Res. Council Canada
100 Sussex Dr.
Ottawa, Ontario

Dr. Sedat Yoruk
Firat Universitesi
Muhendislik Fakultes
Kimya Muhendisligi Bolumu
Elazig
Turkey

Dr. T.A. Zarvalis
Aristlian University
of Thessaloniki
Laboratory of Biochemistry
Thessaloniki
Greece 54006